AMERICAN ENVIRONMENTAL STUDIES

# *Metallic Wealth Of The United States*

JOSIAH DWIGHT WHITNEY

ARNO &
THE NEW YORK TIMES

Collection Created and Selected
by Charles Gregg of Gregg Press

Reprint Edition 1970 by Arno Press Inc.

LC# 74-125766
ISBN 0-405-02692-7

*American Environmental Studies*
ISBN for complete set: 0-405-02650-1

Manufactured in the United States of America

# THE
# METALLIC WEALTH
# OF
# THE UNITED STATES.

THE

# METALLIC WEALTH

OF

# THE UNITED STATES,

DESCRIBED AND COMPARED WITH THAT OF OTHER COUNTRIES.

BY

J. D. WHITNEY.



---

PHILADELPHIA:
LIPPINCOTT, GRAMBO & CO.
LONDON: TRÜBNER & CO.
1854.

# LIST OF ILLUSTRATIONS.

## WOOD-CUTS.

## LITHOGRAPHS.



---

NOTE.—All the sections are drawn on the scale of 200 feet to 1 inch. The stopings are represented by the part in black.

# CONTENTS.

## INTRODUCTION.



## CHAPTER I.

### ON THE NATURE OF THE DEPOSITS OF THE METALS AND THEIR ORES, AND THE GENERAL PRINCIPLES ON WHICH MINING IS CONDUCTED.

## CHAPTER II.

### GOLD, PLATINA, AND SILVER (IN PART).

### SECTION I.

#### MINERALOGICAL OCCURRENCE AND GEOLOGICAL POSITION OF GOLD.



### SECTION II.

#### GENERAL DESCRIPTION OF FOREIGN GOLD REGIONS.



### SECTION III.

#### GEOGRAPHICAL DISTRIBUTION OF GOLD IN THE UNITED STATES.

## SECTION IV.

### PLATINA, AND ITS ASSOCIATED METALS.



# CHAPTER III.

## SILVER.

## SECTION I.

### THE MINERALOGICAL OCCURRENCE AND GEOLOGICAL POSITION OF SILVER.



## SECTION II.

### GEOGRAPHICAL DISTRIBUTION OF SILVER ORE.

## SECTION II.

### GEOGRAPHICAL DISTRIBUTION OF TIN.



# CHAPTER VI.

## COPPER.

## SECTION I.

### MINERALOGICAL OCCURRENCE AND GEOLOGICAL POSITION OF THE ORES OF COPPER.



## SECTION II.

### GEOGRAPHICAL DISTRIBUTION OF COPPER IN FOREIGN COUNTRIES.



## SECTION III.

### GEOGRAPHICAL DISTRIBUTION OF COPPER IN THE UNITED STATES.

## CHAPTER VII.

### ZINC

### SECTION I.

#### MINERALOGICAL OCCURRENCE AND GEOLOGICAL POSITION OF THE ORES OF ZINC.



### SECTION II.

#### GEOGRAPHICAL DISTRIBUTION OF THE ORES OF ZINC IN FOREIGN COUNTRIES.



### SECTION III.

#### DISTRIBUTION OF THE ORES OF ZINC IN THE UNITED STATES.



## CHAPTER VIII.

### LEAD, AND SILVER IN PART.

### SECTION I.

#### MINERALOGICAL OCCURRENCE AND GEOLOGICAL POSITION OF THE ORES OF LEAD.

## SECTION II.

### DISTRIBUTION OF THE ORES OF LEAD IN FOREIGN COUNTRIES.



## SECTION III.

### GEOGRAPHICAL DISTRIBUTION OF THE ORES OF LEAD IN THE UNITED STATES.

## CHAPTER IX.

### IRON.

### SECTION I.

#### MINERALOGICAL OCCURRENCE AND GEOLOGICAL POSITION OF THE ORES OF IRON.



### SECTION II.

#### STATISTICS OF IRON IN FOREIGN COUNTRIES.



### SECTION III.

#### DISTRIBUTION OF THE ORES OF IRON IN THIS COUNTRY.

## CHAPTER X.

### METALS NOT USED IN THEIR SIMPLE METALLIC FORM.



## CHAPTER XI.

### GENERAL SUMMARY.

# INTRODUCTION.

THE records of mining on the American Continent form by no means the least interesting chapter in its history. The expectation of finding a land possessing greater treasures of gold and silver than existed elsewhere, led to the discovery of the New World, and was the predominating motive in its settlement. It was only to the Atlantic coast of the northern half of the Continent, that colonists came who were inspired by any higher aim than that of finding a recompense for the toils and dangers they had undergone, in the rich mines of the precious metals which they were to discover. These golden dreams were not destined to be disappointed, for a land had indeed been reached whose realities of metallic wealth were capable of satisfying an almost boundless cupidity. Under Spanish dominion, the mines of Mexico and South America poured forth their treasures of silver and gold, and deluged Europe with an amount of these metals far beyond anything ever before dreamed of. The history of Spanish mining in America is too well known to require recapitulation here. Confined almost entirely to the metals styled precious, it conferred neither happiness nor prosperity on either country.

The mines of South America had been worked for nearly a century before the first settlements took place upon the less attractive shores of the Northern Atlantic. Mining formed no part of the object of the colonists of that region, and it was only at rare intervals that some adventurous person attempted it. The first to turn his attention in this direction seems to have been Governor Winthrop, of Connecticut, who, from 1650 to 1660, engaged at intervals in examining the metalliferous indications of the Connecticut Valley, in the vicinity of Haddam and Middletown. There is no reason to suppose that any actual mining was ever executed by him.

About the same time, the Jesuit Fathers were exploring the Great Lakes of the far distant Northwest, and in the relations of their journeys for 1659 and 1660, we find the first mention of the copper of Lake Superior, which, occurring as it does in the native state, and not unfrequently found in loose masses along the shore of the Lake, was well known to the Indians, and could not escape the notice of so observing travellers as were Allouez and Marquette. Long previous to their time, at a period of which no record remains, this region had been the scene of extensive mining operations. The copper-bearing rocks had been explored throughout their whole extent, and even on the almost inaccessible island of Isle Royale, and mining excavations had been made at very many places. There is no reason to suppose that these workings were known to the Indians at the time of the first visits of the Jesuits; on the contrary, there seems to have been no tradition of them remaining, and the appearance of the excavations indicates beyond a doubt that they were made long before that period.

Just at the commencement of the eighteenth century, Le Sueur explored the Mississippi as far up as the St. Peter's River, expressly with the view of making discoveries of the metals. The lead deposits of that region did not escape his notice, but his real discoveries are so mixed up in the records of his journey with pretended ones, that the whole account has an almost fabulous air. Although he returned with a cargo of supposed copper ore, of which he fancied he had discovered an immense mountain, he must afterwards have recovered from his delusion, if such it was, since his voyage led to no farther explorations.

About the same time that Le Sueur was thus employed, a great step was taken in New England by the erection of an iron furnace, which was probably the first one on the American Continent. A business was thus commenced which was destined to increase with the growth of the country, and to be a greater aid to its development than any gold or silver mines could have been.

In 1709, the first chartered mining company in the United States came into existence, and the mining of copper was commenced, at Simsbury in Connecticut, where it appears probable that workings were carried on for several years, and some ore obtained. The success at this point seems to have led to the discovery of the same ores in a similar geological position in New Jersey ten years later, as it appears that the Schuyler Mine was discovered in 1719, and that considerable ore was taken from it and sent to England.

In 1719 and 1720, the attention of the French was directed to the Mississippi Valley, and the assumed existence of the precious metals in that region was made the basis of one of the most extensive and wildest schemes of speculation ever started. Both De Lochon and Renault, accompanied by a numerous corps of miners and mineralogists, explored the country near the confluence of the Mississippi and Missouri for the precious metals; but their operations were attended with no practical results beyond the discovery of the lead deposits, which were worked to a very limited extent for a while, and then, on the bursting of the bubble in France, entirely abandoned.

During the 18th century, a number of mining operations were

undertaken and carried on in different parts of the country, apparently with but little success. Between 1750 and 1760, the New Jersey copper mines attracted considerable attention, and were wrought to some extent. In 1762, the cobalt mine in Chatham, Connecticut, was worked; and about the same time, the Southampton lead mine was opened. None of these enterprises appear to have been conducted with much vigor or success.

During the latter part of this century, the western lead region began to be of importance. While Louisiana was in the possession of the Spanish, some mines were opened and wrought, but in a very rude manner, the ore being taken out from mere pits, and smelted on log-heaps. In 1798, improved methods of mining and smelting were introduced, and the business grew rapidly into importance as soon as the territory was ceded to the United States, which event took place in 1803. In 1774, Dubuque had commenced operations in the Upper Mississippi mines, near the town which now bears his name; but it was not until half a century later that the region was opened to general settlement, and the lead business began to be developed. At about the same time, by a remarkable coincidence, the Spanish lead mines of the Sierra de Gador were opened, and produced immensely, so that the price of that metal became very much depressed.

During the early part of the 19th century, pieces of gold were occasionally found in North Carolina, and from 1824 on, the search for this metal was carried on there to some extent. In 1829 and '30, however, it became quite general throughout the Southern States, and no little excitement was raised on the subject. Thousands engaged in the business of gold-washing, and in 1833 and '34 the amount collected in Virginia, North Carolina, South Carolina, and Georgia, was about a million of dollars a year. It afterwards fell off to the half of that sum for a few years, but when the mines in the solid rock began to be worked, it rose again to more than it had been when the washings were most productive.

Up to this time, the attempts at regular mining had been few and far between. Excavations had been made in the rock, but no extensive, permanent, and productive mine had been opened. The lead region of the West was indeed producing largely, but such was the nature of the occurrence of the ores, that systematic working and enlarged plans were not indispensable to mining them with success. And in other parts of the United States, there had never been encouragement enough, or sufficient mining skill, to direct the investment of any considerable amount of capital into that channel.

But in the mean time, the importance of our coal and iron had begun to be appreciated, and their production was increasing with rapidity. In 1820, the first cargo of anthracite was sent to Philadelphia, and in 1847 our annual consumption had nearly reached 3,000,000 tons; while our production of iron was over 500,000 tons, placing us about on a level with France, and only inferior to Great Britain as an iron-manufacturing country.

The mineral resources of most of the different states had been under investigation at the hands of the various state geologists since 1830, and their researches had been of great aid in defining the boundaries of the different geological formations, and thus limiting and directing explorations. As many of the persons thus officially employed were not practically acquainted with the nature of metalliferous deposits, they naturally fell into some errors in describing and recommending them to notice as worthy of working. Still, it must not be supposed that the most practised mine inspector, let it be one even who has spent his whole life in comparing the phenomena of veins and watching their actual development, can always say from mere surface appearances whether there is a prospect of successful working. With regard to the ores of iron, it is quite true that this may be the case; with a sufficient knowledge of the subject, the development of that business need never be a matter of uncertainty, so far as the occurrence, quality, and quantity of the ores is concerned. But in the case of those metals which are worked in deep mines, and on veins, of whatever class, it is often impossible to say before considerable expenditure has been made, what are the chances of success. This is especially the case in a new and unproved region, where the peculiar phenomena of the veins have not been studied, and where the changes of the metalliferous deposits at various depths which are due to local causes have not been satisfactorily made out.

In 1844, however, a new region was opened to the miner upon Lake Superior, to which the name of mining district may properly be applied, since the veins are numerous, well-defined, concentrated within a limited space, and some of them rich in metallic contents. Thither miners and speculators bent their steps, and after a few years of wild excitement and hap-hazard investments, the business became firmly established, and was in most cases prudently conducted. About the same time that the first dividend was paid on any mine, other than those of coal and iron, within the United States, the discovery of the golden treasures of California gave an impetus not only to searching for this metal, but to mining enterprises of all kinds throughout the whole country. The Atlantic States were searched from one end to the other; many long-abandoned mines were taken up; the gold region of the Southern States suddenly became the scene of an unheard-of excitement, and the whole country seemed to swarm with the promoters of mining enterprises. This general movement towards the development of our mineral resources was undoubtedly based in part on a sincere belief in the richness of our mines, and the possibility of new discoveries, which should furnish an opportunity for profitable investment; but it must be acknowledged that a large part of the excitement which sprang up in 1852 and '53, on the subject of mines, was the mere blowing up of a prodigious bubble, a repetition of the old Lake Superior speculations of 1845 and '46, on a new and wider field, and on a greatly enlarged scale.

The facility with which the public allows itself to be deceived, in regard to everything connected with mining, is as remarkable, as the machinery by which the swindling speculation is organized and brought into successful operation is simple. The locality is selected, and visited by some very distinguished scientific geologist, who for a sufficient consideration will write a sufficiently flattering report, and demonstrate the absolute certainty of success. The value of the mine is fixed at an enormous sum, and divided into one or even two hundred thousand shares; the company is organized, and the stock brought into the market. Every means possible is then taken to inflate its value; fictitious sales of ore are announced; the most flattering reports are received from the mine, and published in all the newspapers; the President of the company, who, perhaps, had never seen a mine before in his life, and who may therefore be excused for mistaking iron for copper pyrites, or perhaps even for gold, visits the scene of action, and finds the surface literally "covered with stacks of ore;" a series of dividends is announced as about to be paid, or perhaps, even, the ore or metal from a neighboring mine is purchased with a part of the capital paid in, and sold, and a dividend declared "from the proceeds of the mine;" the whole machinery of fictitious sales of stock is put in motion, the stock rises, and the promoters of the enterprise benevolently allow the public to step in and share with them in the magnificent profits which are certain to accrue. As soon as a sufficient quantity of the stock has been thus disposed of, and the getters-up of the scheme have pocketed the proceeds of their skilful manœuvring, the natural results follow: the stock, no longer artificially kept up, begins to droop; one after another the deceptions which have been practised become suspected; the unfortunate holders rush to dispose of their shares, but it is too late. The property which a few days before was quoted at hundreds of thousands can now hardly be given away; the unfortunate victims having nothing left as the tangible evidence of the brilliant dividends promised but the elegantly engraved stock certificates, and the equally valuable reports by which they were deluded.

And yet the mine, thus made the object of speculation, and perhaps abandoned in disgust, may be really of value, and capable of being worked so as to pay a moderate profit on the capital actually and judiciously invested in its development. But the idea was given out in the beginning of the enterprise that it could be made profitable at once, and because this has not been the case, the holders of the stock lose all confidence, and refuse to furnish the capital, without which hardly any mine, however rich it may be, can be put into a condition in which it can for any length of time be worked with profit. The system which prevails in this country of chartered companies with a large number of shares, seems especially adapted to make the mining business, which contains so much of the lottery element of uncertainty in it, a mere object of stock speculations.

The records of the last few years show almost without exception that companies with large fictitious capital, and an enormous number of shares, have been got up for the purpose of swindling the public, and not for *bona fide* mining purposes. It may be laid down as a universal rule, that the stockholders in a mining enterprise should be kept fully informed in regard to the expenditures and operations of the company. A frank and full publication is the only guarantee of sincerity and good faith. When these things are more generally understood, and the public refuses any longer to be victimized, we may expect to see a less noisy but far more effective development of our mineral resources than we have yet had.

It was with the hope of aiding, in some degree, in this development that this work was commenced, and is, not without some hesitation, given to the public. It seemed as if the time had come, when the history of our mining operations should be taken up, and our capabilities of production in this department be made the subject of a more comprehensive and general investigation than they had yet received. Fully sensible of the importance of the task which I was about to undertake, I have neglected no means to prepare myself to execute it with as much thoroughness as was consistent with the publication of the results within a reasonable time. All of our important mining regions on this side of the Rocky Mountains, and most of the prominent mines, have been visited by me, some of them repeatedly, within the last few years. Having also examined many of the most interesting European mines, I was enabled to make use of the materials derived from my own observations, and those published by others, to compare our own metalliferous deposits with those of Great Britain and the Continent of Europe. This seemed to me to constitute an important feature of a work of the kind projected, since it is chiefly by comparison of new mining districts with those which have been long worked, that light can be thrown on the former.

With these views, the arrangement of the work which has been adopted, is the one deemed best calculated to carry out the plan proposed by its title, namely, a description of our metallic wealth, or a detailed account of the resources of this country in the metals and their ores, and of the present state of the development of our mining interests, as compared with those of other countries. The first chapter is devoted to an explanation of the laws which characterize the deposits of the metals and their ores, and a brief description of the general methods followed in mining operations. This seemed necessary, in order that the technical terms used in the course of the work might not be unintelligible to the general reader, and especially that some clear ideas might be impressed, at the outset, as to the various modes of occurrence of the useful ores, and the importance in all mining operations of carefully recognizing these distinctions in their practical bearing.

In the succeeding chapters each metal is taken up, and treated, usually in separate sections, under the following heads: its mineralogical and geological occurrence; the distribution of its ores in foreign countries; the distribution of its ores in the United States. Under the first head it is not attempted to give a full mineralogical description of all the existing ores; for this information recourse may be had to J. D. Dana's standard treatise on Mineralogy. The principal ores have been noticed, and such information of economical importance given as seemed required in order to throw light on the succeeding sections. Under the head of the geological occurrence, the formations in which the ores of the metal treated of are chiefly found are enumerated, and the most important laws of the veins in their connection with the rocks in which they occur are stated.

In the second section of the chapter, a concise description

is given of the principal foreign districts where the metal under consideration is obtained, and occasionally a more minute account of mines which have an unusual importance from the amount of their produce, or the peculiar circumstances under which they are worked.

Finally, the mines within the limits of our own country are taken up, and described as completely as the materials which could be collected by personal examinations and from official or reliable published accounts would allow, or as was deemed essential to the plan of the work. Having no personal interest in any mining enterprise, beyond that of a sincere wisher that this branch of our national industry may be most speedily and effectually developed with the smallest waste of money, I have not hesitated to give my opinions in regard to the value of some of our metalliferous deposits, in the hope that when favorable they might lead to renewed vigor in their working, and that when unfavorable they might be of some avail in preventing useless expenditures. Whether my judgments will prove in most cases founded in truth (that they should be so in all could not be expected), time only can show. Of the imperfections and omissions inseparable from the first execution of a work somewhat new in its plan and object, no one can be more sensible than the author.

The statistical portion of the work has been carefully compiled from the best accessible data. All the foreign weights have been reduced to the English standard; for this purpose Mr. J. H. Alexander's excellent work has usually been followed as authority. The ton of 2240 lbs. has been used as the unit in the case of all the metals, except gold, silver, and mercury. This is the officially recognized weight of the ton in this country and in England, and it seems hardly worth while to change it, unless the whole absurd system of our weights and measures can be abolished, and one simple and convenient, as that of the French nation, adopted. In the tables of English sales of ores, the ton of 21 cwts. is used, in conformity with the usages of that country.

The sections of the mines are drawn upon the same scale, of 200 feet to the inch, throughout the work, so that a comparison of the amount of ground opened at any two different localities may be easily made. The shafts and levels are shaded with heavy oblique lines, and the ground stoped is indicated by the part in black.

THE

# METALLIC WEALTH

OF

# THE UNITED STATES.

---

## CHAPTER I.

### ON THE NATURE OF THE DEPOSITS OF THE METALS AND THEIR ORES, AND THE GENERAL PRINCIPLES ON WHICH MINING IS CONDUCTED.

Before entering upon a description of the mode of occurrence and geological position of the various metals and their ores, taken singly, it will be well to establish the meaning of some of the more important terms used in treating of subjects of this kind, and to lay down some general principles which the results of mining-experience have shown to be applicable to mineral and metallic deposits. The importance of this preliminary step will be acknowledged, when it is considered that these principles are derived from the combination and classification of facts observed in mining operations all over the world, and that they form the basis on which future enterprises of this kind should be organized. The results which have been obtained in this branch of applied science are drawn from a great variety of sources; and when we reflect on the expense and well-known uncertainty attendant on mining, it will be evident that, amid the multitude of isolated facts, to which every day is adding, some general principles should be recognized which may be relied on as a thread to guide one through the labyrinth of uncertainty.

Great as is the variety of forms under which the metalliferous deposits of different regions appear, and difficult as it might seem, at first sight, to discern any fixed laws in their development, yet a general consideration of their structure will justify the following classification.

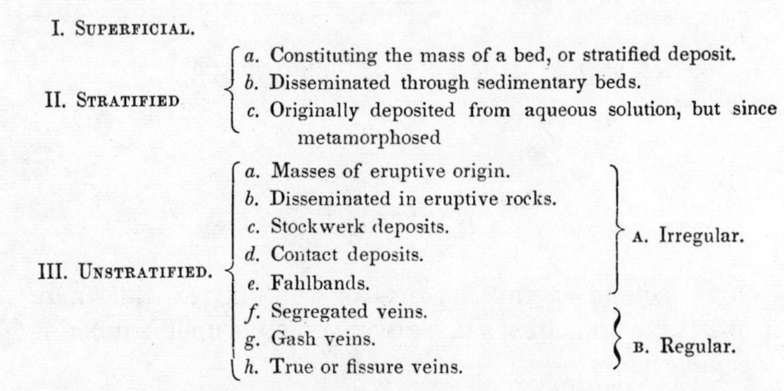

| | | |
|---|---|---|
| I. SUPERFICIAL. | | |
| II. STRATIFIED | *a.* Constituting the mass of a bed, or stratified deposit. | |
| | *b.* Disseminated through sedimentary beds. | |
| | *c.* Originally deposited from aqueous solution, but since metamorphosed | |
| III. UNSTRATIFIED. | *a.* Masses of eruptive origin. | A. Irregular. |
| | *b.* Disseminated in eruptive rocks. | |
| | *c.* Stockwerk deposits. | |
| | *d.* Contact deposits. | |
| | *e.* Fahlbands. | |
| | *f.* Segregated veins. | B. Regular. |
| | *g.* Gash veins. | |
| | *h.* True or fissure veins. | |

I. SUPERFICIAL DEPOSITS.—These comprise all deposits of the metals and their ores, which lie loose upon the surface, or intermingled with the superficial formations, the drift, and the alluvial. They are made up of particles and rounded fragments which have originally formed a part of some bed or vein, and from which they have been detached by various causes similar to those which have produced the great masses of clay, sand, and gravel, which almost everywhere cover the rocks in place. Such metalliferous deposits are usually called "washings," or "stream-works." Most of the gold, and all the platina of commerce, is obtained in this way, as also some of the tin. Their discovery and exploitation is hardly to be classed under the head of mining operations, although they furnish large amounts of the most valuable metals. Lying so near the surface, they can be reached by excavations of the most simple character. The dissemination of the metallic substances through the loose materials in which they are embedded is the result of mechanical causes, especially of currents of water, which have torn them from their original bed and scattered them over the surface.

II. STRATIFIED MINERAL DEPOSITS.—This class of deposits comprehends those masses of ore which are included within rocks of sedimentary origin, and which are in every way identical in their epoch and mode of formation with the strata between which they are accumulated. This is not, however, the mode of occurrence of the more valuable metals. Of all the metalliferous ores, those of iron and manganese only occur in stratified deposits on an extensive scale. The beds of argillaceous ironstone, intercalated among the shales of the coal series, furnish the most striking illustration of this mode of occurrence, as will be fully set forth under the head of the Geological Position and Mode of Occurrence of the Ores of Iron.

There are also strata of slaty and other rocks which are impregnated with ore, or contain it disseminated through them in small particles, so thoroughly intermingled with the rock as to be hardly distinguishable from it. When such deposits occur on an extensive scale, and when the ore forms a persistent portion of the stratum, they become an important source of mineral wealth; although the percentage of ore is usually very small, yet from their great extent, they may be capable of being worked with profit. Such are the cupriferous schists and sandstones of the Mansfeld district in Prussia, and the similar deposits of the government of Perm in Russia. Further on, under the head of Copper, a description of these interesting mining regions will be given.

There is a class of metalliferous deposits which belong to rocks originally stratified, and of aqueous origin, but which have, since their deposition, been metamorphosed so as to have assumed in a great degree the appearance of igneous masses, and which themselves can with difficulty be recognized as possessing distinctly the character of either the stratified or unstratified groups. These are developed in the metamorphic strata of the palæozoic system, but especially in the azoic rocks, as in Sweden and in Northern New York, where the immense deposits of iron ore furnish examples of the class in question. They invariably coincide in their line of greatest linear development with the line of strike of the associated rocks, and are exceedingly irregular in their dimensions, expanding to a great width, and then contracting within very narrow limits. They usually also agree in dip with the planes of the rock enclosing them. Such masses of ore seem to have resulted originally from aqueous deposition, although the materials may in some instances have been derived from the abrasion and destruction of purely igneous masses. It is frequently supposed that these lenticular, or flattened cylinder-shaped deposits have been forced up in a plastic state from beneath, having been driven in between the strata like a wedge, separating them along the line of least resistance. This theory, however, seems hardly tenable, when we consider the great extent of such masses, and the absence of rubbed and polished surfaces along the line of contact of the ore and the adjacent rock. Moreover, such masses are sometimes found thinning out beneath, as if they had occupied a previously-existing depression in the strata, before they had been elevated from their original horizontal position. When such an upheaval of metallic matter has taken place from beneath, we find the result of mechanical displacement and chemical action in the surrounding beds to be most strongly marked. The azoic period having been one of long-continued and violent mechanical action, there is no reason to doubt that many of the masses of which it is made up may have been derived from the ruins of previously-formed rocks of the same age, both sedimentary and igneous. Such abraded material might readily have been swept into the depressions of the surface by strong currents, such as must have existed in those times, and would naturally have a very irregular thickness.

III. UNSTRATIFIED DEPOSITS.—This class comprehends most of the deposits of the ores of the metals, excepting some of those of iron and manganese, as noticed above. Their character is exceedingly varied, and the phenomena which they exhibit frequently of the most complex kind.

In the first place, they may be divided into two great groups, *irregular and regular deposits.*

*Irregular Deposits.*—These include first, igneous eruptive

masses, forming rock aggregations in every respect similar in form and mode of occurrence to the non-metalliferous rocks with which they are associated. This class of deposits consists chiefly of the immense masses of specular and magnetic oxides of iron, so strikingly developed in the azoic system, where they exist in connection with igneous eruptive rocks, and apparently under the same conditions and with the same mode of occurrence. In Europe, the Elba iron ores may be taken as a type of this class of ore deposits; in this country the Iron Mountain of Missouri, and the even more extensive iron-ridges of Lake Superior, furnish striking examples of a metalliferous ore existing in such quantity and under such circumstances as to identify it geologically with the adjacent rock-formations and to cause it to be classed with them, as having originated under the same circumstances. Such bodies of ore usually form ridges parallel with the lines of elevation of the rocky strata in which they occur, or they may take the form of dome-shaped masses which have lifted up the strata on every side, usually filling the fractured envelop with metallic emanations and showing an intense metamorphic action, in the formation of numerous mineral substances not elsewhere found occurring in the formation. As might reasonably be expected, such ore-masses are found developed on the largest scale in the azoic system, having been formed at a period in the geological history of the earth when its crust was thinnest and most easily fractured. If there have been eruptions of metalliferous masses during the later geological periods, they have taken place usually in a region of volcanic fires, or on deeply-seated fissures which have remained open until a late epoch.

As far as the practical application of mining science to this class of deposits is concerned, it is the most simple, for they are developed on so grand a scale that there is no question of exhausting them, and when worked, it is generally by excavations of the nature of the quarry rather than of the mine. In that part of this work which treats of the geological occurrence of the ores of iron, numerous examples will be given of this form of metalliferous deposit, which is almost exclusively confined to that class of metalliferous products.

Closely allied to the last-mentioned class is that of the metals and ores disseminated in the eruptive rocks, when the quantity of the metalliferous substance has not been sufficient to form a special deposit, or when the nature of the circumstances under which it was formed has not permitted its concentration. This is the mode of occurrence of platina and its associated metals, which are disseminated through eruptive trappean or serpentine rocks in fine particles, and rarely masses of a few pounds in weight; but which have never, thus far, been found in regular veins or in any concentrated form of deposit. Serpentine is frequently accompanied by chromic iron, which is disseminated through it in the same way, but much more abundantly. The trappean rocks contain frequently so large a proportion of the magnetic oxide of iron, scattered through the mass, that it may be considered one of the constituent minerals of the rock. Thus, in the mine of Taberg, in Sweden, the trap is so filled with bunches and strings of magnetic iron ore that it can be worked to advantage.

But in general, this class of deposits are of little value, unless nature has performed the operation of separating the metallic contents on a grand scale, by a gradual disintegration and removal of the rocky substance, and the consequent concentration of the much heavier metal or ore. This is the origin of all the platina-washings, which metal was derived from the abrasion of a rock containing it in so small a proportion, that it could never have been separated, by artificial means, with profit.

The tin mines of Europe are, many of them, worked in deposits analogous in their mode of occurrence to those now under consideration. For instance, at Geyer, in Saxony, the oxide of tin is found in small thread-like veins, and disseminated in fine particles through an eruptive dome-shaped mass of granite. It is from an original source resembling this, that the tin of the washings or stream-works of England and the East Indies has been in part derived. Oxide of tin is sometimes disseminated in invisible particles through the granitic rocks, and would probably be frequently found if carefully searched for. I have myself found 0·13 per cent. of this substance in a feldspar from the Riesengebirge, in Silesia, in which not a particle of it was visible to the eye.

Closely allied and passing into the last-mentioned mode of occurrence is that called by the Germans a "Stockwerk," a term which has been adopted into French and English, although not always correctly applied. The stockwerk consists of a series of small veins, interlacing with each other and ramifying through a certain portion of the rock. The ore is not usually entirely confined to the veins or branches, but exists in the adjacent rock in considerable quantity.

Fig. 1.

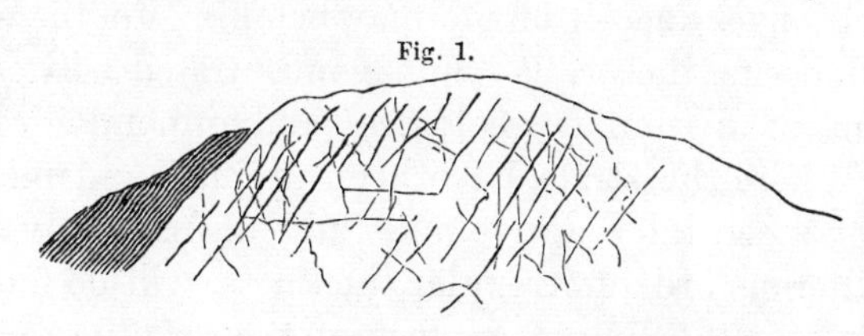

The annexed section (Fig. 1) represents such a network of veins, and the manner in which they may be imagined to traverse the rock.

Such reticulations of mineral matter are frequently called in England, *floors*, and in Germany "Trümerstocke," or "Stockwerke," in allusion to the fact that they are worked in different stages or stories, one above the other, owing to their great extent. Such deposits are usually proportionally poor in metallic contents, in an average of the whole quantity of rock which has to be worked.

The celebrated Carclase tin mine, near St. Austell, is a striking example of the stockwerk form of deposit. It was worked, until recently, in a decomposed feldspathic granite, as an open quarry. The ore occurs in a network of small veins interlacing with each other, the whole of which, together with the rock included between them, had to be removed and the ore afterwards separated by hand-picking, stamping, and washing.

There are other stockwerk mines in Cornwall, most of which have now been worked out and abandoned. The Altenberg stockwerk, in Saxony, furnishes another good example of this mode of occurrence. The rock is a grayish, silicious, feldspathic porphyry. The stanniferous mass is about 1400 feet long, and 900 wide, and like that described above, consists of a great number of small veins, which cross each other in every direction. Those which run nearly east and west, are usually the widest and richest in ore. The enclosing rock is also rich enough to be stamped and washed for a distance of several feet from the network of veins.

In the phenomena of *contact deposits*, *segregated veins*, and *true veins*, there is a gradual passage from one class of forms to the other, so that it is not always easy to say to which of these divisions a deposit of ore may belong. In the class of contact deposits, the ore is found concentrated between two formations of dissimilar geological and mineralogical character. Where the strata have been uplifted and metamorphosed by a central eruptive mass, not unfrequently a band of metalliferous ore, of irregular thickness, will be found extending along the line of contact of the eruptive with the metamorphosed rock. Or, if not always exactly upon the dividing line between the two formations, it may be found perhaps at an inconsiderable distance, and preserving a general parallelism with it. The annexed figure (2), illustrates, by an ideal section, this mode of occurrence.

Fig. 2.

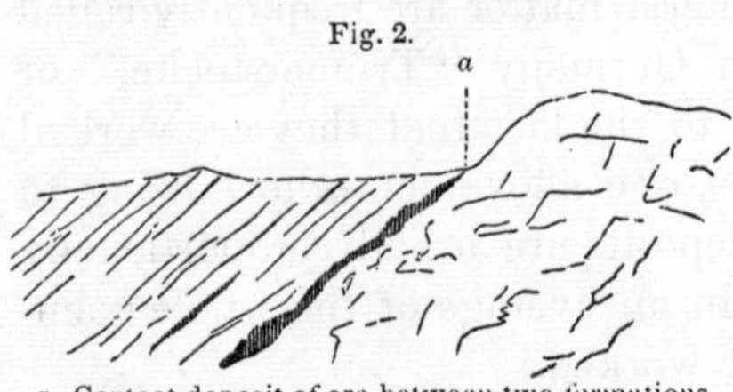


*a.* Contact deposit of ore between two formations.

In the same manner, beds of ore may be found between two successive overflows of an igneous rock, or metallic substances may occur disseminated through that portion of two beds of rock which is adjacent to the plane of separation between them; in such a case, the mineral masses on each side of the ore-deposit may have the same geological position, but will usually be found to differ in mineralogical character. Farther on, interesting examples of this mode of occurrence in the Lake Superior region, will be given. The iron ores of the Harz have a similar position in relation to the eruptive rocks of that district, since they follow the contact-planes of these igneous masses with the uplifted slates. The deposits of ore pass gradually into the trappean rock, showing that there has been a concentration of metallic matter at the junction of two dissimilar formations. The same phenomena exhibit themselves in the Vosges, near Framont, where masses of specular iron surround a central nucleus of quartzose porphyry, which has tilted up the stratified rocks, while the ore has insinuated itself into all the fractures and cavities thus produced, lining them with beautiful crystals and penetrating even into the solid rock itself. At this locality, and others resembling it, the action may be considered to have been principally of an igneous character, the upheaval of the strata opening a passage or chimney for the passage of metallic vapors which have become condensed in the fissures thus formed. In other cases, the phenomena are more satisfactorily explained, by supposing a segregation of the mineral substance along the line of junction of two rocks of dissimilar character, under the influence of electro-chemical agencies. The copper ores of Monte Catini, in Tuscany, belong, according to Burat, to the class of contact deposits, being developed along the line of outcrop of the gabbro, a rock resulting from the metamorphic action of serpentine upon the strata of the cretaceous formation.

Deposits of the character thus indicated are exceedingly irregular in their outline, and are often so combined with each other as to present the features of two or three forms at once, but, in general, in their essential characteristics, they are allied to the rocks in which they occur, in structure and composition. They are often developed at certain points on so large a scale as to be of great importance, but they have not that persistence in depth which is the characteristic of true veins. The ores which they furnish are less crystalline in their texture, and are often quite compact. The gangues are also less distinguishable from the adjoining rocks, and frequently there is no proper veinstone at all. In every mining district, one or the other of these modes of occurrence usually predominates and gives a character to the workings; if the deposits are of the irregular class, they are usually wrought with less skill and capital, although frequently on a very extensive scale; if, on the other hand, they are regular in their character, they require an enlightened system of working, and may be relied on as permanent sources of metallic wealth. Of this description are the mines of Cornwall, Saxony, and the Harz, which have been for hundreds of years the classic regions of mining, and in which the theory and practice of this branch of applied geology have been carried to the highest perfection.

Intermediate between the regular and irregular deposits of the metalliferous ores, are those forms of occurrence designated by the German word "Fahlband." These are more strikingly exhibited in Norway than elsewhere, although the same character has been recognized in other regions on a less extensive scale; and being in some respects peculiar, this mode of occurrence should not be overlooked in a review of the forms of metalliferous deposits.

The "Fahlbands" (German, Fahlband; plural, Fahlbänder), as developed in the district of the Kongsberg silver mines, consist of parallel belts of rock of very considerable length and breadth, which are impregnated with the sulphurets of iron, copper, and zinc, together with a little lead and silver, disseminated through the rock in such fine particles as to be hardly visible, and only to be recognized by their tendency to decompose, and thus to give to the rock in which they are contained a peculiar rotten and disintegrated appearance at the surface; hence the name "Fahlband," or, *rotten belt*, the word *fahl* being a corruption of *faul*, the miner's term for a rotten or decomposed rock.

In general these ore-bearing belts are irregular in their dimensions, although constantly preserving a certain degree of parallelism with each other. They may be traced in the Kongsberg silver-mining district for several miles, and the greatest breadth of any one is about a thousand feet. The quantity of ore contained in them is usually too small to be worth working; but occasionally it is sufficiently concentrated to become the object of mining enterprise. There are seven of these Fahlbands in the vicinity of Kongsberg, and they are parallel in strike and inclination with the gneissoid and schistose strata in which they occur, and have the same local character in relation to disturbances of stratification, schistose structure, and other external forms.

The fahlbands, however, are themselves traversed by fissure-veins bearing argentiferous ores, and the results of extensive mining have shown that these veins are only productive when they intersect the fahlbands, demonstrating that the impregnation of the lode with mineral matter was entirely dependent on the nature of the adjacent rock, and furnishing sufficient evidence that the metalliferous particles in the veins were originally derived from the fahlbands, and probably concentrated there by electro-chemical action.

This mode of occurrence will be recognized as peculiar in its character, being a combination of two distinct forms of metalliferous deposits; but the fahlbands themselves could hardly be considered of much importance, were it not for their enriching action on the lodes which traverse them. The same fact is observed with regard to the enriching of fissure-veins, when they traverse different beds of rock, even in cases where no perceptible metalliferous particles can be observed in any of them, and when the fahlband structure may be supposed to be wholly absent. In such cases, it is not unreasonable to suppose that a chemical examination of those beds, in which a lode shows itself as better filled with ore than elsewhere, might reveal the existence of metalliferous particles the presence of which had been previously unsuspected, because they were too finely disseminated through the rock, or were not of a nature to be easily decomposed, and so failed to give a marked external character to the stratum in which they occurred. The fahlbands may be considered as approaching nearest to the class of segregated veins, which comes next in the classification of metalliferous deposits given above, and to these, in connection with the other modes of occurrence to which the term "vein" is applied, we now turn our attention.

*Regular Deposits.—Segregated Veins, Gash-veins, Fissure-veins.*—The line of demarcation between the three forms of veins indicated above cannot always be easily drawn, although in most cases the difference is very apparent, and in all is of great importance in judging of the value of a metalliferous deposit. But in some instances, there is so gradual a passage from one form to the other, that surface examinations are not sufficient to enable one to decide the question of the actual existence of a fissure or true vein, and it is only by the indication obtained at some depth below the surface, that the vein can be placed in its proper class.

By a vein, as a geological and mining term, in general, is understood *an aggregation of mineral matter of indefinite length and breadth and comparatively small thickness, differing in character from, and posterior in formation to, the rocks which enclose it.*

Werner, the great Saxon geologist, defined veins as "mineral repositories of a flat or tabular shape, which traverse the strata without regard to stratification, having the appearance of rents or fissures formed in the rocks, and afterwards filled up with mineral matter differing more or less from the rocks themselves." This definition would exclude many veins which do not traverse, but run parallel with the strata, and others which occur in unstratified rocks. The definition given above, of course, includes veins of mineral matter not metalliferous, which are of frequent occurrence; but, of course, of no importance in the present connection.

Weissenbach, in a paper published in Cotta's "Contributions to the Knowledge of Mineral Veins," which is especially devoted to the subject of the metalliferous lodes of Saxony, has given a general classification of all the modes of occurrence which can be brought under the head of veins. He divides them as follows:—

1. *Veins of Sedimentary Origin.*—Where a fissure has been filled from above by deposition of mineral matter, in a manner similar to that in which the sedimentary rocks themselves have been formed. The origin of these is exceedingly simple, since it is evident that in the process of deposition of any of the stratified masses, as of sandstone or limestone for instance, if this operation were going on in a region where open fissures existed in the subjacent rock, they would be filled by the sedimentary matter, which would assume a stratified appearance in the fissure exactly as above it. The matter thus deposited would not properly be a metalliferous ore, since this class of products have a chemical and not a mechanical origin. This is perfectly evident, and it cannot

fail to strike one with surprise that a theory so inconsistent with facts should have been adopted by Werner to account for the formation of metalliferous veins. The whole class of veins of sedimentary origin is of little importance.

2. *Veins of Attrition.*—Fissures filled with matter introduced by purely mechanical means, such as by fragments of the wall-rock falling from above, or produced by friction of their sides against each other. These phenomena are exhibited in many ore-bearing veins, in the formation of which a twofold action was concerned.

3. *Veins of Infiltration, or Stalactitic Veins.*—These result from the filling of fissures by incrustation of the sides with calcareous matter deposited from aqueous solution, in the manner of the stalactites so common in caverns.

4. *Plutonic Veins.*—Fissures filled with mineral matter identical with that of rock formations, or mountain masses, and supposed to have been introduced by injection, or pressure from beneath upwards, while in a plastic state. Such are the common granite veins in granite or slate.

5. *Segregated Veins.*

6. *Metalliferous Veins, proper.*

These two latter classes are included in the class of regular, unstratified mineral deposits, as previously shown in the table of mineral formations; the latter division including both gash and fissure veins; and they are the only ones which are of importance in the consideration of the metalliferous veins.

*Segregated Veins.*—Under this class of metalliferous deposits are included those vein-like masses which have a crystalline structure, or, at least, a gangue differing from the adjacent mass, but which do not seem to occupy a previously existing fissure in the rock, being so enveloped and limited on all sides within it as to show that the metalliferous and mineral substances of which they are made up could not have been introduced into their present position in any other way than by a gradual elimination of their component particles from the surrounding formation. This process seems to have been of a chemical nature, and one by which materials of similar character were collected together from all directions, or segregated, as it is termed. Of the conditions under which the adjacent rocks must have been when such an elimination of their metallic contents took place, we know little with certainty. We see, however, examples of segregation in masses of lava as they cool from a state of igneous fluidity, when crystals of the different mineral species found in such rocks are found to have crystallized out into distinct individuals, from what was before an apparently homogeneous paste. The same is true of granite and the trappean rocks. In the former, the single crystals sometimes attain a length of several feet. If circumstances cause the segregating crystals to imitate an elongated mass, we have at once the rudimentary form of a veinlike mass, which may continue to develop itself, and acquire considerable dimensions. A tendency to this separation of pure quartzose material will be noticed in almost all of the so-called metamorphic rocks, which frequently show bands of pure quartz, parallel with each other, and lying in the plane of stratification of the enclosing rock; and it should be noticed that quartz is almost universally the veinstone of ores occurring in this form of deposit.

These segregated veins differ from true veins in some important respects. In the first place, they usually lie parallel with the cleavage planes of the formation in which they occur, which is the gneissic and schistose portion of the metamorphic palæozoic rocks. The annexed wood-cut (Fig. 3), represents an ideal section of the form of deposit here under consideration. The ore-bearing mass may or may not appear at the surface, but its extent downwards in one plane is not to be relied upon as in the case of a fissure-vein, since the accumulation is liable, at any point, to be found thinning out in depth, and transferred to another plane as represented in the figure.

Fig. 3.

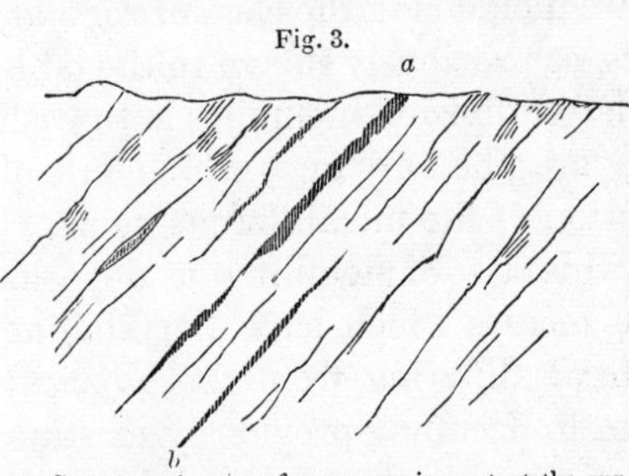


*a*. Segregated mass of ore cropping out at the surface.
*b*. Parallel layer not extending upwards so far.

A striking illustration of this form of deposit may be seen in the Rammelsberg, one of the most celebrated localities of the Harz, a section of which is given in the annexed cut (Fig. 4). The principal mass of ore (*a*, *c*), lies in the direction of the argillaceous slates of which the mountain is made up. At the depth of about four hundred feet, it sends off a branch (*b*), which dips at a considerably less angle than the slates themselves. The greatest thickness of the mass of ore is over one hundred and fifty feet; and its length is about nineteen hundred; but the dimensions decrease in depth, and at eight hundred feet its thickness is about twenty, and its length seven hundred and fifty feet; so that there can be little doubt that the mass will terminate entirely at a certain not very great depth. The ores of this singular mass are chiefly sulphurets of iron, zinc, lead, and copper, intimately blended together, and almost entirely destitute of gangue.

Fig. 4.

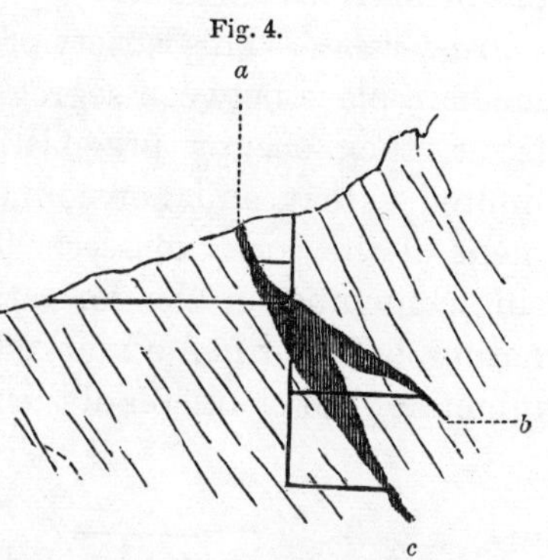


Section of the Rammelsberg.

The auriferous quartz-veins of most gold regions belong to this class of deposits. They consist of belts of quartzose matter, with sulphuret of iron, which near the surface is decomposed into a hydrated oxide, and contain gold disseminated through these substances, and sometimes in the adjoining rock, in fine particles, or, occasionally, large lumps. These belts run with the strata and dip with them, and in other respects exhibit the phenomena of segregated rather than of fissure-veins.

Practically, the most important feature of this class of deposits is that they cannot be depended on in depth as true veins; as they seem almost always to be richest near the surface, and frequently terminate altogether at no very considerable depth. Nor is the ore or metallic matter distributed through them with as much regularity as in the true veins, forming often a series of nests and pockets ranged in a general linear direction, and connected by mere threads of ore or barren veinstone.

*Gash-veins.*—This variety of mineral deposit holds an intermediate place between segregated and true veins. Like the latter, they occupy pre-existing fissures; but these are of limited extent, and not connected with any extensive movement of the rocky masses. They are usually confined to a single member of the formation in which they occur, terminating below, when a marked change in the lithological or mineralogical character of the rock takes place. The annexed cut (Fig. 5) represents, in an ideal section, the mode of occurrence to which the name of gash-veins is applied. The stratum *b*, included between *a* and *c*, cuts off the veins in *a* entirely, the fissures not having extended through that bed at all. Should circumstances favor, and the bed *c* resemble *a*, similar fissures may be found again below *b*, as represented in the figure, but it can by no means be asserted that they are the continuations of the identical fissures which were found in *a;* on the contrary, they are a new set, originating in and comprised in the bed *c*, as those above were in *a*. Lateral branches will usually be found in connection with the main fissures, which may or may not be nearly vertical, according to circumstances; but, whatever their position, the two sets of cracks will be nearly at right-angles with each other, and will possess the same character in regard to their mineral contents, although one set will generally predominate over the other greatly in extent.

Fig. 5.

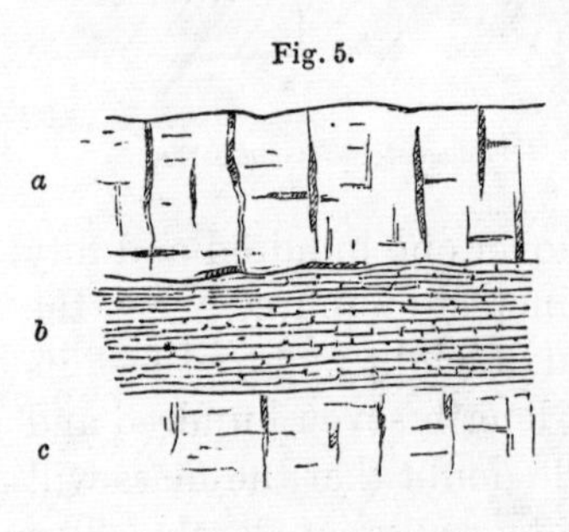


Gash-veins.

The origin of this class of fissures must, in all probability, be referred to the contraction of the rock caused by shrinkage, either while gradually undergoing consolidation, or from the effect of long exposure to a somewhat elevated temperature, after a cessation of which, certain strata might, on cooling, under peculiar circumstances of texture and thickness, be more liable to be fissured than others.

The filling of fissures thus originating with mineral substances may have taken place in various ways. Most of them would naturally be the recipients of the sedimentary matter in process of deposition, if they occurred in strata still accumulating. In a rock impregnated with metalliferous combinations, segregation would more readily take place from the walls of such cavities, and their contents would then have the characters of those of segregated veins. From true fissure-veins they differ especially in not showing as distinctly marked selvages, and in having a less crystalline and comby structure of the veinstone. They are especially to be found in the unmetamorphosed sedimentary rocks, where these have undergone so little change as not to have assumed a thoroughly crystalline texture, and have retained the original lines of stratification, which, in the fully altered rocks, are almost obliterated.

The difference between segregated and gash-veins may be sometimes hardly perceptible, but their origin is sufficiently distinct to justify their separation from each other; and there are mining regions where the peculiarities of one form or the other may be seen with great distinctness. The latter are still less reliable than the former, and are usually soon worked out in depth; but frequently their number makes up for a want of continuous extent in any one of them, so that a region where they abound may furnish, for a time, a large amount of ore.

*True Veins.*—A true vein may be defined as a fissure in the solid crust of the earth, of indefinite length or depth, which has been filled more or less perfectly with mineral substances; or, in other words, an aggregation of mineral matter, accompanied by metalliferous ores, within a crevice or fissure which had its origin in some deep-seated cause, and may be presumed to extend for an indefinite distance downwards.

True veins are almost universally admitted by geologists to have originated in "faults" or dislocations caused by great dynamical agencies connected with extensive movements of the earth's crust, and for this reason they are believed to extend indefinitely downwards; an assumption which is supported by facts, since no well-developed and defined vein has ever been found entirely terminating in depth. Gash-veins, on the other hand, as before remarked, occupying fissures which have resulted from shrinkage of the rock, cannot be expected to extend into strata of different character from that of the bed in which they originated.

Among all the forms in which the metalliferous ores occur, that of true veins is of much the greatest interest, since they are the principal repositories of the ores of the useful metals, and their exploitation is a matter of lasting importance, involving the employment of both skill and capital; hence their principal features require to be described somewhat in detail.

The linear extent of true veins is very various in different instances. Some of the longest known have been traced many miles; but, usually, even if they extend for so considerable a distance, they are not found to be impregnated with ore through the whole of their course. The longer the vein, as a general rule, the more likely it is to be, in some part of its course, rich in ores: thus, some of the great veins of Mexico, which have produced such enormous quantities of ore, have been followed for more than six miles, and have been opened and worked in a great number of places. The width of a vein is not necessarily in relation to its length; some, which are well-defined and traceable for a great distance longitudinally, are quite narrow. From the nature of their origin, their direction and dimensions must be somewhat irregular, since the two portions of the rock on each side of the original crack having been moved in relation to each other, the uneven sides of the fissure will have given rise to cavities of unequal width. That such motion has taken place, is abundantly proved, in numerous instances, by the actual displacement of strata or mineral masses once evidently continuous and now removed to a greater or less distance from each other both in a vertical and a horizontal direction. Such cracks frequently exist in the stratified rocks, and have not been filled up with mineral and metallic substances; in such cases, when there has been a perceptible vertical movement of the two sides in regard to each other, the break of continuity is called a "fault."

Fissures thus formed may have been the receptacles of mineral matter unconnected with any metalliferous ores, as in the case of dykes of trap or veins of granite, which are usually considered to have assumed their present position while in a plastic state. The filling of cracks in such cases was a process of short duration compared with the time required for the formation of veins containing metallic matter. In other instances there may be regular veinstone accumulated between the walls of a fissure, and yet ore be entirely wanting; but it is rare that this is the case, since in almost every vein in which mineral matter has slowly gathered, there have been metallic substances present in some part of its course, during some part of the time of its formation.

In regard to the occurrence of true metalliferous veins, there is no fact more striking than that they are rarely found singly, but rather in groups, often in a complicated network crowded into a comparatively narrow space. The great mining regions occupy but a small part of the earth's surface; while very extensive tracts are almost wholly destitute of metalliferous indications. Let any one take a map of Europe, and color upon it the best-known mining districts, which furnish the larger portion of the metals to commerce, and he will be astonished at the small space which they cover. The groups of veins of the Cornish and Saxon mining districts are so complicated in their number and variety of relations to each other, that centuries of working upon them have not yet fully developed even the more important facts with regard to them.

The larger portion of the vein-fissure is occupied almost invariably by the "gangue" or "veinstone." This is the earthy or non-metallic portion of the lode or vein, consisting of mineral substances, of which a few are of almost universal occurrence as associates of valuable ores. The principal one of these is quartz, which may be said to be almost never absent entirely from any vein. It occurs in a great variety of forms, usually more or less crystalline, sometimes beautifully so, especially in "vugs" or cavities of the vein. Next to quartz, carbonate of lime is most common, in the form of calcareous spar, sometimes compact, and sometimes crystallized, and often passing into brown-spar and dolomite. Fluor-spar and heavy spar are also minerals of frequent occurrence as veinstones. Sometimes these substances form the entire mass of the vein for some distance, either singly or together, and are completely destitute of metallic ores; but this is not usually the case.

The various minerals which make up the body of the lode are frequently arranged in a succession of plates parallel to its walls. Usually, these plates, or *combs*, as they are called, are made up of aggregations of crystalline matter, the separate crystals of which have their axes at right-angles to the wall of the lode, and are developed on the side turned towards its centre. The annexed figure (Fig. 6) represents part of a lode at Wheal Julia, near Binner Downs, in Cornwall.* In this, which furnishes a good example of a comby structure, we see a central predominating mass of crystalline quartz (*e*), on each side of which are seams of copper ore (*f*, *f*). This constitutes the main body of the lode. Between this central comb and the walls are two smaller combs of the same material, on each side, separated, in one instance, as seen at *c*, by a partition of earthy matter. Each comb consists of two corresponding portions, whose crystalline faces meet and interlock towards its centre.

Fig. 6.

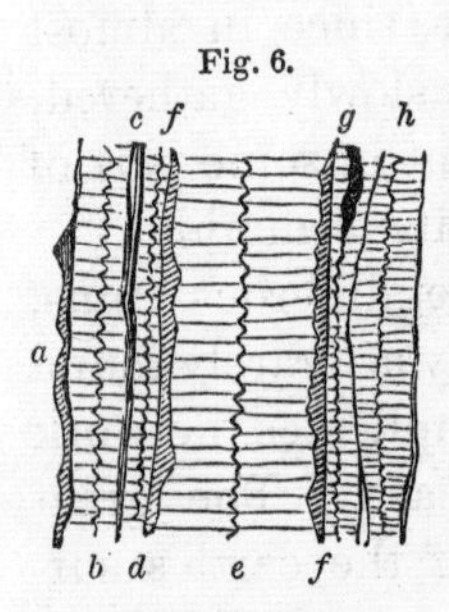


*a*, Bisulphuret of copper and sulphuret of zinc; *b*, comb of quartz; *c*, wall of indurated argillaceous matter; *d*, comb of quartz; *e*, larger comb of quartz, with blende and copper ore (*f*, *f*) on both sides; *g*, cavity, or vug, in another comb of quartz; *h*, more solid comb of quartz.

These appearances indicate a long-continued chemical action, occasionally broken off, and then renewed; the materials held in solution within the space of the vein, or gradually segregating from its walls, varying in character at different times. Sometimes a fissure may have been reopened, after having been once filled up with mineral matter, and

* De La Beche's Geology of Cornwall, p. 340.

may thus have afforded space for another deposition of veinstone. There are cases where this enlarging of the cavity of the vein seems to have been repeated, at intervals, several times in succession, the thickness of each comb indicating the width of the fissure at the time of its deposition, crystallization commencing on the walls as bases, and developing towards the centre, until the whole space was filled up.

In other comby lodes the width of the fissure remained the same, and a succession of deposits took place against its sides, the nature of the material varying at different periods, so that a section of the lode shows a series of various mineral substances arranged in corresponding parallel layers on each side of the centre. A beautiful instance of this mode of filling a vein-fissure may be seen in the annexed figure (Fig. 7),

Fig. 7.

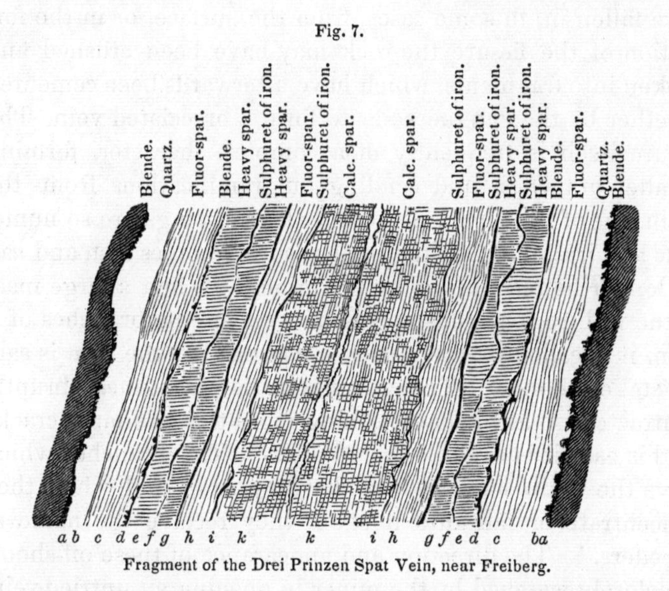


Fragment of the Drei Prinzen Spat Vein, near Freiberg.

from Weissenbach's "Illustrations of Remarkable Vein-phenomena," which represents a fragment of the "Drei Prinzen Spat" Vein, near Freiberg in Saxony. Next to the walls on each side is a crystallized deposit of blende (*a a*); to this succeed layers of quartz, followed by others of fluor-spar, sulphuret of iron and heavy spar, as indicated in the figure, each comb on one side having one exactly corresponding on the other, while the middle portion is occupied by crystallized calc. spar, with a cavity in the centre, the whole showing eleven symmetrical repetitions of six different mineral substances. Such perfect symmetry is, however, not often met with, and frequently the whole mass of the vein seems to have been formed by one uninterrupted process.

Every mineral district seems to have certain veinstones, as well as ores, in a measure peculiar to itself, and the locality of an ore may frequently be recognized by a simple inspection of a fragment of the accompanying gangue.

Besides the veinstones proper of a lode, there are frequently found enclosed in it fragments of the adjacent strata or wall-rock, which have been introduced mechanically. These may have fallen in, in some cases, from the surface, or in the formation of the fissure the rock may have been crushed and broken into fragments, which have afterwards been cemented together by the gangue, so as to form a brecciated vein. The fissure itself is frequently of a complex character, forming parallel branches, and sending out ramifications from the main line of fracture. Sometimes these strings are so numerous and irregular that the main fissure becomes lost, and can no longer be recognized by the miner. When a large mass of the wall-rock is thus enclosed between the branches of a vein, it is called by the miners a "horse," or the vein is said to "take a horse." The vein-fissure is sometimes abruptly contracted, so as to show nothing more than a simple crack; in this case it is said to be "nipped." The branches which leave the main lode are called "droppers," and when they concentrate or fall into it again they receive the name of "feeders." The direction and appearance of these off-shoots are closely watched by the miner in opening an untried vein, and from them he forms an opinion as to what parts of the lode are likely to prove richest in ore.

The mass of the veinstone is usually separated from the wall, by what are called "selvages" (French, *salbandes;* German, *saalbänder*). These are usually thin bands of clayey matter, and are of importance, since they prevent the adherence of the lode to the wall-rock, and of course facilitate its removal. The walls of the vein, themselves, are frequently smoothed and striated, as if there had been motion of the lode on its wall, accompanied by pressure: these polished surfaces are called "slickensides."

The annexed ideal section of a part of a vein (Fig. 8), will

Fig. 8.

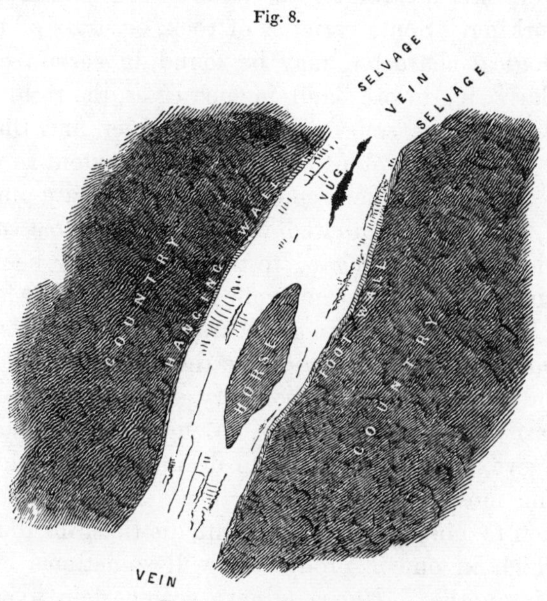


Transverse section of a vein.

serve to convey an idea of the various technical terms used in speaking of a vein, and its connection with the rock. Cornish miners having been first and most extensively employed in mining in this country, their terms have been generally adopted, as they are in England, from the fact that Cornwall is the most extensive mining district in that country.

It has already been noticed, that the metalliferous portion of a lode, or the really valuable ore, makes up, usually, but a small portion of its contents. The barren or worthless mineral substances extracted from the lode, and from the adjacent rock, in the process of removing the vein, are called *deads*, or *attle*. But few veins are sufficiently rich throughout the whole extent of the mine in which they are worked to allow of their being removed entirely. Frequently a large part of the lode is left standing, or unexcavated, while only the rich bunches of ore are taken out and raised to the surface. The occurrence of these rich deposits of ore in a vein is a matter of great uncertainty, and only to be judged of by the miner, after a careful observation of the district in which he is working. Some varieties of rock, or strata of peculiar mineralogical character, may be found, in certain districts, particularly rich in ore; and, in such cases, the right kind of "country" will be sought for by the miner, and the workings on the vein developed to as great an extent as possible in that rock. In other regions, when there are numerous parallel, or similarly situated veins, the *run of the courses of ore* (their situation in regard to the vein) having been determined in one case, the same rule may be found to apply to others. The differences in the character of two rocks, one of which may be highly favorable to the development of a rich lode, and the other just the contrary, are often not of a kind to be easily expressed in words. Long practice, and a well-trained eye, may be required, to decide at once which presents the most favorable appearance. Nor does the same rule seem to hold good in different districts, for the formation, which in one is productive, will sometimes be found barren in another. Hence miners, with certain fixed ideas, derived from the observation of a limited district, are apt to be led entirely astray, when they insist on applying the same rules to another mining region.

Since the crevices occupied by veins have resulted from dislocations of the rocks in which they occur, and since their formation has not been confined to any particular geological epoch, it follows that there may be, in any one formation, a variety of fissures of different ages and directions. This is found to be the case in many important mining regions, and the phenomena resulting from their influence on each other are often of great interest. The direction of a vein may sometimes be altered suddenly by a change in the course of the fissure, without being heaved out of its course, as it is termed, by the intersection of another. Thus, in the annexed cut (Fig. 9), which represents a transverse section of a vein at Holzappel, in Baden, where the vein coincides in dip with the strata, and has been shifted from one plane to another, a distance of twenty-five or thirty feet, without any dislocation having taken place in the rocks. The fissure still connected the two portions of the vein, but was so indistinctly marked that the level was driven through it, as shown in the figure, without its having been perceived; and it was only found again by sinking, when it was struck, having the same characters as before.

Fig. 9.

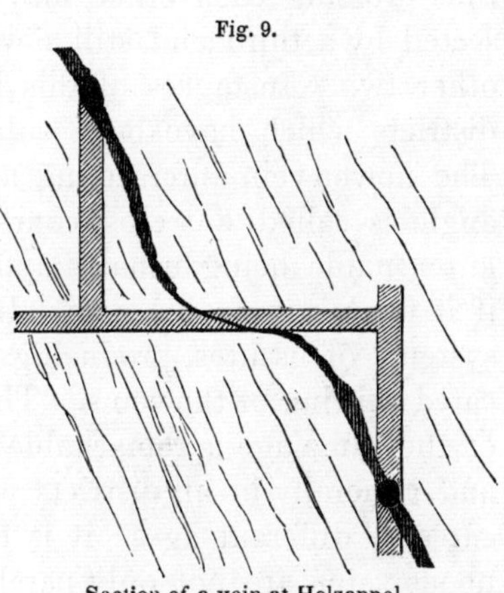

Section of a vein at Holzappel.

In a similar manner, a true fissure-vein may at some part of its course seem to coincide with the dip of the strata and actually send out branches which follow the lines of dip or cleavage; and yet the main fissure may pursue its course across the lines of bedding, although only cutting them at an acute angle, as represented in the section (Fig. 10). In such a case it may not be possible, without sinking upon the lode for some distance, to ascertain its real character as a true vein. The branches which are parallel with the stratification may be regarded as occupying fissures subordinate to the main one, which have been opened along the line of easiest fracture.

Fig. 10.

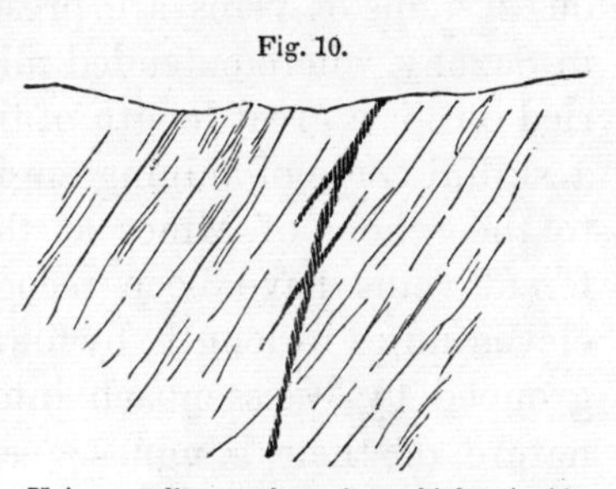

Veins, sending out branches which coincide with the bedding of the rock.

The derangements caused by veins of different ages and directions crossing are of great interest. If a vein traverses an older one it interrupts its continuity, and frequently "heaves" it to one side or the other. Two systems of veins thus crossing each other, may in their turn be both intersected by a third and still newer one, which may heave the other two. Instances of this kind are common in mining districts which have been subject to frequent dislocations. The newer vein intersecting an older one at a considerable angle is called a "cross-course," or a "contra-lode;" the latter name being usually applied to the intersecting vein if it contain valuable ores. In Cornwall, there are several systems of fissures, producing in many instances a complicated shifting of the veins. The general parallelism of veins of the same age is remarkable in many parts of that region, and generally in all districts where there are numerous fissures of different ages: it is also noticed that contemporaneous veins are not only parallel with each other, but that they usually contain the same varieties of ore associated with similar veinstones. In the Harz, there are two principal directions of fracture; the Samson, the most powerful lode of the district, cutting and heaving the Gnade Gottes and Bergmann's Trost, two other important veins. The most interesting and complicated phenomena of the intersection of numerous systems of veins are presented by the Freiberg district, in Saxony, where extended mining operations have been carried on for a great length of time, and have been studied by a skilful corps of mining engineers and professors attached to the School of Mines at that place. More than 900 different veins have been recognized within a space ten or eleven miles in length, by four or five in breadth. They are grouped by Weissenbach into four classes, according to the nature of their gangues; each of which is also in some measure characterized by certain ores, and by a constant direction.

In this country there have been no such complicated phenomena of veins of different ages observed. In the region to which most properly the term mining district may be applied, namely, that of Lake Superior, there are numerous veins, but they do not afford evidence of more than one epoch of formation. In any limited district, as for instance that of Keweenaw Point, the veins are nearly parallel with each other, and are not intersected by cross-courses which heave them from their course. In a few instances there seem to have been slips of the different beds of rock upon each other, which have shifted the veins at the junction for a few feet to one side or the other. In the Southern States the metalliferous deposits usually lie in the direction of the stratification, belonging to the class of segregated veins, and are not crossed by any fissure-veins. It will not therefore be necessary to go into any lengthened description of the particular appearances attendant on such intersections, especially as they differ much in different districts. In general, however, the crossing of two veins is wont to be considered as advantageous to the production of ore at that spot, and rich bunches of mineral are looked for by the miner at their intersection, especially if the angle at which they meet is a small one. Sometimes one lode impoverishes the other for some distance, carrying all the ore with it. As an instance of this appearance presented by two lodes intersecting each other, the annexed cut (Fig. 11) is given, representing one lode crossing another as sketched by Weissenbach.

Fig. 11.

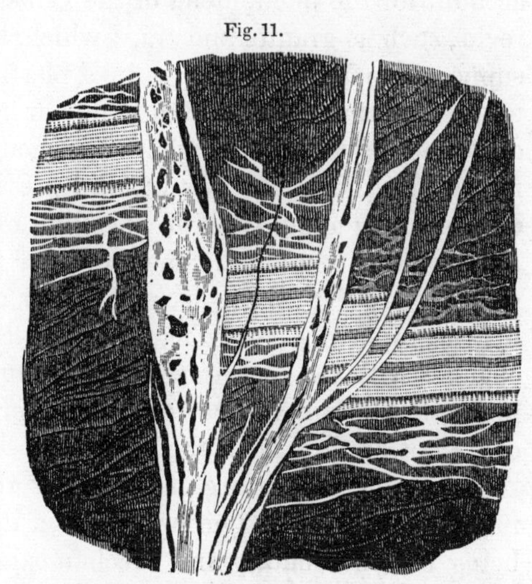

Ground plan of the intersection of veins in the Himmelfahrt mine, near Freiberg.

*Theories of the Formation of Mineral Veins.*—Although a full discussion of the various theories of the formation of metalliferous deposits would occupy much more space than can be given to the subject in the present work, yet it will not be out of place to notice briefly some of the most important theoretical considerations which the study of vein-phenomena suggests. The following are the principal heads under which the theories which have been proposed may be arranged.

1. The veins originated contemporaneously with the rock in which they are contained, and are, so to speak, a mere accidental phenomenon, not governed by any fixed laws of formation. This theory may be dismissed at once, as entirely at variance with all the facts, and as unworthy of consideration.

2. Veins have originated in the filling of fissures, by injection of metallic and mineral matter in a state of igneous fluidity from below. This is the theory usually adopted to account for the phenomena of the veins of so called igneous rocks, such as granite and trap; which, like modern lava, are supposed to have been once in a plastic or semi-fluid state, under the influence of a high temperature, and in such a condition to have invaded the superincumbent rocks, being forced into the crevices by upward pressure. However such a theory may adapt itself to the Plutonic veins, it cannot be considered as explaining the modes of formation of metalliferous lodes. It fails to account for, or rather is contradicted by, the often observed fact, that the character of the lode changes with the rock in which it is found, being rich in ore in one formation, and barren in another adjacent one. This could not be the case if the vein had been forced up through the strata, as the nature of the rocks through which it was raised could have had no influence on its contents, the action being but momentary and mechanical. Besides, if we consider the immense force which must have been required for such an upward motion as this theory supposes, it will be apparent that, had it really taken place, evidence of its existence must have remained in the widening out of fissures in depth, and in the shattered condition of their walls; while there would have been a constant tendency in the more valuable metalliferous substances, being heavier than the veinstone itself, to occupy the lowest position in the vein. Such phenomena, however, have only been observed in isolated cases, while usually the appearance of the walls and the distribution of the mineral matter and ore between them

in true metalliferous veins, is such as to make this hypothesis of their formation entirely untenable.

3. The theory of formation by sublimation, according to which vein-fissures were filled by the volatilization of metallic matter from the great centre of chemical action beneath, namely, the ignited interior of the earth. That such may have been the origin of some metalliferous deposits, and that this agency may have contributed in some degree to the filling of veins, cannot be denied. The fact of the volatility of some metallic combinations is well known, and can be observed at the present day in the products of volcanic ejections. Evidence of the same character is afforded, in some instances, by the position of metalliferous particles on the under side of crystals lining the walls of a lode; as, for instance, at Nagyàg, in Transylvania, where metallic arsenic is seen to have been sublimed and deposited on those faces of crystals of manganese-spar which were turned downwards. Specular iron is found sublimed into the fissures of volcanic craters, and sometimes carried to a considerable distance, and deposited. But these phenomena are of limited extent, and not by any means sufficient to account for the existence of the masses of ore and of earthy minerals filling the body of a large vein. Neither would such a theory account for the variation in the character of lodes in passing from one kind of rock to another, nor for the presence in them of substances not volatile in their nature, nor for any of the complicated phenomena exhibited by veins in their intersections with each other. Hence we must conceive that the agency of sublimation was of very secondary importance in the formation of regular metalliferous veins. In contact deposits, and some other irregular forms of occurrence, where the whole mass of a bed seems to have been impregnated equally throughout by metallic particles, as especially exemplified in some mercury mines, we can conceive of no theory more probable than that of the diffusion of the metallic matter through them by sublimation. Thus originated the extensive beds worked at Almaden, so rich in mercury; and they offer the most striking example which can be given of the class of deposits to which this theory may be applied.

4. The theory proposed by Werner, which may be called that of aqueous deposition, presupposes a chemical solution covering the region in which the veins are found, from which solution, by chemical precipitation from above downwards, the vein-matter was accumulated in the fissures existing in the rocks below. This theory is in direct opposition to that of igneous injection, since, according to its principles, the origin of the contents of veins was a superficial one, their introduction into the fissures from above instead of from below, and the action a chemical instead of a mechanical one. But, in the sense in which this mode of formation was understood by Werner, but little importance can be attached to it. If any such fluid holding metalliferous substances in solution had actually covered the surface, we can conceive of no reason why it should have deposited its contents in the fissures rather than on the surface adjacent; and we ought, in accordance with his ideas, to find every vein connected with a flat sheet of metalliferous ore somewhere along its course, at the place which the solution occupied in the series of formations at the time of the filling of the vein-fissure. Such, however, is not the case, at least in regard to true veins; although there may be a limited class of mineral deposits to which this theory will apply. Besides, if deposition in veins took place in this manner, we should expect to find more or less matter introduced at the same time, mechanically, and showing its origin by its stratified condition. There is nothing of this kind, however, observed in true veins. The deposits all took place in a direction parallel to the walls, and not horizontally, as they would have been, in part, under the circumstances required by this theory. There are many other reasons, equally conclusive, against the ideas of Werner; but it is not necessary to enter into them at length, since his theoretical views with regard to the origin of veins have ceased to have the weight which was once attached to them.

5. The theory of lateral secretion. The views which are at present most generally adopted, assume a somewhat complicated series of phenomena as concurring in the formation of mineral veins. It cannot be doubted that the process has been a complex one, and one requiring a long period of time

for its development. No one simple cause can be considered sufficient to account for all the facts, but the main idea is that of *lateral secretion*, or segregation of the mineral and metalliferous particles from the adjoining rocks in a state of chemical solution, and their deposition upon the sides of a previously-formed fissure under the influence of electro-chemical forces. This is the only theory which will account for the often noticed fact of the change in character of a lode in passing from one geological or mineralogical formation to another of a different character, and it seems to be more in accordance with the other phenomena of veins than any other one yet proposed.

The metallic appearance of the most common ores and their resemblance to certain well-known furnace products, are apt to lead to the supposition that they must be exclusively of igneous origin, especially in view of the fact that the mineralizing substances with which the metals are generally found combined are such as have the effect of rendering them more volatile. Nor can it be denied that some ores and metals do occur in rocks of undoubted igneous character in such a way that they must have had an origin contemporaneous with that of the mass in which they are disseminated. But these deposits do not belong to the class of regular veins, in general, but are to be classed, as already shown, among the irregular modes of occurrence. The metalliferous substances produced artificially, and found filling cracks and cavities in connection with furnaces, and in the chimneys of smelting-works, do not seem to possess that comby structure, or parallel arrangement so characteristic of true veins. If metalliferous particles were introduced into a vein-fissure by sublimation, they would naturally rise and be condensed together in accumulated masses, which might have a crystalline structure, but would be distributed in such a manner as to show their igneous origin.

On the other hand, we know that the earthy substances forming the gangue of metalliferous lodes, and usually far predominating in quantity over the ores themselves, are not the products of igneous action. The minerals which form the mass of the trappean rocks and of volcanic lavas, such as the feldspar family, hornblende, augite, olivine, and the like, are not the usual components of the veinstones, as we should suppose would have been the case, were the veins in which they occur the result of igneous action. On the contrary, we know that the usual gangue minerals, quartz, calcareous spar, and heavy spar, according to the researches of Bischof and other eminent chemists, cannot possibly have been introduced into vein-fissures by injection while in a state of igneous fluidity, or by sublimation. We know also that some of the ores accompanying these veinstones undergo decomposition at a high temperature; and many of the most common, such as the sulphurets of iron and lead, we see in constant process of formation in aqueous solutions, at the present time. The sulphuret of iron is one of the most common products, where ferruginous waters are brought into contact with decaying organic matter; and it is also one of the precipitates from hot springs. It has been shown by Murchison that the copper ores of the Permian strata of Russia must have originated in a similar way, since they are accumulated around and on the surface of the carbonized remains of the stems and branches of plants.

Fissures being opened far down into the heated interior of the earth, they would become filled with water charged with various kinds of metallic and mineral substances. In their passage through rocks of various characters these solutions would be variously acted on, and would, under hydrostatic pressure, penetrate more or less deep into the surrounding strata, according to their texture and other circumstances. If the water was acidiferous, it might thus combine directly with metalliferous particles, previously existing there. There can be no doubt that the contents of a vein-fissure have often acted chemically upon the adjacent rocks, since it is very frequently found that the walls have undergone a great change for a considerable distance from the lode; often the "country" is found decomposed and rotten, its texture being entirely destroyed; or it is impregnated with silicious matter, so as to be much harder and more quartzose than at some distance from the lode. These hardened silicified masses are called by the Cornish miners the *capels* of the lode.

Under the presupposed circumstances, it is evident that a fissure being filled with solutions thus originating, there would be ample reason why there should be deposits of ore within the limits of certain strata, in preference to others. Either those beds might themselves contain more metalliferous substances, which would enter into solution, to be again deposited within the walls of the fissure, or there might be something in their structure which would enable them to act with more efficiency in decomposing the solutions brought from other sources, by circulation of the column of water filling the cavity of the vein.

We cannot doubt that electro-chemical action is, and has been going on in mineral veins, and that the explanation of many of the more obscure facts with regard to the distribution of the ores within them is to be found in this all-pervading agency.

Mr. Robert Were Fox seems to have been the first to be impressed with the idea that electric currents might, "not only contribute to produce the extraordinary aggregation, and position of homogeneous minerals in veins," but that the action was still kept up in them, and he instituted a series of experiments in the Cornish mines, for the purpose of determining this fact. He found the mine-waters to be impregnated with a variety of salts, but in very different proportions in different parts of the same mine. He also found by numerous experiments that the reaction of these solutions on each other, and their contact with extensive surfaces of rocks of different characters, gave rise to electrical currents, the effect of which was to promote decomposition and deposition of various mineral substances, similar to those found occurring in veins. He thus was enabled to account for some of the most curious facts in relation to the position of the ores in the Cornish mines. Professor Reich, in the Freiberg mines, arrived at similar results, and Mr. Robert Hunt also made numerous experiments, determining the existence of galvano-electric currents in the veins of Cornwall, and producing chemical decomposition by their aid. He arrived at the conclusion that these currents were purely local in their nature, originating in the decompositions going on in the vein itself, and not connected with the great magnetic forces which sweep around the earth.

Whether or not it be allowed that these electro-chemical currents performed a very important part in the original formation of veins, it cannot be denied that in the chemical changes which have taken place in the upper portions of metalliferous lodes, and which are evidently still going on, they were and are conspicuous agents. These changes of character are well recognized in all mining districts. They are usually indicated on the surface by a discoloration of the lode by oxide of iron, and a general rotten appearance of the veinstone. The Germans call this the "iron hat" of the vein, the Cornish give to it the name of "gossan." In the cupriferous lodes of Cornwall, the normal ore, below the point to which decomposition has reached, is copper pyrites, or sulphuret of iron, and copper, and other sulphurets of copper. But near the surface, the metalliferous ores, instead of the sulphurets, are oxides and oxidized compounds, such as silicates, carbonates, and phosphates. Often a large portion or all of the copper near the surface has been removed, having been oxidized to sulphate of copper, and dissolved out and washed away, leaving the iron mixed with the quartzose veinstone, in the form of a hydrated oxide, or gossan. Similar appearances are presented by lodes containing lead ores, the sulphuret of lead, or galena, which is almost the only ore found in any considerable quantity at some depth, being, near the surface, converted into a great variety of oxidized combinations, of which the carbonate, sulphate, and phosphate are the most common.

In general, this zone of decomposition does not reach much below one hundred feet, but in some instances it penetrates to the depth of three hundred, and there are, on the other hand, veins in which no such chemical changes have taken place, the original sulphurets existing undecomposed almost at the very surface. The causes which have operated in accelerating and retarding these transformations are not quite clearly understood. Bischof has shown that heated steam has been one of the most powerful agents in the decomposition of the contents of veins, and that it has been most effectually aided by carbonic acid. In order to prove this, he subjected sulphuret of lead, while heated, to the influence of a current of steam, which reduced the galena to metallic lead, the sulphur passing off in the form of sulphydric and sulphurous acid gases. Sulphuret of silver is reduced in the same way, the resulting metal taking that dendritic form which it commonly has in its associations with the ores of lead and silver. The reduced metals thus set free, if they belong to the oxidizable class, are then ready to enter into combination with the various acids everywhere being liberated under electro-chemical agencies. No doubt the effect of atmospheric influence is to be largely taken into account also, since these changes are principally superficial only, compared with the total depth to which the ores are found to extend. The penetration of water charged with oxygen into the fissures and minute pores of the rocks must be in some degree essential to these processes. To give any idea of the complicated changes thus slowly taking place in many lodes, would require far more space than the object and scope of the present work could permit; it is sufficient here to have indicated some of the causes at work in the formation and transformation of mineral veins.*

The variations of the metalliferous portions of lodes at different stages are of the highest importance to the miner, not only in relation to the changes which the ores have undergone near the surface, but as viewed in connection with their persistence in mines worked to great depths. In this country we have hardly more than begun to open our mines, so that we have but little more than superficial indications to judge from; but in the mining districts of Europe, which have been wrought for centuries, there are abundant data accumulated to enable us to draw some important conclusions in regard to the termination of veins in depth. The distinction between true veins and all other forms of occurrence of ores must be carefully studied in reference to this question, or otherwise it will be difficult to obtain a clear

* For further information in regard to this class of subjects, the reader is referred to G. Bischof's "Lehrbuch der Chemischen und Physikalischen Geologie," which has been in course of publication since 1847.

conception of the truth. The facts may be briefly summed up thus:—

1. True fissure-veins are continuous in depth, and their metalliferous contents have not been found to be exhausted, or to have sensibly and permanently decreased, at any depth which has yet been attained by mining.

2. Segregated and gash-veins, and the irregular deposits of ore not included under the head of veins and not occurring in masses as part of the formation, cannot be depended on as persistent, and they generally thin out and disappear at a not inconsiderable depth; at the same time they are often richer for a certain distance, and contain larger accumulations of ore, than true veins, so that they may be worked for a considerable time with greater profit than these, although not to be considered as of the same permanent value. Of these assertions, many illustrations will be given in the course of this work, so that a clear idea may be formed of the distinction which is here intended to be drawn between deposits of a persistent and those of a temporary character.

#### THE GENERAL PRINCIPLES ON WHICH THE CONTENTS OF A VEIN OF MODERATE WIDTH ARE REMOVED.

In order to render the descriptions of mines and mining operations in the succeeding chapters more intelligible to those who have never given their attention to these subjects, it is proposed to add a few pages on the general principles of *exploitation;* by which term is understood the process by which ores and minerals of value are won from their natural position, often at great depths below the surface, and brought where they can be rendered available. Those who are not acquainted with the nature of such operations, and who have never had occasion to reason on the many circumstances which must be taken into consideration in laying them out, frequently suppose that a mine is a mere excavation in the ground without law or rule, and that this, which would seem to them to be the simplest method, would be also the best for such a work. Nothing can be farther from the truth; for to open a mine, if on a vein or regular deposit of ore, and work it as an excavation open to the air, or as an *open-cut,* is quite impossible, if it is necessary to go below the most moderate depth. And such is the case with metalliferous lodes, the superficial portion of which, for some distance down, is often less valuable than that below, and sometimes of no value at all. Moreover, the richest part of a vein is usually of limited extent, and its treasures can only be won by penetrating into the earth to great depths. To raise the ore from such a distance beneath the surface, requires powerful machinery which must be placed on firm ground directly over the vein. The surface-water and rain must be kept from running into the excavations, which could not be done were they open, throughout, to the day; and as the freeing of a mine from water is frequently one of the most expensive operations connected with its working, the greatest care should be taken to get rid of as much as possible in the simplest and least expensive manner. Again, it is necessary that mining should be carried on uninterruptedly by night and day, and in all weathers, since only a limited number of persons can be usefully employed at one time; also that a system of ventilation should be kept up; all of which would be impossible, were the excavations exposed to the weather, or the mine made to consist of a single open cavity.

The first operation in opening a mine is usually to remove the surface-water by driving an *adit-level* or horizontal gallery, which shall commence in an adjacent valley, at the lowest point not liable to inundation and conveniently accessible, and thus drain all that part of the proposed work which lies above. Figures 12 and 13 illustrate the position and object of the adit-level. In extensive mines large sums are frequently expended in thus getting rid of the water; sometimes a number of mines are drained by a simple adit of great length driven at their joint expense, or that of the state, when the work is of great magnitude. So long as the exca-

Fig. 12.

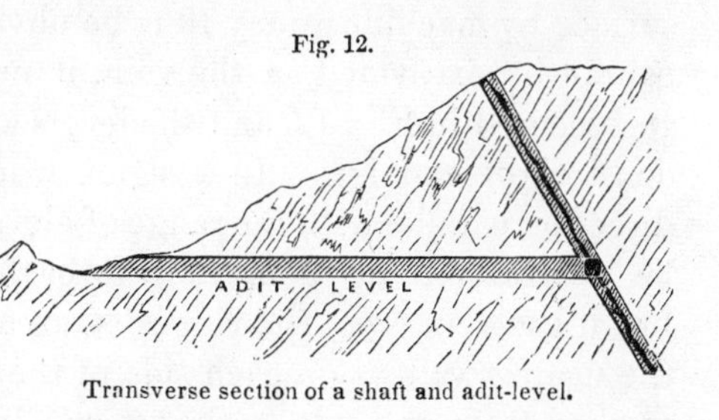


Transverse section of a shaft and adit-level.

vations in the mine are kept above the level of the adit, no machinery for pumping is required; and when extended in depth below it, the water is elevated by the pump or other contrivances employed, only up to that point; thus, in a very extensive mine, every foot through which a large body of water is to be raised, which can be saved, is of importance, and large sums of money may be economically expended in obtaining even a small reduction of the distance.

Frequently only a few feet of "back," as it is called, or of perpendicular height above the adit-level can be obtained; but, in other cases, by an adit planned with judgment, several hundred feet of drainage may be obtained, so that mining may be carried on for a long time without the expense of costly machinery for pumping. If possible, the adit should be driven on the vein, or by the side of it, so that its character and richness may be ascertained by breaking into it occasionally as the work progresses. A large number of the Lake Superior mines are thus favorably situated; and extensive drainage may be had at the same time that the vein is proved in depth.

If sufficient advantages present themselves for the location of machinery for washing and dressing the ores at the mouth of the adit-level, they may be brought out on a tram-road laid in the level, and the necessity of hoisting them to the surface by machinery may thus be obviated. If the adit-level cannot be excavated *on* the vein, it must be driven *to* it, as represented in Fig. 12, and the levels are then to be extended on it right and left. In a region where the veins run in a direction parallel with the range of elevations, it will generally be necessary to drive the adit-level through the "country;" but if the vein cross the ridges at an angle, it will also cross the nearest valleys on each side of the elevation, and may be worked from one side or the other. In the majority of cases, however, the adit-level is only used to remove the surface-water, but few mines being so favorably situated as to have a very large portion of their work above it.

As the necessity of ventilation would not allow the adit-level to be driven to an indefinite distance without some communication with the surface, so as to establish a current of air through it, it becomes necessary to commence excavations on or near the line of the veins, which may intersect the adit, or a cross-cut or level leading from it, and which are carried down to a still greater depth for the purpose of opening and working the mine. These excavations, whether vertical or inclined, are called *shafts*. Their inclination depends on a variety of circumstances, such as the underlay of the vein, or the angle which it makes with the perpendicular, its width, or the nature of the rock in which it occurs. If the vein approach a vertical position the shafts are usually sunk upon it, unless its underlay is so irregular as to spoil the shaft by making it too crooked, in which case it is sunk by the side of the vein, and connected with it by cross-cuts. In Fig. 12, a shaft thus sunk on a vein of generally regular underlay is represented. Whether a shaft shall be sunk on the vein, so as to include the vein within it, or by the side of it, is also dependent on some other circumstances. If the lode is wide and the veinstone such that it can be wrought with as much facility as the adjoining rock, it is best to sink in the vein itself, especially when its dip is nearly vertical, and the country solid, so that but little ground need be left standing in the lode to support the walls, and preserve the shaft. In the mines of Lake Superior this method is generally followed.

If the vein has an underlay as great as 45°, it is a matter of some importance to know whether it is best to work the mine through shafts inclining with the vein, or by vertical shafts connected with it by cross-cuts. The latter method is illustrated by the annexed ideal section (Fig. 13). In such a case a shaft is commenced at some distance from the lode, in the direction from its outcrop towards which it dips, and sunk perpendicularly until it cuts the lode, at a point the depth of which can be calculated if its dip is known; after intersecting the lode it may be continued below it, and cross-cuts must be driven as before at suitable distances, so as to open a sufficient number of points for working. If there are several different lodes within a small space, as represented in the figure, or if it is desirable to make underground explorations for veins not supposed to appear at the surface, the method of sinking vertical shafts is to be recommended. In the English mines the practice of sinking shafts inclining with the lodes has not generally been adopted; although,

Fig. 13.

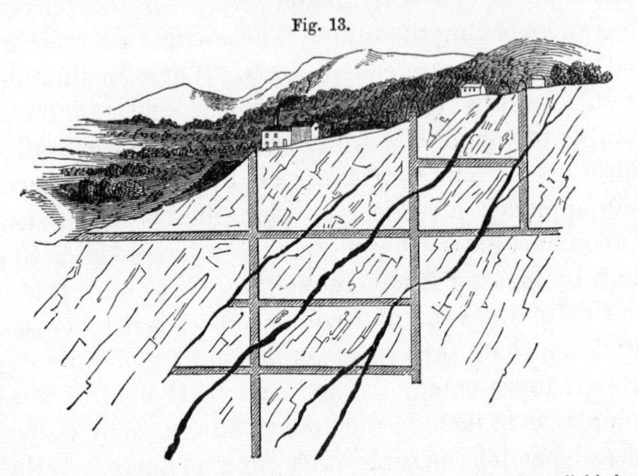

Ideal section of a system of shafts for opening a mine on several parallel lodes.

when it can be done, it presents some great advantages over the other method, a large amount of unprofitable excavation in the rock, or *dead-work* as it is called, in cross-cutting from the shaft to the lode, being thus saved. This method may be recommended when the dip of the lode is tolerably regular and the rock solid, and where there is only one vein to be worked through the same shaft. In order to raise the ore in such a case with advantage, a double tram-road should be laid in the shaft, and the ore hauled up in cars which run from the levels directly on to small platforms, mounted on wheels, on which they are raised to the surface. In the Lake Superior region, where there are numerous veins underlaying from 40° to 50°, the shafts are invariably carried down on the lodes, and no practical difficulty experienced; on the contrary, it has proved the most economical and judicious method under the circumstances. In the Southern gold mines, where the rocks are frequently decomposed and softened, so that for the first fifty to one hundred feet sinking costs but little, the shafts are generally vertical. If the underlay of the lode is very irregular, it is better not to sink upon it, as crooked shafts present many disadvantages in fitting up the machinery for pumping, tram-roads, and other requisites.

The dimensions of a shaft are variable, and dependent on the work to be done by its aid; when it is to be used for winding up the kibbles, or buckets of ore, and at the same time as an "engine-shaft," or the shaft in which the pumping machinery is placed, it should be from twelve to fourteen feet long, and is usually from six to eight feet in width. Shafts which have to be timbered up for the whole, or a portion, of their length, are square or rectangular; those which are built up with stonework are usually round. In this country, wood is universally used for this purpose, timber being usually cheap and abundant in our mining districts. In the Lake Superior region, the shafts only require to be supported for a few feet, or until solid rock is reached. The timbers are squared roughly, and laid upon each other, forming a substantial framework, which rests in a bed-place cut in the rock where it is sufficiently solid to obviate any danger of its giving way.

For a distance of fifty feet, or even one hundred, if desirable, the ore and rubbish may be raised to the surface by the simple windlass, worked by hand, on which a rope is so wound that one bucket or kibble descends while the other ascends. The same means are used in sinking a *winze*, or excavation from one level to another, not extended up to the open day. As soon, however, as the depth of the shaft becomes more considerable, it is necessary to resort to horse or steam-power for raising the ore. The common machine used for this purpose is called a whim; as usually constructed when worked by horse-power, it is represented in the annexed cut (Fig. 14). In the steam-whim, which is generally used when the shaft has a depth of more than two hundred feet, the cage or drum on which the rope or chain is wound is usually placed horizontally, instead of vertically as in the common horse-whim. Frequently, in opening a mine, before the excavations have reached any considerable depth, the pump-rod is attached to the whim, and worked by the same power which raises the ore.

The steam-engine is not usually applied to working a

mine, until there is good reason to suppose that it may become of permanent importance. For hoisting ores, no very great power being required, small horizontal high-pressure

Fig. 14.

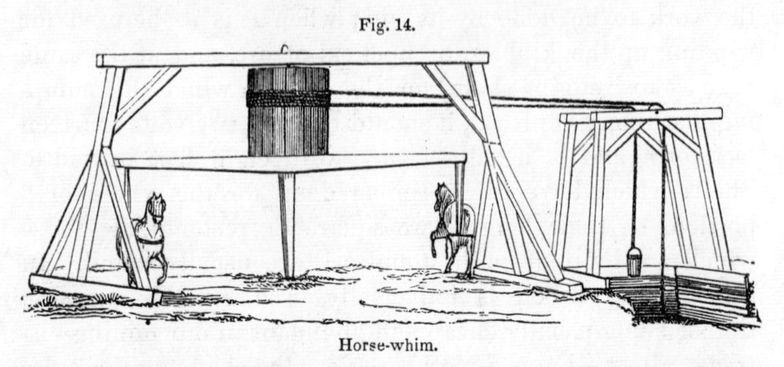

Horse-whim.

engines, working full steam, are usually employed. The machinery must admit of instantaneous reversal, and should be as simple as possible. Water-power, if at hand, can be advantageously applied to the same purpose. The pumping-engines in extensive mines are usually low-pressure ones, and every effort has been made in Cornwall and other mining regions to reduce the consumption of fuel to the smallest possible amount, by perfecting the form of the boiler and machinery.

Most mines, when worked to any considerable extent, have several shafts; two, at least, are almost necessary for ventilation. The distances from each other at which they are located depend on a variety of circumstances, such as the nature of the surface, the distribution of the ore in the vein, and the extent of the sett or length of the vein which can be worked. Two shafts should not be too near each other, if it can be avoided, as in that case the blocks of ground would be too short, causing a wasteful expenditure in opening the ground. If far apart, winzes must be sunk between them, in order to afford ventilation and the means of attacking the vein at a sufficient number of points. In general, three hundred feet distance between two shafts, in a vein of moderate width and richness, may be considered reasonable.

After the shaft has been sunk to a proper depth, if on the vein, drifts or levels are commenced from it in each direction, so as to open blocks of ground, as they are called, or portions of the vein, which are thus rendered accessible so that their contents can be removed. The Cornish miners usually arrange the drifts so as to leave between them a back of sixty feet or ten fathoms; that is, one level having been started, the shaft is continued sixty-six feet (six feet being allowed for the height of the drift), and then another level is commenced, leaving a portion of the vein between them sixty feet in height. Like the shafts, the levels may be driven in the vein itself, or by the side of it, according to its width, texture, and other circumstances.

The contents of a block of ground thus prepared are removed by what is called *stoping*, or working in steps; the object of which is to take out all that portion of the vein

Fig. 15.

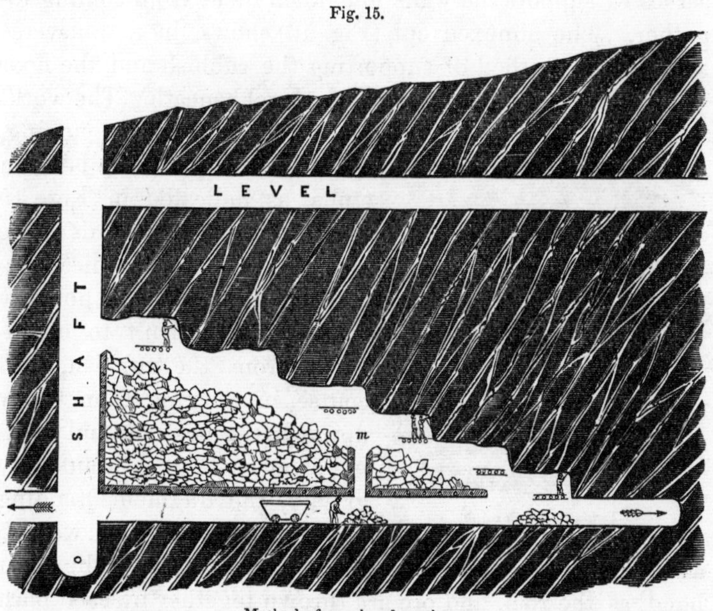


Method of overhand stoping.

which is worth reserving for its ore. There are two methods of doing this, one of which is called overhand and the other underhand stoping, the difference being, that by the former method the vein is taken down by working from below upwards, in the other by excavations from above downwards.

Overhand stoping is almost universally practised in this country. The annexed section (Fig. 15) will illustrate this method. It represents a supposed longitudinal section, on the vein, of a shaft and two levels, between which stoping is going on. The rubbish which is necessarily blasted down with the valuable part of the vein, is piled up back of the miner, as his work proceeds, on a scaffolding of stout timbers, called *stulls.* At suitable distances, openings are left through the rubbish (one of which is indicated in the figure by the letter *m*) called *mills* or *passes*, through which the ore is shot down to the level below, to be loaded on to a wheelbarrow or car, and conveyed to the mouth of the drift, or to the shaft, to be hoisted to the surface. The rubbish accumulated in the space from which the vein has been removed serves to support the walls and keep them from coming together. The annexed cut (Fig. 16) shows, in a transverse section, the method of supporting the rubbish and the floor of the level beneath. The workman, while engaged at stoping, stands on a platform laid on stulls, as indicated in Fig. 15 above. Stoping upwards is the most rapid and economical method, when the ores are not very valuable, or difficult to distinguish from the rubbish. Of course, in working from below upwards, the looseness and working away of the rock is aided by gravity, the fragments tending downwards by their own weight, and it is therefore easier for the miner; but, on the other hand, as the rock and ore are thrown together by each blast upon the pile of rubbish below, it requires care, on the part of the miner, to select out the valuable portion of the lode without sending up too much of the worthless rock, which, apart from the cost of hoisting it to the surface, is needed to fill up the space left by the previous excavations, in order to help to support the walls.

Fig. 16.

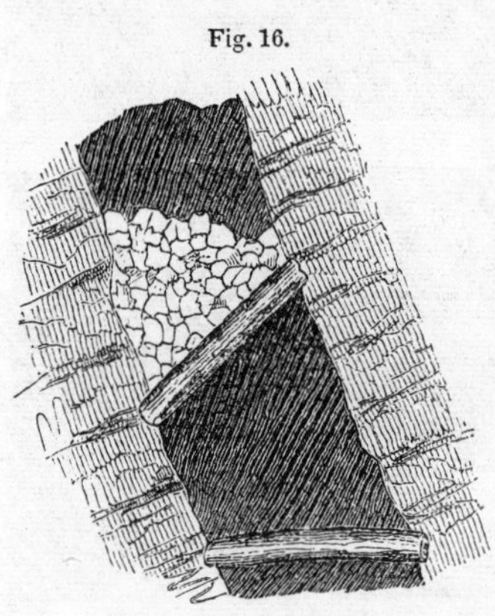

Timbering in a mine.

The proper timbering of mines is a matter of great importance in their management, and is often an item of considerable expense. The amount of wood consumed in the necessary operations about a mine is very large, and care should always be taken to secure a good supply of this indispensable material. In some mining regions, where it is scarce, its cost becomes one of the most considerable items of expense to be incurred, as for instance in the Australian copper mines. Cornwall is mostly supplied from Norway; and the timber cut from hundreds of square miles of Norwegian forests is now buried deep beneath the surface, in the Cornish mines. In Germany, large tracts of forest are set apart for the use of the great mining districts; and the wood is cut from a certain portion every year, so that a constant supply is kept up.

Fig. 17.

Timbering in a level in a yielding rock.

Fig. 18.

Timbering in a level with solid rock on one side.

The above figures (Figs. 17 and 18) show two modes of timbering in levels, and will serve to give an idea of the manner in which such work is executed. In Fig. 17, the rock being soft and liable to inward crushing, the side-timbers require to be kept apart by pieces laid across above and below. The other figure represents a drift of which one side is of firm rock, while the other requires a lining of plank behind the upright timbers.

*Exploitation of Thick Veins and Masses.*—When the deposit of ore to be mined is in masses or veins of extraordinary width, the process of removing their contents becomes a more difficult one, especially if the surrounding rocks are not firm and unyielding. The general method, however, is the same: the work must be subterranean and not open to the day. One or more shafts must be sunk into or near the mass to be removed, and the work commenced at as low a point as possible. Levels are then driven into the ore, and as fast as the valuable mineral is removed, the rubbish which is made is piled up in its place, so as to support the portion left overhead. If there is not enough material for this, stones must be sent down from the surface. In this way the whole contents of a mass may be removed without loss. The methods pursued in such workings must be specially adapted to the shape of the deposit of ore, the degree of solidity of it and of the adjoining rock, and a variety of other circumstances; the general principles, however, are those above indicated.

# CHAPTER II.

## GOLD, PLATINA, AND SILVER (IN PART).

### SECTION I.

#### MINERALOGICAL OCCURRENCE AND GEOLOGICAL POSITION OF GOLD.

MINERALOGICAL OCCURRENCE.—Gold occurs, in nature, in the following forms:—

*Alloys.*

*Native Gold.*—An alloy of gold and silver, with traces of iron, copper, and other metals.

*Porpezite.*—Gold and palladium. (*Ouro poudre.*)

*Rhodium Gold.*—Gold and rhodium.

*Amalgam.*

*Gold Amalgam.*—Gold and mercury, with a little silver.

*Ores.*

*Graphic Tellurium.*—Telluride of silver and gold.

*Aurotellurite.*—Telluride and antimonide of gold, silver, and lead.

Of the above-named substances, the first, native gold, is that form of combination in which almost the whole amount of this metal obtained in the world is found. Of the others, graphic tellurium is the only one which exists in sufficient quantity to be of economical importance, and that only in one district; the others are exceedingly rare substances.

Native gold is invariably found alloyed with silver; no analysis has ever been made of it which has not given a small amount of this metal: besides the silver, there are traces of other metals, among which iron and copper are rarely wanting, although generally present only in minute quantities. The following table will exhibit some of the analyses of native gold from different localities.

| Locality. | Analyst. | Gold. | Silver. | Copper. | Iron. | Sp. Gr. |
|---|---|---|---|---|---|---|
| RUSSIAN EMPIRE. | | | | | | |
| Schabrowskoi washings, | G. Rose, | 98·96 | ·16 | ·35 | ·05 | 19·099 |
| Boruschkoi washings, | G. Rose, | 94·41 | 5·23 | ·36 with iron and loss | | 18·440 |
| Beresowsk mine, in brown hematite, | G. Rose, | 93·78 | 5·94 | ·08 | ·04 | |
| Béresowsk, in quartz, . | G. Rose, | 91·88 | 8·03 | ·09 | trace | |
| Zarewo-Nicolajewsk, near Miask, washings, | G. Rose, | 89·35 | 10·65 | trace | trace | 17·484 |
| Alexander-Andrejewsk, near Miask, washings, | G. Rose, | 87·40 | 12·07 | ·09 | trace | 17·402 |
| Petropawlowsk washings, | G. Rose, | 86·81 | 13·19 | trace | trace | 16·869 |
| Boruschkoi washings, . | G. Rose, | 83·85 | 16·15 | trace | trace | 17·061 |
| TRANSYLVANIA. | | | | | | |
| Vöröspatak, in porphyry with quartz, | G. Rose, | 60·49 | 38·74 | ·77 with iron | | |
| Sta. Barbara mine, at Füses, scales in porphyry with quartz, | G. Rose, | 84·89 | 14·68 | ·04 | ·13 | |
| AFRICA. | | | | | | |
| Senegal, . . . . . . | D'Arcet. | 86·97 | 10·53 | | | |
| AUSTRALIA. | | | | | | |
| Bathurst, . . . . . . | J. H. Henry, | 95·69 | 3·92 | | ·16 | |
| Locality not given, . . | A. B. Northcote, | 99·283 | ·437 | ·069 | ·203* | |
| South Australia, . . . | A. S. Thomas, | 87·78 | 6·07 | | 6·15 | |
| SOUTH AMERICA. | | | | | | |
| Antioquia, New Granada, washings, | Boussingault, | 64·93 | 35·07 | | | 14·149 |
| Brazil, . . . . . . | D'Arcet, | 94·00 | 5·85 | | | |
| Marmato, . . . . . | Boussingault, | 73·45 | 26·48 | | | 12·666 |
| CANADA AND UNITED STATES. | | | | | | |
| Rivière du Loup, Canada, | T. S. Hunt, | 86·40 | 13·60 | trace | trace | 15·761 |
| Georgia, . . . . . . | W. W. Mather, | 95·579 | 4·421 | trace | trace | |
| North Carolina, Lewis mine, | F. A. Genth, | 65·03 | 34·18 | | | 14·90 |
| American Fork, scales, . | Rivot, | 90·90 | 8·70 | | 0·20 | 15·70 |
| Feather River, scales, . | Rivot, | 89·10 | 10·50 | | 0·20 | 17·55 |
| Locality not given, . . | Henry, | 90·01 | 9·01 | ·86 | | 15·96 |

* Bismuth, ·008.

It was clearly shown by G. Rose that gold and silver are not combined in atomic proportions in native gold, as had been suggested by Boussingault; but that the two metals occur alloyed, so that the silver forms from one-half to less than one-hundredth of the mass. The gold of California is remarkably constant in its composition, yielding about ninety per cent. of the pure metal. The usual range of fineness is from 875 to 905 thousandths; the average is 885 to 890.* That of Australia is very pure, as shown by the analyses above.

Gold and platina form an exception in their mode of occurrence to all the other metals in common use. Silver, tin, copper, lead, zinc, and iron are obtained almost exclusively in the form of *ores*, that is, in combination with a *mineralizer*, of which the most common one is sulphur; while gold is found, all over the world, in the native state, alloyed with silver; its combination with tellurium, an exceedingly rare substance, being confined to one or two localities. Silver is not unfrequently found in the native state, though much the larger portion of the produce of this metal is obtained from the sulphuret. Native copper has never been very rare, but had furnished no noticeable part of the copper of commerce, prior to the discovery of the Lake Superior mines. Lead, zinc, tin, and iron are found only in a mineralized state, with the exception of the masses of meteoric iron, alloyed with nickel, which have an extra-terrestrial origin. Though there is a great variety of ores of lead, yet almost the whole of the lead of commerce is produced from the combination of that metal with sulphur. The sulphuret of zinc is of almost universal occurrence, but it is not worked to any considerable extent; the combinations of the oxide of this metal with carbonic acid, and with silica, being its chief ores. The principal ore of tin, on the other hand, is an oxide. Iron occurs in abundance in combination with sulphur; but in this form it has but little value. The oxides, the magnetic and specular, together with the combination of the protoxide with carbonic acid, form the great bulk of the workable ores.

* Eckfeldt and Dubois, Manual of Coins, p. 234.

Gold and iron are almost as intimately associated as gold and silver. There is hardly a specimen of sulphuret of iron (iron pyrites) in which a minute trace of gold might not be detected by sufficiently delicate manipulation. When the iron pyrites has undergone decomposition and become converted into a hydrated oxide, the gold may often be separated with advantage, and the auriferous quartz of most of the gold workings in the solid rock is associated with iron ores. When gold is contained in iron or copper pyrites which remains undecomposed, there is a considerable loss in the process of amalgamation, which has hitherto prevented its profitable separation; but several attempts have been recently made to contrive machinery for effecting this economically, although it does not seem, as yet, to be determined whether any one of them is likely to be successful.

The varieties of form of this metal as it occurs in nature are not very great. It is most usually found in fine particles, and scales or flattened grains. Sometimes, however, it exhibits a crystalline form, usually that of the octohedron, but the crystals rarely attain any considerable size. The finest have been obtained in California. The lumps of larger size, which are not frequent, are called "nuggets" (French, *pépites*). They rarely exceed a few pounds in weight, and are usually accompanied by more or less of the quartzose gangue in which they were originally embedded. The largest of them ever found is said to be the great Australian nugget, which weighed 1615 ounces before melting, and yielded 1319 ounces of fine gold. The finest pure mass now in existence is that preserved in the Russian School of Mines, which weighs over 97 pounds troy.

*Geological Position and Mode of Occurrence of Gold.*—In general, it may be said that the older the geological formation, the greater the probability of its containing valuable ores and metals. On examining the table of geological formations, it will be seen that gold is confined entirely to the two lowest groups, the azoic and the palæozoic. Unfortunately, it is not possible in all cases to distinguish between these two groups, since in localities where the formation is metalliferous, the rocks, if of palæozoic age, have in all cases been so changed from their original character, that their place in the geological series is with difficulty to be recognized. For this reason, the azoic and the lower palæozoic rocks have been generally confounded together under the names of primary, primitive, primary fossiliferous, Cambrian, Cumbrian, and the like. In but few countries is the distinction so clearly marked that it could not be overlooked by a careful observer, as, for instance, in Sweden or on Lake Superior.* In the latter region, the strata of the Potsdam sandstone, the lowest fossiliferous rock thus far known to exist, are found resting, nearly in their original position and unchanged in character, on the upturned edges of a series of slaty and quartzose rocks, which were once deposited from water, as is evidenced by the ripple-marks still preserved on the faces of the quartzose strata.

The general facts in regard to the mineralogical character of the gold-bearing rocks, are very nearly the same the world over; whatever were their original structure or composition, they have, by the agency of a long chain of similar geological events, been brought to exhibit a striking resemblance to each other. They consist most frequently of slaty rocks, more generally talcose, although occasionally chloritic and argillaceous. It is in these rocks that the gold-bearing quartz, which forms, almost invariably, the gangue or accompanying mineral of this metal, is found to be most productive. Those veins which occur in the hypogene, or eruptive rocks, are rarely of much value. The auriferous quartz veins, which are themselves worked for gold in various parts of the world, and in which that obtained from washings originated, seem almost invariably to belong to the class of "segregated veins." The masses of quartz rock of which they are composed have the same dip and strike as the slaty rocks in which they are enclosed, and they exhibit no appearance of occupying a pre-existing fissure. Their width is variable, extending from a mere thread up to a hundred feet, and their richness in gold is equally uncertain. There are frequently, however, in some gold districts, small

* See Foster and Whitney's Report on the Geology of Lake Superior, Part II.

veinlike masses, which appear to traverse the strata, and to be analogous to true veins; but they seem, generally, to be merely subordinate to larger segregated deposits, and not to fill fissures which originated in any fracturing of the rocks by a deep-seated cause. The characteristic phenomena of the auriferous quartz veins will be fully illustrated in the discussion of the most important gold regions.

When the palæozoic rocks remain in nearly the same condition in which they were originally deposited, as, for instance, in the valley of the Mississippi, there is but little probability of finding gold, although there seem to be pretty well-authenticated accounts of small quantities of this metal having been found in a region underlaid by unaltered Silurian rocks, and far from any metamorphic action; where, on the other hand, the strata have been invaded by igneous masses, broken up, and raised upon their edges, and rendered crystalline in their structure, there is good reason to expect its presence.

There is room for doubt whether the great gold deposits of the world did not originate exclusively in the palæozoic strata, since we are not aware that the rocks which have been proved to be of azoic age have been found to be auriferous. In the Ural Mountains, the metamorphosed strata represent the whole palæozoic series, from the lower Silurian up to, and including, the carboniferous. The Australian rocks associated with auriferous quartz, contain fossils of Silurian age, while the flanks of the Sierra Nevada in California may probably be referred to the palæozoic epoch, though farther evidence is needed on this point. In regard to the gold of the Appalachian chain, in the United States, the question remains yet undetermined; but the evidence, thus far, preponderates in favor of its occurrence in the metamorphic palæozoic strata. Certainly the azoic centres of New York, Lake Superior, and Missouri, have given at most only doubtful indications of gold as yet, while the metamorphic rocks of Vermont and Canada, which are known to be of Silurian age, have furnished this metal in not inconsiderable quantity. In the southern gold region, the distinction between the azoic and palæozoic has not yet been attempted to be drawn.

As the undisturbed and unaltered Silurian rocks exhibit, at most, only traces of gold, it appears that this metal only becomes evident after the strata have been metamorphosed, or invaded by igneous and eruptive rocks. The period at which the segregation of the auriferous veins took place, and that of their impregnation with gold, still remain open questions, even after the age of the formation in which they are included has been determined. In general, the quartz veins may be presumed to have originated at the time of the metamorphic action on the strata themselves; and where there are igneous rocks in the immediate vicinity, the development of the metallic contents of the adjacent veins is usually ascribed to their presence, since it is so often found that the metalliferous deposits are intimately associated with eruptive masses. It is not necessary to suppose, however, that these phenomena throughout the world were confined to any particular geological epoch; on the contrary, there may have been a repetition of similar conditions at periods of time very distant from each other.

Murchison has shown that the impregnation of the rocks of the Ural with gold took place at a very recent geological period, as late even as the drift-epoch: this is one of the most striking facts developed by this distinguished geologist in his great work on Russia. In other important gold regions we have not sufficient data for fixing with much definiteness the time of the concentration of the metal into veins. In the Southern United States, it appears that no very great change in the character of the strata has taken place since the epoch of the new red sandstone; and the date of the gold-bearing veins may with probability be assigned to a period between that formation and the carboniferous. In Australia, the elevation of the auriferous strata and their probable impregnation with gold seem to have been later than the epoch of the coal. In regard to the age of the quartz veins of California, but little positive information can be given. Analogy and all the facts thus far obtained, lead us to suppose that the auriferous masses are included in slates of palæozoic age, highly metamorphosed; but of the period of the igneous action which may be supposed to have been the cause of their impregnation with gold, we know but little. In the Cordilleras of South America, the period of metalliferous emanations seems to have been after the deposition of the cretaceous strata; and disturbances of the rocks, which may have been attended with phenomena of this character, have evidently continued down to a very recent geological period, as well here as in California.

It is not, however, from workings in the solid rock that the principal portion of the gold of commerce is derived. Probably nine-tenths of it, at least, are obtained from *gold-washings*, or the separation of the metal from the superficial detritus which lies upon the rock in place, and which is included by geologists among the drift and alluvial deposits. Nature has performed the washing and concentrating processes herself on a large scale, and accumulated the precious metal in a position from which it can be obtained with facility, without any considerable outlay of skill or capital. It is this circumstance which causes such extraordinary fluctuations in the production of gold, since an almost unlimited quantity of unskilled labor can be at once applied to its collection, while the same amount of metal, were it remaining in its original position, could only be acquired by the application of a vast amount of capital. Indeed, the experience of the past has shown very clearly, that had not nature effected this concentration, the larger portion of the auriferous veins could not be worked, since the metal is too much scattered through them to be separated with profit.

The separation of the gold from its original matrix and its deposition among the strata of gravel, sand, and clay, or beneath them upon the surface of the rock, has been the result of causes acting through an immense period of time; and which have not yet ceased to operate, although their energy seems no longer equal to what it must have been at a former epoch. The rocky strata of the earth are constantly undergoing abrasion from the combined action of various meteorological causes; of which one of the most powerful at present is the alternate freezing and thawing of water in fissures and cavities, which tends to wear away and disintegrate the most elevated portions, especially of the slaty beds, and to carry down the abraded and loosened materials, and spread them out in the adjacent valleys. In lofty and rugged mountain chains, where torrents of rain frequently fall, and the streams, suddenly swollen to a great volume, rush with tremendous violence down rapidly declining valleys, their force becomes capable of wearing away the rocks with great rapidity. This mechanical action is frequently aided by a chemical one; the strata undergoing a molecular change which softens them and renders their abrasion easy. As the enclosing rocks are thus worn away, the quartz-veins become disaggregated by the oxidation of the iron they contain, and are themselves crushed into fragments and borne down into the valleys, where the metallic particles, having by far the highest specific gravity, are first deposited and sink to the bottom, while the lighter earthy portions are carried farther.

The exact identity in origin of the auriferous gravel and sand of various gold regions and the modern alluvial formations is not admitted by all geologists. Murchison, especially, insists strongly that the superficial deposits which contain gold are in no way to be confounded with detritus formed by present atmospheric action, but rather that they were the result of diluvial currents connected with and originating in physical changes in the surface of the globe, such as the elevation of mountain chains. A source as vast as this seems to be required to account for the accumulated masses of loose material which are scattered over the Siberian plains, or at the base of the Cordilleras; and it seems reasonable to admit that the elevation of the Ural and the Andes may have been connected with the deposition of the auriferous deposits which lie upon their flanks. But until geologists shall have more satisfactorily settled the relations between the drift and alluvium, and the transition from one to the other, and until a much more thorough exploration shall have been made of some of the greatest gold-producing districts, it will be a hazardous matter to enter into any discussion of the subject. It is most probable that the causes by which the auriferous detritus was accumulated must have been similar in quality, if not in quantity, to those now in action; and when we take into account the immense period through which they must have been at work, perhaps a less degree of intensity will be thought necessary to account for their effects.

It will be indeed an interesting fact, if it should be proved that the largest amount of gold is to be obtained in regions where the rocks have been most recently and extensively invaded and uplifted by igneous masses; and it might furnish a clue, among other things, to the fact of the vastly greater accumulation of gold in the superficial débris of California than in those of the Southern Atlantic States, although the richness of the auriferous veins in place does not seem to be much, if any, greater in one region than in the other.

---

## SECTION II.

### GENERAL DESCRIPTION OF FOREIGN GOLD REGIONS.

Having prefaced by some general remarks on the occurrence of gold, we proceed to illustrate the subject more fully before entering upon the gold-regions of the United States, by giving a condensed account of some of the principal auriferous districts of other parts of the world. The geographical order which will be followed in this, as in the succeeding chapters, will be: to commence with Northern Europe, including the northern part of Asia embraced in the Russian Empire; Scandinavia; England; Central Europe; France; Spain; Southern Europe; Northern Africa; Central Africa; Central Asia; Southern Asia, and the East India Islands; Australia; South America; Central America, and Mexico; Canada, and the United States from the northeast towards the southwest.

Gold Region of the Ural and Siberia.—The gold of the Russian Empire is obtained almost entirely from the eastern slope of the Ural Mountains, from Siberia, and in the Caucasus.* In Russia in Europe a very small quantity only is obtained, from the western slope of the Ural. A small amount of this metal was formerly obtained in the government of Archangel, but this locality has been abandoned

* Consult Tschewkin in Russian Journal des Mines; and G. Rose, "Reise nach dem Ural."

since the commencement of the present century. In Asiatic Russia, the auriferous districts are principally included in the governments of Perm, Oremburg, Tomsk, Yenisseisk, Irkoutsk, and the district of the Kirghese. The first discovery of gold was made near Ekatherinenburg, in 1743; the first workings were in the solid rock, and were commenced in 1752, at the mines of Beresow, which are still productive, though in a very diminished degree; their yield, in 1850, being less than 100 pounds. In 1823, there were sixty-six localities in the Ural where gold had been mined from the solid rock; but they had, with the exception of eight, all been abandoned. The veins are numerous at the Beresow mines; they are contained in granite, which itself forms veins in the talcose, chloritic, and micaceous slates.* The productive ones are of quartz, and they cut the granitic masses at right angles, having a nearly vertical dip. They do not generally extend beyond the limits of the granite. The working seems to have demonstrated that the amount of metal decreases in depth. During the latter part of the last century, the Beresow mines yielded from 600 to 800 pounds of gold annually; and the stamped ore yielded from five to eight zolotniks of fine gold to the hundred poods; equal to ·0013 to ·00208 per cent.

The proper gold washings of the Ural, which have produced such large amounts of this metal, and which had acquired such a high celebrity before they were eclipsed by those of California and Australia, were commenced by the crown in 1814. Those of Western Siberia were established in 1829; of Eastern Siberia in 1838. The following table shows the produce of these washings from the commencement.†

| | lbs. troy. |
|---|---|
| 1814 to 1820, product of the crown washings, . . | 1,085 |
| 1820 to 1830, crown and private, " . . | 73,200 |
| 1830 to 1840, " " " " . . | 175,460 |
| 1840 to 1850, " " " " . . | 553,955 |
| And of gold obtained from mining in the rock, from 1752 to 1850, . . . . . . . . . | 128,570 |
| Total, . . . | 932,270 |

* G. Rose, "Reise nach dem Ural," i. 175.

† Tschewkin, Jour. des Mines, quoted in Ann. des Mines, (5) iii. 805.

Since 1847, when the produce of the Russian mines seems to have reached its maximum, there has been a decided diminution; few new localities have been discovered, and in the Siberian washings the yield of the sand is lessening, and also the amount of gold obtained; in the Ural, on the other hand, owing to the perfection of the apparatus employed, and the skill with which the works are conducted, the produce of gold is increasing, although the tenor of the sands has diminished to ¼ zol. per hundred poods, equal to ·00006 per cent. It may be safely inferred that the Russian produce of gold has reached its maximum, and that it will continue slowly to decline.

The washing machinery of the Ural mines is remarkably efficacious, much time and money having been expended on its perfection; were it not such it would be impossible to wash the auriferous sands with profit, since their tenor in gold is very low, the average not being more than from 4 to 6 parts in a million, equal to about 3½ grains per bushel of 100 lbs. av.

GREAT BRITAIN.—The number of localities in England where gold has been found is very considerable, but there is no reason to suppose that it occurs in large quantities. In South Wales the Romans mined extensively in the Silurian rocks, with but little return of metal. In Dumfriesshire, in the Lead Hills, several hundred men were employed in washing gold sands during the reign of James V. The occurrence of this metal in the tin stream-works of Cornwall and Devon has been observed for hundreds of years, and a few ounces are still obtained yearly by the miners,* which are mostly preserved as a mineralogical curiosity. Borlase mentions a piece weighing 15 dwts. 3 grs.

The Wicklow Mountains, in Ireland, are made up of rocks which might, from their geological nature, be presumed to contain gold; and, in fact, this metal has been obtained there in small quantity; one piece weighed twenty-two ounces, and quite an excitement was raised on the subject.†

* De la Beche, "Survey of Cornwall," p. 614.

† E. Hopkins, in Mg. Almanac for 1849, p. 195.

A large capital was wasted here in an endeavor to find the *lodes* in which the precious metal originated.

The yield of gold of Great Britain amounted to about four ounces per annum, for the few years previous to 1853. Recently, quite an excitement has been raised with regard to this metal, and the most extraordinary stories are told of vast treasures hidden in the gossans of the Cornish mines. This excitement may be traced, in considerable degree, to the appearance of a work on "The Gold Rocks of Great Britain and Ireland," by John Calvert, in which some astounding stories are told with regard to the former productiveness of England and Scotland in this metal. For some time past, the papers have been filled with accounts of the large yield of gold obtained in crushing and amalgamating various Cornish and other gossans, which are said to be the ores richest in this metal, by the aid of Berdan's machine. This is one of the numerous inventions recently introduced for pulverizing and amalgamating, at the same time, auriferous rocks. As it has been but a short time in operation, and as it appears thus far to have been but little used except for experiments on a small scale, it is impossible to say what its ultimate value may prove to be; but it is certain that the results obtained by it in England have created much excitement. Many of the samples of quartz and gossan tried by its aid, have yielded several ounces of gold per ton, and some even as high as ten to sixteen ounces. It has heretofore been the opinion of those best informed on these subjects, that grinding and amalgamating simultaneously is not an economical method, the Cornish stamps having always proved more satisfactory in their operation than any machine contrived on this principle; and it will require considerable evidence, based on actual running, to convince many mining engineers of the contrary. At the St. John del Rey mine, the only very successful gold mine wrought in the solid rock, where a rock poor in gold is worked, it has been found that nothing answered the purpose, and proved, in the long run, so economical as the Cornish stamps. The wear and tear in machines of this kind being enormous, it is necessary, before all things, that they should be capable of being repaired, or having their worn-out parts exchanged for others, without much delay or cost. In the construction of the latest invented "gold-quartz crushing machines," it seems as if this circumstance had not been sufficiently attended to.

If these machines are capable of separating the gold with as much economy as they profess, there can be no doubt that there are numerous mines in England which furnish some ore which will pay for working; but, in general, the gossan, which is supposed to be the richest gold ore, does not continue to a very great depth, usually not more than twenty fathoms; and unless the lodes are very wide, the quantity which they would furnish would be small in comparison with that which would require to be stamped in order to furnish a large supply of gold.

Mr. John Taylor, Jr., whose experience in these matters is very great, remarks on this point* as follows: "I have seen evidence to make me believe that British gold ores, in moderate quantities, can be obtained, and that, if they are skilfully treated, they can be made to yield a moderate profit; but, beyond this, I cannot persuade myself that producing gold in England will ever be a large or very lucrative branch of industry."

GERMANY.—*Austrian Empire.*—The amount of gold obtained in Germany is very small indeed, although in some places mining or washing has been pursued almost uninterruptedly since the days of the Romans.

*Hungary, Transylvania.*—The exploitation of the mines of gold, silver, lead, and copper, of Hungary has been carried on almost uninterruptedly since the eighth century; and the results of these long-continued labors present one of the most interesting fields of investigation for the mining engineer which can be found in the world. At Schemnitz, Kremnitz, Neusohl, and Libethen, in Lower Hungary, the visitor may study the results of centuries of experience in the slow development of the processes for the treatment of auriferous and argentiferous ores.

There are mines of gold and silver, as well as copper and

* Eng. Mining Journal, Dec. 1853.

lead, at Nagybánya, Kapnik, and Felsöbánya, on the western border of Transylvania, in which the ancient works are on a gigantic scale; and there are also gold mines at Zalathna. In the Lower Hungarian mines, which are grouped around the towns of Schemnitz, Kremnitz, and Neusohl, the ores are chiefly auro-argentiferous; galena is also obtained in sufficient quantity to afford the necessary lead for separating the silver. These mines are much less productive than formerly; but works are in progress, on a gigantic scale, to open them at a great depth. The quantity of the precious metals in all this region has, however, been found to decrease as the veins were worked downwards. The Transylvanian mines afford those interesting combinations of gold and tellurium which are so rare and curious. The present annual produce of the Hungarian mines is about 3800 marks of gold = 2850 lbs. troy, and 68,000 marks of silver = 57,000 lbs. troy. These mines could not be worked except under the two conditions of the highest development of mining and metallurgic skill, and a low price of labor.

*Tyrol and Salzburg.*—A very small quantity of gold has been obtained, during many centuries, from these districts, by the application of great skill and labor to ores of the smallest possible richness. The whole production in 1847 amounted to only 109 marks = 78 lbs. troy.

At Zell the average yield of the stamp-work was, in 1845, only from 11 to 12 loth in 1000 centner, equal to 4 parts in 1,000,000.* This is the smallest yield of any gold ore known to be actually mined. The production amounted, according to the last official accounts, to 20 marks only per annum.

*Bohemia.*—At a very early historical period, the mines of Bohemia were extensively wrought. It is said that as early as 734 the gold mine of Eula was so productive that golden images were manufactured from its produce. From the eleventh to the fifteenth century, the produce of the Bohemian washings helped in some degree to furnish a circulating medium to Europe. The present total produce of gold in Bohemia, however, does not much exceed $500 in value per annum.

* Russegger, Der Aufbereitungs Prozess Gold und Silberhaltiger Pocherze, p. 3.

The different provinces of the Austrian Empire furnished each the proportional amount of gold specified in the following table, according to the average of the official returns of the years 1840 to 1847 :—

| | |
|---|---|
| Transylvania, | 53·3 per cent. |
| Hungary, | 45·6 " |
| Salzburg, | 0·6 " |
| Tyrol, | 0·25 " |
| Styria, Bohemia, &c., | 0·25 " |
| | 100·00 |

The total production of the empire amounted, in 1848, the last year for which returns have been published, to 5645 lbs., having steadily, although slowly, increased from 2682 lbs. in 1820. Its present yield may be estimated at 6000 lbs.

FRANCE.—*The Rhine.*—The sands of the Rhine are still washed, on a small scale; formerly, there is good reason to believe, the production was very considerable. Historical documents show that, in 667, the right of washing the sands of the bed of the Rhine was considered to be of value. The present yield (1846) is estimated by Daubrée at 45,000 francs. The auriferous portion of the bed of the river is included between Basle and Mannheim, but the washings have been especially numerous in the vicinity of Strasburg.* The sands recently washed yield from thirteen to fifteen hundred-millionths; they rarely contain more than seven ten-millionths. The washer makes, on an average, from one and a half to two francs a day, and sometimes as much as ten or fifteen francs. Daubrée considers the average yield of the sands of the Rhine, Siberia, and Chili, to be in the proportion of 1 : 10 : 37.

There are several localities in France where gold has been produced in small quantity. Diodorus speaks of the richness of the Gallic rivers in golden sands, but nothing at present existing would lead to the opinion that the amount obtained was ever very great. The river Ariège derived its name from its yield of auriferous sands (*aurigera*), and up to

* Daubrée, Comptes Rendus, xxii. 640.

the close of the fifteenth century it produced something like a hundred pounds of the metal annually.

The only quartz vein bearing gold known to exist in France is that of La Gardette, in the department of the Isère. It is from one to three feet wide, and enclosed in gneiss. Native gold was discovered in it in 1700, and it was worked at intervals up to 1841, yielding but a small amount.

Many of the plombiferous veins contain a little gold.

SPAIN.—Although the amount of gold at present obtained from Spain is very insignificant, that country was at one time highly productive in this metal. Diodorus, Strabo, Pliny, and other ancient writers, allude to Spain as the land of golden treasures. Adrien Paillette* has investigated the subject of ancient mining in the Peninsula, and has arrived at the conclusion, that gold was produced in large quantity by Spain and Portugal at the most remote periods, and that this metal was obtained from workings in the solid rock, as well as from washings of the auriferous sands of the Douro and Tagus. So far as the working of the mines was concerned, it seems that 1800 years ago, at least, a system of adits, levels, and shafts was adopted. In the time of Pliny, Gallicia and the Asturias furnished 20,000 pounds of gold per annum, and were the richest gold-field known. Pliny not only speaks of the shafts and levels, but describes the mode of timbering them up, the preparation for opening the ground, and the system employed in smelting the ore; and it appears evident, that in those remote times the arts of cupelling and amalgamating were fully understood and practised. It is interesting to learn that, even in the time of Diodorus, these processes were spoken of as being of the most ancient origin, an opinion corroborated by Pliny.

It appears evident from Paillette's investigations, that the auriferous region must have been pretty thoroughly exhausted, for in all their examinations of the old workings, which are of immense extent, even by carefully washing portions of the detritus, they only obtained a few traces of gold.

The present production of gold is almost nothing, though

* Bull. Soc. Geol. de France, (2) ix. 482.

there are reports of the recent discovery of rich gold sands. The washings at present are principally confined to the rivers Sil and Salor, of which the whole produce may perhaps amount in value to $8,000 per annum.

ITALY.—Quite a number of localities were known to the ancients as producing gold. The Po is still one of the gold-bearing streams. The only works of any consequence at all are in Piedmont and Savoy. In the Anzasca Valley, near Monte Rosa, an auriferous pyrites was worked until quite recently, which yielded only from 2 to 85 francs per cwt. of ore, or 8 dwts. of gold to the ton. The amalgamation works were scattered up and down the valleys on the small streams. The whole amount produced in the province of Ossola, to which these works belong, was, in 1829, about 250 lbs. troy, with a profit of not much over $15,000. These mines were extensively worked in the time of Pliny; and, according to him, the Senate fixed 5,000 as the number of slaves who were to be allowed to work in them, lest the price of the precious metal should be reduced.

A French mining company was also recently engaged in working a gold mine in serpentine, near Genoa, on the flanks of the Col Bochetta.

CENTRAL ASIA.—Golden sands had frequently been discovered in the Caucasus; and, in 1851, the Russian government took steps to establish washings there.* In that year the deposits of auriferous sands occurring on the branches of the Koura, which originate in the great trans-Caucasian chain, were examined, and the region was found to be in all respects identical in character with the gold-bearing districts of Siberia. Evidence was obtained that these sands had been worked previous to the Christian era. On the River Akstafa, the auriferous bed consists of rolled pebbles and boulders, mixed with quartz, brown iron ore, and clay, the whole covered with six feet of unproductive alluvium.

Thibet is generally supposed to be rich in gold, as well as other metals. The rivers of the western portion of that country are especially referred to as abounding in auriferous

* Ann. des Mines, (5) iii. 830.

sands. Jacobs, who gives the only estimate of the amount produced which I have been able to obtain, fixes it at 10,000 ounces.

Southern Asia.—There can be little doubt that the portion of the Asiatic continent south of the great Himalayan chain has, in former times, yielded vast amounts of gold. The much vexed question of the locality of Solomon's Ophir has not been definitely settled; but it seems not improbable that it was on the Malayan peninsula or some of the adjacent islands. Ritter places it in Hindostan; by others it is supposed to have been somewhere in Thibet. Gold is still obtained, in small quantities, at many localities in India; but it appears that they have long since lost the productiveness which history leads us to believe they must once have had. What metal is now obtained is mostly used for personal ornaments. The washings of the Burrampooter are estimated to yield from 30,000 to 40,000 ounces yearly, by Jacobs and others.*

In the Burman Empire, many streams have gold-washings upon them, which are irregularly worked by the natives, who are represented as especially fond of decorating their persons with golden ornaments; and, their own produce not being sufficient, they are said to import an additional quantity from China for that purpose. According to Jacobs, no reliable data are to be obtained in regard to the amount produced; Mr. Birkmyre, however, puts that of Ava at about 2000 pounds.

Authentic information represents the Malayan peninsula as having been rich in gold, but the amount produced at the present time is small. The character of the inhabitants forbids any extended researches on the part of foreigners, although such have been attempted recently with some success.†

China and Japan.—Of the mining statistics of these countries we have little or no exact knowledge. We know, however, even after making allowance for exaggerations of travellers,

* Jacobs, Historical Inquiry into the Production and Consumption of the Precious Metals, ii. 330.

† Ann. des Mines (5), iii. 516.

that Japan is rich in gold as well as copper and other metals. It is stated that the Japanese foresaw the gradual impoverishment of the gold-washings, and, for that reason, checked the exportation of that metal, which formerly took place on a large scale. At present none of it enters into the commerce of the world. The Siberian gold-bearing formations extend into China, and have been worked to some extent; but, according to Murchison, operations were suspended, in conformity with Chinese politico-economical theories, in order to preserve the balance of circulation.

The East India Islands furnish gold at many points, and in considerable quantity; but the statistical information of the amount produced is very indefinite.

Borneo has extensive gold-washings, which are principally on the western coast. The auriferous beds, on the authority of James Brooke, Rajah of Sarawak,* consist of coarse sand and gravel from one to four feet thick, overlying a bed of clay of about ten feet in thickness. A large quantity of platina is also obtained in washing out the gold, although it is only quite recently that its value has become known, and that it has been preserved. According to Mr. Brooke, there are 5,000 persons, mostly Chinese, employed on the western coast, who obtain over $5,000,000 of gold annually. Mr. Jacobs, however, estimates the produce at only one half of this.

Sumatra, Celebes, Timor, and the Philippine Islands are all stated, in various authorities, to furnish considerable amounts of this metal.

The whole produce of Southern Asia, including the East India Islands, may be estimated at 25,000 pounds, making allowance for the small quantity which finds its way out of China. Other estimates vary from 10,000 to 30,000 pounds.

Africa.—There can be no doubt that this continent is rich in gold. The few travellers who have penetrated into the interior have all coincided in their accounts of golden wealth. The attempts to establish European mines there have had a most disastrous termination. Kavelowski, a Russian mining

* Narrative of Events in Borneo and Celebes, &c., London, 1848.

engineer, reported the existence of extensive deposits of auriferous sands on the river Somat, near Kassan; but neither his attempts, nor those of the Pasha of Egypt, have led to any systematic mining operations under the direction of European miners. The climate alone is a sufficient hindrance.

The gold-washings in Senegambia and in Bambuk, are said to be of importance. In Nubia, the negroes wash the auriferous earth in wooden bowls or in gourds, and in all probability lose the larger portion of the precious metal.

Russegger, who travelled through Nubia in 1838, gives the following as the results of his investigations in regard to the occurrence of gold in Central Africa.*

1. In the interior of Africa, in the so-called primitive chain which stretches from east-northeast to west-southwest, and in the alluvions of the rivers flowing from it, there exists a large amount of gold, though not equal to the fabulous quantity imagined by some. 2. The gold is found in scales and dust in Sennaar and the south of Abyssinia, in the gneiss and chlorite slates, and also in granite. In the latter rock, it occurs in quartz veins, with iron pyrites and hematite; in the chlorite slates it is found in immense quartz beds, with various ores of iron. 3. In the gold-bearing alluvial strata, those beds are richest in the precious metal which are of an ochrey character, with coarse detritus intermixed, or of a fine clay, with roots of plants scattered through it. 4. The swifter and rockier the stream, so much the richer is it in gold. 5. The gold is remarkably pure and of a deep yellow color.

The whole amount of gold obtained from Africa at present is estimated by Mr. Dusgate, a gentleman spoken of by Murchison as familiar with the subject, as not over $\frac{1}{17}$ of the produce of the Siberian mines in 1850, which would be 3,744 lbs. Mr. Birkmyre puts it at 4000 lbs. The physical difficulties are too great to render it probable that Africa, however rich its mines may be, will add much to our metallic wealth for a long time to come.

AUSTRALIA.—The southeastern corner of Australia is bordered by a series of chains of mountains, of which the culminating ridge is from fifty to one hundred miles distant from

* Karsten and Dechen's Archiv. xii. 153.

the ocean.* Its highest point, Mount Kosciusko, is 6,500 feet above the sea. West of Sydney, in New South Wales, the line of the chain is nearly north and south, but in Victoria the watershed has an east and west direction. This Australian Cordillera, as it has been called by Murchison, is made up principally of schists; namely, micaceous, argillaceous, and silicious slates, interlaminated with granite. The slates, like almost all other schistose rocks occurring over large districts, have a general north and south strike, and stand nearly vertical; they appear to be of Silurian age, but are highly metamorphosed. Silurian fossils have been found on the flanks of the dividing range of New South Wales, but exactly under what conditions we have been unable to learn. These slates have been broken up and invaded by igneous rocks, syenite, porphyry, basalt, and trap; as, for instance, on the Macquairie and the Turon. In short, the resemblance between the geological position and mineralogical character of the rocks of the Australian and Uralian Cordilleras is so striking, that Murchison, in an examination of the specimens and maps of Count Strzelecki, in 1844, was led to declare, that analogy would justify him in predicting the discovery of gold in Australia. Two years afterwards he received from that distant region specimens of auriferous quartz, and, on the strength of these indications, he advised the Cornish miners to make an attempt to develop the supposed riches of the Southern Continent. This advice was printed and circulated abroad, as well as at home, and, it may be presumed, had some influence in determining a search for the precious metal. The discovery of the golden wealth of California, however, was the immediate cause of the opening of the Australian gold region, having the same stimulating effect there on the search for gold that it had everywhere else.

It is worthy of being recorded that Earl Grey, who was addressed on this subject by Murchison, took no steps whatever to promote the discovery of gold, on the ground that

* G. H. Wathen, in Qu. Jour. Geol. Soc. ix. 74. See also Delesse, in Ann. des Mines (5), iii. 185.

the production of this metal would interfere with sheep-growing! Luckily, such men have not the power of long restraining Anglo-Saxon energy. A writer in the Quarterly Review* has well said, "We were quite unprepared for such pastoral predilection in the colonial office under Lord Grey's presidency. To realize Arcady in New South Wales, and convert convicts into Strephons, might be a very amiable conception, but would hardly justify the minister of a great commercial empire—and, above all, a zealot of *Free Trade*—in an attempt to *cushion* rich sources of mineral wealth opened in a colony under the watch of his intelligence."

Thus things went on while California was raining her golden shower, a "Mr. Smith," in the meantime, having applied to government for a reward for the discovery of auriferous deposits, which he was willing to disclose on being paid for it, till a "returned Californian," of the name of Hargraves, urged on by the strong resemblance of the rocks of Australia to those of the gold region in California, without waiting for the slow motion of government in reply to his application, and notification that gold would be found, commenced "prospecting" on Macquairie River, and on the 8th of May, 1851, the Commissioner of Crown Lands received a notification from him that several ounces of gold *had* been found on a branch of the Macquairie, and, soon after, that a piece weighing thirteen ounces had been dug up. It need hardly be added that the excitement grew amain, and that the whole neighboring population rushed to the second California.

As soon as the discovery became an established fact, the government began to stir in the matter, and laid claim to the region, granting licenses to dig for gold for £1 10*s.* per month, and also instituted a geological commission to explore the country, whose reports have been published in the government blue-books.

The gold discoveries extend over a space of at least nine degrees of latitude, and occupy a breadth of fifty miles, or more, along the line of junction of the palæozoic strata and the eruptive rocks. The first diggings on the Summerhill

* Quart. Rev. xci. 511.

were christened "Ophir," and soon after, the Turon Valley was found to be auriferous for a distance of one hundred and thirty miles; twenty miles north of the Turon, on the Meroo, a native shepherd discovered blocks of quartz, rich in gold, lying on the surface, from one of which 60 lbs. of the metal were taken. The next great discovery was at Araluen, two hundred miles south of the Turon; and between these localities numerous others have since been found, extending over a distance of seven hundred miles in length, north and south.

The name of Victoria was given, in 1850, to a district previously almost unknown, which lies around Port Philip, and of which Melbourne was the principal settlement. In this colony the discovery of gold was announced in August, 1851, and the Ballarat diggings, about fifty-five miles north-west of Geelong, at the junction of the slates and the trappean rocks, were soon found to surpass all that had been discovered in New South Wales. This, and the Mount Alexander gold-field, a few miles north, have yielded almost fabulous quantities. In October, 1851, about 3,000 persons were at work at the Ballarat diggings; but, in December, it was computed that at least 12,000 persons were collected at the Mount Alexander gold-field, within a space of fifteen square miles. It is not necessary to give particulars of the large nuggets found, or the excitement which followed on each successive discovery; the statistics of the yield of these gold-fields are sufficient to enable every one to judge of the effect which such revelations of golden wealth must have had on the astonished Australians.

The Australian gold is of remarkable fineness, the different analyses giving from three to seven per cent. of silver. The whole amount, almost without exception, has been thus far obtained from washings; and although numerous quartz-mining companies have been organized, they have, it is believed, without exception, failed to accomplish anything. Mr. Calvert, in his work, which draws so heavily on the credulity of the reader, declares that he had discovered nearly two hundred and thirty-eight gold-bearing quartz veins, producing on the average 2½ dwts. to the ton. He speaks of one, in particular, the Macquairie vein, as having been traced forty miles, running north and south. This vein, he admits, could not probably be worked with profit at the present time. From another vein, Mr. Calvert mentions that he broke off 3 cwts. of rock, which yielded 76 lbs. of the pure metal. The investigations now going on in London, with regard to the gold-quartz mining companies of Australia, would indicate that they are mostly swindling concerns; and also that the quartz veins are not rich in gold, or, at least, have not yet been proved to be.

The auriferous deposits present the most striking analogy with those of California. There are innumerable "diggings," which are sometimes quite superficial, and sometimes extend through the detritus and loose materials to the "bottom rock" or "ledge," namely the rock in place. In the surface-workings, which are exclusively on the flanks of the hills, the gold is diffused through the gravelly soil to the depth of six to twelve inches, beneath which there is a stiff, red clay, which contains little or no gold. In the deeper workings, it is necessary sometimes to sink twenty-five or thirty feet, or even more, before reaching the auriferous deposits. These are of varying character. The channels of the small streams coming down from the mountains, and often entirely dry for most of the year, are usually rich in gold, which is found accumulated against the "bars" or projecting ledges of rock, and forced into the chinks between the strata. The more precipitous the torrent, the larger the nuggets which are found; where the valley expands out, the golden particles are smaller. These rich deposits are frequently found at the entrance of a lateral into the main valley, where there has been an eddy in the current, caused by the contraction of the mouth of the lateral valley.

In the valleys of the Bendoc and Delegete, according to Rev. W. B. Clarke, the superficial deposits are in the following order:—

1. Gold-bearing detritus, made up of fragments of slate and quartz, cemented by argillaceous matter.
2. Pipe-clay.
3. Erratic blocks and pebbles of quartz containing gold.
4. Rock in place.

The minerals and rocks associated with the gold deposits are similar to those of other great gold-bearing districts. Quartz is exceedingly abundant, and evidently the gangue of the gold. Magnetic iron sand, sometimes titaniferous, is rarely wanting. The rarer minerals are topaz, garnet, zircon, corundum, rutile, and the diamond, of which one crystal, at least, is said to have been found on the Turon. Platina is not wanting; but, thus far, seems to be quite rare. The quartz and oxide of iron were evidently originally associated with the gold; the precious gems may or may not have occurred in the same connection; their hardness, indestructibility, and high specific gravity, would cause them to be found under the same circumstances as the gold, and in company with it, in the superficial detritus, even if derived from rocks at a considerable distance from the gold-bearing quartz veins.

The question, whether the accumulations of superficial detrital matter containing gold, such as pebbles, fine clays, and sands, are due to the long-continued and gradual action of causes which are still in operation, such as currents of water, rain, and the like, or whether they are the result of cataclysmal action, namely, of a drift-period similar to that of our "Northern drift," is one of great interest both scientifically and practically; but the data are too incomplete, at present, to allow any positive opinion to be formed on that point.

Mr. Stutchbury, one of the government geologists in New South Wales, considers that the evidence is decidedly in favor of the gold-bearing detritus having been accumulated by the slow disintegration and consequent denudation of the rocks, under the action of causes not differing from those now in operation.

The apparatus used by the miners for separating the gold is of the greatest simplicity, consisting of the cradle and pan. In the richest fields of Victoria, up to a very recent period certainly, thousands of miners have collected large quantities of gold without even so simple a machine as the cradle, by simple "panning."

Imperfect as are the statistics of the total yield of Australia, they are far more satisfactory than the accounts of the average contents of the auriferous earth washed. Mr. Calvert estimates it at $\frac{9}{10}$ grain in a cubic foot, which he says will seem very low. Considering the average depth of the auriferous beds to be thirty-nine inches, and the area covered by them to be equal to 68,700 square miles, the total amount of gold would be 434,191 tons, worth, at £3 19*s*. per ounce, the trifling sum of £46,100,571,660. To this Mr. Calvert adds, that this amount is exclusive of the gold contained in the solid rock and the quartz veins.

The difficulty of arriving at an accurate knowledge of the amount of gold produced in Australia seems to be very great. The following estimates are selected from those supposed to be most reliable.

The gold produced in the colony of Victoria is mostly brought, under government or private escort, to Melbourne, Geelong, or Adelaide; but a portion of it is carried by private hand across to Sydney. That of New South Wales, which is much less in quantity, goes to Sydney almost exclusively. Hence, we have a first approximation to the total produce in the amount carried under escort to these sea-ports. But this is but a portion of the gold, since a considerable quantity finds its way to the cities through private hands. The amount exported and manifested is a better guide; but to this must be added what is taken by passengers privately, and what remains in the country, in circulation, in the hands of bankers, and in transitu.

The returns of the exports of native gold for the different years are believed to be approximately as follows:—

| Year. | Victoria. | New South Wales. | Total of Australia. |
|---|---|---|---|
| | lbs. troy. | lbs. troy. | lbs. troy. |
| 1851, . . . . . . . | 12,176 | 11,910 | 24,086 |
| 1852, . . . . . . . | 164,580 | 70,636 | 235,216 |
| 1853, . . . . . . . | 212,105 | 63,800 | 275,905 |
| Total since the discovery of the gold to end of 1853, . | 388,861 | 146,346 | 535,207 |

The production of 1852 is that which can be most accurately determined; and here there are numerous estimates which agree tolerably with each other.

The total production of the colony of Victoria is estimated as follows by different authorities:—

<table>
<tr><th></th><th>1851.</th><th>1852.</th><th>1853 (1st half).</th></tr>
<tr><td>English Mining Journal, .</td><td>224,140 oz.</td><td>4,167,571 oz.</td><td>1,296,059 oz.</td></tr>
<tr><td>Mr. Khull, bullion-broker at Melbourne, . . . . .</td><td></td><td>4,247,657</td><td></td></tr>
<tr><td>William Westgarth, . . .</td><td></td><td>4,545,780</td><td></td></tr>
<tr><td>Melbourne Argus, . . . .</td><td colspan="2">4,000,000</td><td></td></tr>
</table>

The total yield of Victoria and New South Wales for 1852 is estimated by M. Delesse, from data obtained by Murchison from official sources, at 400,000,000 francs = £16,000,000.

Mr. Birkmyre, who was on the spot, and may be supposed to have had good opportunities for judging, has placed his estimates somewhat lower, namely:—

| | |
|---|---|
| 1851, . . . . . . . | 23,667 lbs. |
| 1852, . . . . . . . | 266,178 " |

These estimates, however, seem rather too low, since the quantity actually exported in 1851 seems to have exceeded that given by Mr. Birkmyre.

During the year 1853 the produce diminished considerably. The yield of New South Wales reached its maximum early in 1852, apparently. That of Victoria seems to have been at its highest point from July to November, 1852, when Mount Alexander and Ballarat were producing enormously, as shown by the following account of the gold taken from those fields to Melbourne and other places, according to escort returns:—

| | | |
|---|---|---|
| June, | 1852, . . . . . . . | 162,990 ounces. |
| July, | " . . . . . . . | 353,182 " |
| August, | " . . . . . . . | 350,968 " |
| September, | " . . . . . . . | 366,193 " |
| October, | " . . . . . . . | 264,683 " |

The following figures show the amounts taken from the Victoria gold-fields for each month of the years 1852 and 1853, specifying for the first nine months of each year the number of ounces brought by escort to Melbourne, and for the last three months, to Melbourne and Geelong, thus affording a comparative view of the yield of the two years:—

| | | 1852. | 1853. | |
|---|---|---|---|---|
| Brought to Melbourne, . . . . . | January, . | 53,594 | 186,615 | ounces. |
| " " . . . . . | February, . | 56,142 | 172,329 | " |
| " " . . . . . | March, . . | 62,026 | 169,654 | " |
| " " . . . . . | April, . . | 68,041 | 170,427 | " |
| " " . . . . . | May, . . | 77,247 | 116,812 | " |
| " " . . . . . | June, . . | 116,009 | 122,695 | " |
| " " . . . . . | July, . . | 320,218 | 198,007 | " |
| Brought to Melbourne and Geelong, | October, . | 334,648 | 200,280 | " |
| " " " " | November, | 327,623 | 162,393 | " |
| " " " " | December, | 121,827 | 176,580 | " |

The same inference may be drawn from the following table, which is stated to have been prepared by a gentleman in South Australia, who had given much attention to the statistics of gold.* It shows the amount of gold brought to Melbourne, Geelong, and Adelaide, by government and private expresses, and also gives an estimate of the quantity brought by private hand, together with the shipments for each quarter from the first discovery to the middle of 1853:—

| Year. | Quarter ending | Brought by escort. | Estimated by private hand. | Total. | Shipped. |
|---|---|---|---|---|---|
| 1851, . . | Dec. 31, . . . . | 124,522 | 99,618 | 224,140 | 145,116 |
| 1852, . . | March 31, . . . | 178,006 | 356,012 | 534,018 | 420,445 |
| " . . | June 30, . . . . | 299,615 | 599,230 | 898,845 | 339 729 |
| " . . | Sept. 30, . . . . | 1,017,283 | 508,641 | 1,525,924 | 512,692 |
| " . . | Dec. 31, . . . . | 967,027 | 241,757 | 1,208,784 | 702,110 |
| 1853, . . | March 31, . . . | 576,192 | 144,048 | 720,240 | 661,957 |
| " . . | July 2, . . . . | 460,655 | 115,164 | 575,819 | 572,121 |
| | | 3,623,300 | 2,064,470 | 5,687,770 | 3,354,170 |

The principal yield of Victoria, up to the middle of 1853, was from the Ballarat and Mount Alexander gold-fields; several others have been discovered, but none of them have compared in richness with these two. As might reasonably have been expected, the large number of persons congregated there, and employed in digging and washing, amounting to over 100,000 probably, caused these limited areas of unpre-

* Eng. Mg. Journal, Nov. 5, 1853.

cedented richness to be soon worked out. New localities may be discovered, and the old ones will continue to be worked with more moderate returns; but it seems probable that there will be a gradual falling off in the yield. This is the opinion of Mr. E. Hopkins, an English mining engineer well known to the scientific world, who also gives it as the result of his observations in Australia, that there are no quartz veins there worthy of being worked.

The amount of gold imported into England from Australia is stated as follows:—

| | |
|---|---|
| 1852, | £7,282,635 |
| 1853, | 14,972,743 |

Of course the amount received in England in any particular year, would not be a guide to that actually obtained during the same year in Australia, the length of time required for the passage being several months.

On the whole, after a careful consideration of the various returns and estimates, some of which evidently bear the marks of exaggeration, while others appear to have been carefully compiled, I have adopted as the production of Victoria and New South Wales together, the following amounts of pure gold, which probably nearly approximate to the truth.

| Year. | lbs. troy. |
|---|---|
| 1851, | 30,000 |
| 1852, | 330,000 |
| 1853, | 210,000 |

SOUTH AMERICA.—Although the silver mines of South America were at one time pouring forth a stream of wealth unparalleled in the history of the world, and causing a general rise in prices, similar to that which we are now witnessing, as the result of an unprecedented yield of gold; yet, of this latter metal, the quantity furnished by that country has never been very large. In 1800, the whole produce of the Southern American continent was estimated by Humboldt at 33,524 lbs. of gold; of which 9,900 lbs. were the yield of Brazil; while at the same time the annual proceeds of the silver mines, according to the same authority, amounted to 691,625 lbs., gold and silver being as 1 to 21 nearly by weight. In 1850, the amount of gold had fallen off to about 24,000 lbs., or three-fourths of what it was at the commencement of the century. According to Chevalier, the total yield of the precious metals from the Andes of Peru and Bolivia was, up to 1846, in the ratio of 1 part of gold by weight, to 170 of silver.

*New Grenada.*—The ancient vice-royalty of New Grenada was, from 1819 to 1831, united with Venezuela, and part of the time with Equador, forming the republic of Columbia. Each of these states is now nominally an independent republic. This part of South America, as also Brazil, produces almost exclusively gold. Of the New Grenada mines we have ample accounts, although not of recent date, from Boussingault, Chevalier, and other travellers and mining engineers, sent out to explore that region; the principal mines and washings are in the provinces of Antioquia and Veraguas. In the former there is said to be hardly a river which does not flow over auriferous sands, and there are also numerous quartz-veins in the granite which are worked to some extent. The principal mines of this character, worked in 1850, were on the river Porce. The veins are of quartz, carrying auriferous pyrites, which, near the surface, is much decomposed. They resemble in every respect the gold-bearing quartz-veins of other parts of the world. In the provinces of Panama and Veraguas the veins are also numerous; but their yield of gold is small, not amounting, according to Boucard,* to over 0·002 per cent., on the average.

The yield of the province of Antioquia, in the year 1847–8, was supposed to be 12,500 lbs. Chevalier estimated the amount annually produced in the republic at 13,276 lbs., and the total produce as follows:—

| | |
|---|---|
| Previous to 1810, . . . | $295,000,000 |
| From 1810 to 1846, . . . | 81,500,000 |
| | $376,500,000 |

Equal to 556,840 kilos. pure gold, or 1,492,331 lbs. troy.

Mr. Danson,† from information based on the returns of

* Ann. des Mines (4), xvi. 377.

† Jour. Stat. Soc. of London, xiv. 40.

the English consuls, estimates the total amount produced, from 1804 to 1848, at $204,085,328.

There is abundant evidence that New Grenada is still rich in gold; under a different climate, and with more energetic inhabitants, its produce of this metal would be greatly increased.

There are several English companies working gold mines in this country.

*West Grenada, or Veraguas Mining Company.*—This property has been very favorably reported on by a gentleman formerly connected with the St. John del Rey mines. The gold-bearing quartz is said to be very rich. Preparations were making, in 1853, to work it on an extensive scale.

*Mariquita and New Grenada.*—This company owns the Marmato mines, among others, and is working them with considerable profit.

*The New Grenada Mining Company,* in the district of Antioquia, has recently purchased the mines of Frontino, nine leagues west of the city of Antioquia. The produce of this mine, as worked for several years by the inhabitants, has been from 15 to 25 lbs. of gold per month. The workings have been confined to one quartz vein, which is mixed with iron pyrites, and varies in thickness from two inches to five feet.

*Venezuela.*—That part of the ancient viceroyalty of New Grenada which is now Venezuela had produced very little or no gold up to very recent times;* but lately this metal has been obtained in considerable quantity in the canton of Upata, province of Guyana, and also in the province of Cumana, at Campano.†

*Brazil.*—Although Mexico and Peru have furnished the larger part of the metallic treasures of the New World, their yield has been chiefly of silver. Brazil, on the other hand, has, for a great length of time, been famous for its gold alluvia, which have been worked since the beginning of the last century, and have produced an enormous amount. There is considerable and more precise information than is usually to be had in regard to South American mines, in the works of Von Eschwege, formerly Director-General of the Brazilian mines, Burat, and others.

It was chiefly from the washings of the auriferous alluvia of Minas Geraes, that the large quantity of gold which

* Ann. des Mines (4), xviii. 107.

† Ann. des Mines (5), i. 600.

flowed from Brazil in the eighteenth century was obtained; but, at present, the principal source of produce are the veins in the solid rock. Burat has given an excellent account of the mining region of Brazil,* which shows that the auriferous deposits are somewhat different in character from those in other parts of the world. The rocks are supposed to be of palæozoic age, but are so metamorphosed as to be not referable to any precise epoch. They have not been elevated into Cordilleras, but form a series of disconnected elevations; the trappean and other eruptive rocks appearing rather in dome-like protuberances and dykes than in mountain châins. It seems pretty certain that there have been no geological disturbances of so late a period as the elevation of the Andes. The rocks most developed are gneiss, and those varieties of quartz-rock which are known in Brazil by the names of itacolumite, jacotinga, and itabirite.

These rocks are characteristic of Brazil. The itacolumite is a quartz mixed with chlorite, and is sometimes of enormous thickness. When it takes into its composition specular iron, it becomes the rock known as itabirite or jacotinga, according as it is crystalline or compact. The gold is disseminated through the whole of the metalliferous bed, there being no regular veins; but it is especially concentrated in the vicinity of masses of specular iron, and in connection with quartz.

The great importance of the Brazilian gold mines, as being the only ones both extensively and profitably worked in the solid rock, induces me to present a somewhat detailed description of one of the most important, that of St. John del Rey. This, as well as the other principal companies working in the neighborhood, is owned and managed entirely by English capitalists.

The St. John del Rey Company is divided into 11,000 shares, on each of which £15 have been paid up. The operations were commenced on the estate of Morro Velho in 1834, having been previously carried on in other localities with considerable loss. In 1838 the mine was first worked with profit; and since that period, up to the present time, it has been steadily increasing in value.

The gold is found in a heavy bed of jacotinga, which is intercalated in

* Comptes Rendus, xii. 252.

a very argillaceous itacolumite slate of a reddish-blue color. It had been worked for a hundred years before coming into the possession of the present Company, and was considered exhausted. The auriferous mass averages about 44 feet in width, and dips with the rocks of the vicinity at an angle of about 45° to the southeast. It consists mainly of specular iron mixed with sulphuret of iron, magnetic pyrites, and quartz. Its average yield of gold for the last four years has been as follows:—

| 1849. | 1850. | 1851. | 1852. | |
|---|---|---|---|---|
| 3·89 | 4·07 | 3·89 | 4·25 | oitavas per ton. |
| ·00137 | ·00143 | ·00137 | ·0015 | per cent. |

In numerous places there are cavities and "vugs," lined with fine crystals of calc. spar and spathic iron, which, themselves, are frequently covered with smaller crystals of specular iron, magnetic pyrites, &c. When these drusy cavities are frequent, the yield of gold diminishes, and this metal is most abundant in the compact, uncrystallized, specular iron. There are three mines known as Bahu, Cachoeira, and Gambu. The principal shafts are sunk inclining about 45° with the auriferous bed, and the ore is raised on a tram-road by cars. The Bahu mine is about 1200 feet deep. As the ore comes up, it is broken by negro women, and then carried to the stamps, of which there are one hundred and twenty heads in operation, moved by three or four large water-wheels. The coarse gold is caught on cow-skins, which are changed every two hours, while the slimes are amalgamated in barrels. In each barrel are placed 80 lbs. of mercury to 16 cubic feet of slime, and the whole is allowed to revolve for thirty hours. The Tyrolian mills have been tried here and found not to succeed.

The Company own 1000 slaves, and employ about 80 or 90 Europeans as overseers, captains, mechanics, and head-men, in every department. Some 300 Brazilians are hired by the day as surface-men. The annexed table shows the quantity of rock stamped, and the amount of gold produced, for a few years past, together with the profits:—

| | 1846. | 1847. | 1848. | 1849. | 1850. | 1851. | 1852. |
|---|---|---|---|---|---|---|---|
| Ore stamped (tons), . . | 34,935 | 40,234 | 58,122 | 67,336 | 67,106 | 79.810 | 82,642 |
| Gold produced (lbs. troy), | 1,465 | 1,638 | 2,108 | 2,473 | 2,541 | 3,000 | 3,233 |
| Net profit, . . . . . | £14,820 | 21,536 | 32,269 | 38,136 | 35,880 | 51,586 | 55,391 |

The average number of stamp-heads working, during the year 1852-3, was 118·58. The quantity of stamp-sand or slimes amalgamated, was 19,709·59 cubic feet, which yielded 17·02 oitavas of gold (or 2 oz. troy, nearly) to the cubic foot. The loss of mercury was 728 lbs., or 0·037 lb. per cubic foot of slime amalgamated. The net profit in the same year amounted to £55,390 16*s.*, and the total profits of the Company, from 1838 to 1852, to £312,621; a result due mainly to economy and skill in working an ore occurring in abundance, but very poor in gold. The gold contains about 20 per ct. of silver.

*Imperial Brazilian Mining Association.*—This Company was formed in 1825, for the purpose of working gold mines in the province of Minas Geraes. They commenced by purchasing the estate of Gongo Soco, which in fifteen years produced nearly a million pounds sterling. From 1840, the returns were less satisfactory, the mine for some years not paying its expenses. Recently, however, the accounts are more encouraging, valuable new discoveries having been made, and old workings re-opened.

This Company has 10,000 shares, and £25 per share paid in; it had paid to its stockholders in dividends, up to December 1844, £380,000, having produced in ten years 35,000 lbs. of gold. The Gongo Soco mine was, in 1840, 378 feet deep. The jacotinga at this locality is 50 fathoms in thickness, and softer than that of the Morro Velho mine. The Camara lode, at present worked, is 10 fathoms in width, and composed of a series of quartz layers, or branches, containing iron pyrites, oxide of iron, and oxide of manganese.

*National Brazilian Mining Company.*—This Company was wound up in 1853, the mine having been ruined by culpable neglect on the part of the manager, who had been superseded. It had produced, in 1850, 120 lbs. of gold, containing 14 per cent. of silver.

The greatest yield of Brazil in gold was about the middle of the eighteenth century. Between 1752 and 1761 the amount on which the royal quint was paid varied from 17,000 to 21,500 lbs. yearly. From that time it gradually fell off, and was in 1822 less than 1000 lbs. The mean annual production from 1810 to 1817 of Minas Geraes, the principal mining district of the country, is given by Humboldt at 4288 lbs. At present the English companies mining in the rock furnish almost all the gold obtained. The washings have nearly ceased. The present annual produce is probably about 6000 lbs.

Chevalier calculates the grand total of gold produced by Brazil, from the earliest period up to 1845, at 3,576,192 lbs. troy. The data on which any such conclusions with regard to the former yield of Brazil are based are of very doubtful character. From 1800 up to 1850 the average seems to have been about $2,000,000 per annum.

MEXICO.—This country is pre-eminently that of silver veins. Gold, however, occurs, hardly in any other way than as a constituent of small percentage of the argentiferous ores. Yet so large is the quantity of silver produced that the accompanying gold becomes a matter of considerable importance. The silver of Guanaxuato and Guadalupe y Calvo is remarkably rich in gold, while that of Tasco, Catorce, and Zacatecas, is poor. The ores are in some instances ground

8

under the arrastras with the addition of mercury, and the silver thus obtained by amalgamation yields from 4 to 6 per cent. of gold. M. Duport calculates that in 1840, or thereabouts, the value of all the gold produced in Mexico, both from the washings and by parting from silver, was equal in weight to about $\frac{1}{135}$ of the latter metal, and, in value, to $\frac{1}{8}$. A considerable amount of gold is also obtained from the washings of Sonora. In the next chapter, when speaking of the silver of Mexico, some statistics of the yield of both the precious metals will be given.

There are gold mines in Oaxaca which have been worked in the solid rock to some extent, and which, in the opinion of Chevalier, are destined to be of importance at some future time.

Central America.—Nothing definite can be ascertained in regard to the gold-washings of Central America. Costa Rica is known to produce a small quantity; but it is mostly smuggled out of the country, so that no estimate can be made of its amount.

---

## SECTION III.

### GEOGRAPHICAL DISTRIBUTION OF GOLD IN THE UNITED STATES.

The gold regions of the United States are at least two in number, but of very unequal importance, although similar in extent. The one, that of the Atlantic slope, the "Appalachian gold-field," has been worked to a moderate extent for over thirty years; the other, that of California, "the Sierra Nevada gold-field," has, however, in the six years since it was first discovered, produced more than twelve times as much as has hitherto been obtained on the Atlantic side.

The development of the riches of California reacted upon the gold-bearing districts already known all over the world, causing a general search for the precious metal in many almost abandoned auriferous localities, and stimulating its production in an incredible degree. Among the great discoveries which were the immediate result of this excitement, the greatest was that of the Australian gold-fields, which rival with those of California itself in productiveness. On this side of the American continent the influence of California has been felt in the renewed attempts to develop the Appalachian gold deposits, which are now attracting a greater share of attention than ever before. These we will now proceed to describe, so far as the materials collected by a personal examination of some of the most important points, and those published by others, will allow. It must be premised, however, that descriptions of our gold mines are at best but unsatisfactory. The workings are, thus far, chiefly superficial, and do not possess much interest as specimens of the mining art; neither can their probable value in most cases be ascertained by inspection, since the precious metal usually lies in invisible particles within its rocky matrix. Assays of numerous samples of the ore are necessary to determine its value; and the safest guide to enable one to judge of the character of a gold mine is a correct statement of what it is producing.

The first notice of gold in the Southern States which I can find, is in Jefferson's "Notes on Virginia," in which it is stated that a lump of this metal, weighing 17 dwts., was found near the Rappahannock. Drayton's "View of South Carolina," published in 1802, also mentions the finding of a small piece in Greenville District, on Paris's Mountain. Cabarrus County, in North Carolina, was the first district where it was obtained in any noticeable quantity. In 1799, a lump of gold, said to have been of the size "of a small smoothing-iron," was found by the son of a Mr. Reed, who kept it for several years without knowing what it was or suspecting its value. It was finally disposed of by him for $3 50. Soon afterwards, gold was discovered in Montgomery County, and washings were carried on, on a small scale, for several years in these two counties, near the small streams, and nuggets of considerable size were found; one of these, from Cabarrus, weighed 28 lbs. avoirdupois, and a number of others from 4 to 16 lbs. This was at the Reed mine, and the proprietor estimated that, before 1830, 100 lbs. of gold had been obtained in pieces of over a pound in weight. At a mine in Montgomery, a number of pieces over one pound in weight had been found previous to 1830; one weighed 4 lbs. 11 oz.; another, 3 lbs. In Anson County, in 1829, one piece was obtained of 10 lbs.; one of 4 lbs.; and a number of smaller ones. The first native gold of this county was coined at the Mint in 1825; and, up to 1830, four-fifths of our gold coinage was from native metal. There had been small amounts brought to the Mint previous to 1824; but from that time the quantity increased rapidly. From 1804 to 1827, North Carolina had furnished all the gold produced in this country, amounting to $110,000. In 1829, $2,500 was deposited from Virginia, and $3,500 from South Carolina, the first which found its way from these states to the Mint. In 1830, the first deposit of Georgia gold took place, to the amount of $212,000.

Up to 1825, all the gold procured in North Carolina was from washings; about that time, however, the gold-bearing rock was discovered in place by Mathias Barringer, of Montgomery County, who obtained from an excavation 30 to 40 feet long, and not more than 15 or 18 feet deep, over 15,000 dwts. of gold. This discovery turned the attention of those searching for the precious metal from the "branch" or "deposit mines," as they were called, to the "vein mines;" and soon after Mr. Barringer's discovery some valuable quartz-veins were found in Mecklenburg County. The product of these was so great as to excite general attention, although they were worked in the rudest possible manner, and in consequence this county was pretty thoroughly explored and extensively occupied by the gold miners. The developments here were followed by others in Guilford, Cabarrus, and Davidson Counties.

About the year 1824, Prof. Olmsted made a geological reconnoisance of North Carolina, and his observations on the gold region of that state are published in Silliman's Journal for 1825.* At the time of his survey, no other than the deposit mines were known to exist, and the limits of the gold-bearing region were considered by him to embrace an area

* Vol. ix. p. 5.

of 1000 square miles, included in a circle, whose diameter would be about thirty-six miles, and whose centre was near a point eight miles west by south of the mouth of the Uwhare, and including a great part of Montgomery County, the northern part of Anson, the northeastern corner of Mecklenburg, Cabarrus, to a little beyond Concord on the west, and a corner of Rowan and Randolph. Of this region, he says, that gold may be found in almost any part of it in greater or less abundance. Its true bed, however, was considered by him to be a thin stratum of gravel, which, in low grounds, was covered by alluvium to the depth of eight feet; but when no cause had operated to alter its original depth, it was about three feet below the surface. He remarks that the gold is not confined to the beds of the rivers, but is found in the superficial formations, on the summit of elevations sometimes 100 or 200 feet above the nearest valleys.

At the time of Prof. Olmsted's survey, there were three principal localities where washings were carried on, called Anson, Reed's, and Parker's mines. At Reed's, where the large 28 lb. lump was found, the diggings occupied the bed of Meadow Creek, a branch of Rocky River, and the process by which the gold was obtained was as follows: during the dry season, the workman excavated through three or four feet of dark-colored mud full of angular fragments, until he met with the peculiar bed of gravel and clay, recognized as the gold-bearing stratum. This, which was only a few inches thick, was carefully taken up with the spade and first cradled, and then panned. The average produce per hand, at Reed's mine, was only sixty cents a day. According to C. E. Rothe,* in 1827 no gold washing was considered worth working which did not yield at least 1 dwt. per day for each hand employed. At that time the rocker was universally used for separating the gold from the sands. It was generally considered that but a small part of the gold found its way to the Mint; it was mostly worked by the jewellers, who were anxious to get it, on account of its fineness being above that of coin.

In 1829, Prof. Mitchell published a map of the gold region

* Sill. Am. Jour. xiii. 201.

of North Carolina,* on which are noted nine different mining localities, three of which are included by him in the "primary," and the remainder in the "transition or slate."

It was in the summer of 1829 that the first discoveries were made in Georgia; and in 1830 the first gold was sent to the Mint, to the amount, in that year, of $212,000. The first locality known was in Habersham County, and very soon the explorations were extended into Hall and Carroll Counties. The excitement, or "gold-fever," produced by these discoveries was extraordinary. As many as six or seven thousand persons were, soon after, engaged in washing for gold in that region. The researches were carried on, until it was ascertained that the whole of that part of the state lying along the base of the Blue Ridge was more or less auriferous.

South Carolina sent to the mint, in 1829, $3,500; this was the first arrival of gold from that state. Deposits were worked, in 1830, in Chesterfield, Lancaster, and Kershaw Districts. Brewer's mine, in Chesterfield, was one of the most productive localities, there being from 100 to 200 persons employed in 1830 and 1831, who were supposed to average from $1 50 to $3 00 each per day.

The Georgia gold excitement did not last very long; but there, as in the other Southern States where this metal was found, gold-washing continued to be followed by the majority of the washers, not as a regular pursuit, but as one to be taken up and dropped again, as circumstances directed. In 1831, the subject of establishing a Mint in the gold region was agitated, and, as each state claimed one as necessary to the development of its metallic wealth, three were at length built, and commenced operations in 1838; one at Charlotte, N. C.; one at Dahlonega, Ga.; and a third at New Orleans, which latter, however, had, previously to the Californian discoveries, hardly coined any bullion of native production. The production of gold seems to have reached its maximum in the period from 1828 to 1845. From the last-named year it was evidently falling off rapidly, previous to 1853.

* Sill. Am. Jour. Science, xvi. 1.

In 1836, Prof. Silliman* visited the gold region of Virginia, and from him we have an account of what was doing there at that time. The mines examined by him were principally in Goochland, Louisa, and Culpeper Counties.

*Moss and Busby's Mines*, Goochland County, are about fifty miles from Richmond. At Busby's mine, a shaft had been sunk fifty-seven feet, and a steam-engine erected. The auriferous quartz is coarsely granular in texture, but firm, and quite free from any foreign substances except the gold, which is mostly invisible, before being separated from it. The average yield, according to experiments made under Prof. Silliman's direction, was $8 16 per bushel of 100 lbs. The earth around the mine also yields a considerable quantity of gold.

*Moss's Mine* is three-fourths of a mile from Busby's. At the surface, the rock is entirely decomposed into a red clay. Inclination of the bedding of the rock and included quartz masses, 45°; average width of the latter about twenty-four inches. The quartz breaks readily into tabular masses, and does not adhere to the adjacent rock, hence it is easily mined. Average yield, $7 39 per bushel. Average yield at Fisher's mine, $3 15 per bushel. At the *Walton Mine*, $5 92. This mine is in Louisa County, about forty miles southwest of Fredericksburg. A large number of workings exist in this neighborhood. Some of the ore of the Walton Mine yielded as high as $133 73 per bushel. Prof. S. remarks, "That the Walton gold mine, and many others in Virginia, may be profitably wrought, admits of no doubt—provided, that in all cases, good judgment, sound economy, competent skill, adequate machinery, and strict fidelity, combine their salutary influence."

*The Culpeper Mine* is situated on the river Rapidan, eighteen miles west of Fredericksburg. At the time of Prof. Silliman's visit, operations were commencing here on an extensive scale. The main vein is said by him to be divided into a number of parallel branches, in which the gold is found to be more abundant than when they unite into a single main vein. There is a large quantity of iron in the quartz, mostly decomposed iron pyrites. The strata stand nearly vertical, and have a direction of about north 10° east. The results of a number of trials of the ore gave, as an average, a yield of $2 13 to the bushel.

The following remarks by Prof. Silliman deserve to be read with attention by all interested in this kind of property: "In my judgment, nothing could be more inauspicious to the mining interest and the welfare of the country, than a spirit of speculation in these concerns. In an excited state of the public mind, it is rare that facts are correctly reported, or correctly viewed. The speculator, who buys merely that he may sell again, is, too frequently, ignorant of the facts, and reckless of the consequences, in regard to those who may succeed him in his obligations; flattering gains from sales of stocks are reported from day to day; the property rapidly changes hands; the public mind, being morbidly excited, is of course blinded, and, at no distant period, accumulated ruin falls heavily on the last in the train."

---

* Sill. Am. Jour. xxxii. 98.

Professor Tuomey's State Geological Report enables us to form a good idea of the state of the gold mining interest in South Carolina about 1848, the time of its publication. Although a great number of localities were worked, and called mines, they must have been either on a very small scale, or else worked with loss, since the yearly produce of the state, at that time, could not have much exceeded $50,000. The following mines are noticed as the most important ones by M. Tuomey.

*Brewer's Mine,* Chesterfield District. A bed of quartz of immense thickness, 800 yards in its widest part, is here worked and found to be more or less auriferous throughout its whole extent. The upper part of this bed was so disintegrated, that it was worked for some time under the impression that it formed a part of the superficial deposit of clay, gravel, and pebbles, which was itself rich in gold, and which had been worked from 1838 to 1843. Two hundred persons were employed here in 1848, but the workings were conducted in the most shiftless manner. Other mines were worked on a small scale in this neighborhood.

*Catawba and Lynch's Creek Mines,* Lancaster, Chesterfield, and Kershaw Districts. The following mines are enumerated here: Lawson's Mine (in North Carolina, near the line), Ezell's Mine, Blackman's, Hale's, Cureton's, Belk's, Perry's, and Stevens's.

*Lawson's Mine* had been opened over a space of one mile in length, and in some places, to a depth of 40 feet.

*Blackman's Mine.* A talcose slate furnishes the gold here. The productive portions are enclosed in the barren in lenticular masses. Very extensive workings have been made.

*Hale's Mine* is of the same character. The mines of this neighborhood had been steadily worked from 1828 up to 1848. They are, however, but little more than a series of open cuts from twenty to forty feet deep.

*Fair Forest Mines,* Spartanburg and Union Districts. Among these are Nott's, Harman, Fair Forest, West's, and Bogan Mines.

*Nott's Mine* is opened in an enormous bed of quartz, which is in some places forty feet thick. This is considered by M. Tuomey to be a true vein, and to cross the slates at a small angle. A large quantity of ore has been taken out from an open cut fifteen to twenty feet deep. A perpendicular shaft ninety feet deep was sunk, and connected with the vein by a short cross-cut. Three thousand dwts. of gold were taken from eleven bushels of decomposed ferruginous matter.

*Harman's Mine.* Copper pyrites is said to form a considerable portion of the quartz bed worked here.

*Fair Forest and West's Mines.* These are in talcose and micaceous slates, which are decomposed to a depth of from 90 to 100 feet. Below that point they cease to be worked with profit. At West's Mine a shaft 115 feet deep was, in 1848, the deepest mine-work in the State.

*The Bogan Mine* belongs to this group, and has the same characters.

*King's Creek Mine; Nuckols and Norris Mine; Lockhart's Mine.* These are considered by Tuomey as being true vein-mines. They have been worked only to a trifling extent.

Having thus noticed with sufficient detail what had been done during the earlier periods of working the Southern gold mines, an attempt will be made to give a more connected and complete account of the present state and prospects of the Appalachian auriferous district, under which name it is intended to include all the gold deposits of the eastern side of the North American continent, from Canada to Georgia. As discovery after discovery has been made, the limits of the gold region have been extended far to the north, although it is not yet demonstrated that any deposits of sufficient extent and richness to be profitably worked exist north of Virginia, the results thus far showing pretty conclusively that the greatest concentration of the gold is in the State of North Carolina. Throughout the whole of this extensive region, the circumstances under which this metal occurs seem to be nearly identical, the modifications in the form of the deposits being due to local causes. The geological structure is the same from one extreme to the other.

The Appalachian chain takes its origin in Canada, southeast of the St. Lawrence, and forms a broad belt of mountain ridges, extending in a southwesterly direction to Alabama. The entire length of the chain is about 1300 miles; its breadth is variable, gradually expanding towards its centre, and contracting at each extremity. The most striking feature of this mountain system is the fact that it is made up of a series of parallel ridges, very numerous, especially in Pennsylvania and Virginia, no one of which can be considered as being the main or central chain to which the others are subordinate, but the whole forming a system of flexures which gradually open out from the southeast to the northwest, as has been made evident from the results of the geological surveys of Pennsylvania and Virginia, under the direction of Professors H. D. and W. B. Rogers. Along the southeastern edge of this great Appalachian system is a relatively narrow, undulating range, known under different names in the different States. In Vermont it is called the Green Mountains; in New York, the Highlands; in Pennsylvania, the South Mountains; in Virginia, the Blue Ridge; in North Carolina, the Smoky Mountains. The rocks of this belt, which has a width of ten or fifteen miles, are of the lower palæozoic age, but highly metamorphosed, and, for the most part, having their organic remains entirely obliterated. Still farther to the southeast lies the great auriferous belt, nearly parallel with the Blue Ridge, and not easily separated from it in geological age, either lithologically or by palæontological characters. The central axis of this belt has a direction in Virginia of about north 32° east; towards the north it assumes a more nearly north and south direction, and to the south it approaches an east and west line. Its width, where most developed, does not exceed seventy miles. This is about its extent on the borders of North and South Carolina. In Virginia it does not exceed fifteen miles. Starting from Georgia, and proceeding northward, we find it developed in the following counties: in Georgia, in Carroll, Cobb, Cherokee, Lumpkin, and Habersham Counties; in South Carolina, through the whole northwestern corner of the state, especially in the following districts: Abbeville, Pickens, Spartanburg, Union, York, Lancaster; in North Carolina, in Mecklenburg, Rutherford, Cabarrus, Rowan, Davidson, Guildford, and Rockingham; thence, through Virginia, in Pittsylvania, Campbell, Buckingham, Fluvanna, Louisa, Spottsylvania, Orange, Culpeper, Fauquier; in Maryland, Montgomery County. Beyond Maryland, to the north, the indications become fainter, and consist only in a few scattered lumps or fine scales occasionally picked up, until we reach Canada, where there is a considerable extent proved to be auriferous.

Throughout this whole extent, the auriferous belt presents rocks of nearly the same character; they are slates of every variety, intermixed with bands of a granite and syenitic character. The predominating kind of slate is talcose, passing into chloritic and argillaceous. The prevailing dip is to the east at a very high angle. In Virginia, they stand nearly vertical.

CANADA.—The auriferous district of Canada, according to Logan,* is found to comprehend an area of between 3000 and 4000 square miles. It appears to occupy nearly the whole of that part of the province which lies on the south-east side of the prolongation of the Green Mountains into Canada, and extends to the boundary between the colony and the United States. The gold has been obtained exclusively from washing the superficial deposits, and is not known to have been found in place. The deposit in which it occurs is considered by Mr. Logan to be part of the ancient drift, the "Laurentian" of Desor, alluded to in various reports of the Canada survey as "tertiary" and "post-tertiary," and containing bones and shells of animals of existing species. In the localities where the gold occurs, the coarse materials of the drift are chiefly fragments and rolled pebbles of the clay-slates and gray sandstones on which it rests; but it contains also pebbles and boulders of talcose slate and serpentine, with magnetic, specular, chromic, and titaniferous iron, and masses of white quartz, which are derived from the mountain-range bounding the district on the northwest.

Gold-washings have been carried on, on the Du Loup and Chaudière Rivers, and about 1900 dwts. were obtained during the season of 1851–52, by fifteen men. The largest nugget weighed 2 oz. On the Touffe des Pins, lumps of greater size were found; one weighed as much as 4 oz. The results, thus far, have not been very satisfactory, as far as profitable working is concerned, although it is demonstrated that the precious metal does occur over a very considerable extent of surface.

VERMONT.—Gold has been known to exist in Vermont for twenty-five years. I have had specimens of native grains from that State in my possession for a long time. According to Rev. Z. Thompson,† a lump was picked up in Newfane, in 1826, which weighed 8½ ozs. It was pure gold, with exception of some small quartz crystals attached to it,

* Geological Survey of Canada, Report of Progress, 1850–51, p. 6.

† Appendix to Thompson's Vermont, p. 48.

weighing perhaps half an ounce. The specific gravity was 16·5. The gold-formation of this state forms, according to the same authority, a narrow and irregular belt extending through the entire length of the state. The rocks which mark the line of the formation are talcose slate, steatite, and serpentine, accompanied by magnetic, specular, chromic, and titaniferous iron, and also the sulphuret and the hydrous peroxide of the same metal. Rock-crystal is common, and is sometimes traversed by rutile.

At Bridgewater, the gold occurs in quartz associated with the sulphurets of iron, copper, and lead. The quartz is in seams or beds, dipping 55° to the east, and of irregular thickness, not exceeding ten or twenty inches. The quantity obtained here seems to have been very small.

There do not seem to be any well-authenticated accounts of native gold between Vermont and Maryland. A systematic search along the line of the proper formation might reveal its presence, but there is no reason to suppose that it would be found in any considerable quantity. In Pennsylvania, the occurrence of a single grain has been noticed,* but of uncertain locality, and under circumstances which render it doubtful whether it was a genuine specimen of native gold. According to F. A. Genth, however, the lead and copper ores of Lancaster County contain distinct traces of both gold and platina.

MARYLAND. — Although gold has been found to some extent in Maryland, the amount produced seems to have been, thus far, very trifling. I am not aware of any mines of this metal worked in that state at present. It has been discovered on the farm of Samuel Elliot, in Montgomery County.† The quartz, which forms the gangue, crops out amidst a decomposed talcose slate, so that it is easily mined. The average yield of a portion assayed at the Mint is said to have been at the rate of $522 per ton.

VIRGINIA.—The product of the Virginia gold mines has been small and pretty constant, from 1830 up to the present time, the amount annually deposited at the Mint being

* C. M. Wetherill, Trans. Am. Phil. Soc. N. S. x. 350.

† Proc. Am. Phil. Soc. 1849, p. 85.

between $50,000 and $100,000. Within the last year or two, however, mining has been commenced, on a considerably extended scale, at a number of localities, the result of which will soon determine whether the auriferous quartz is sufficiently rich to be worked with profit. The general physical and geological features of the auriferous belt of this state have been elaborately described by Clemson and R. C. Taylor,* and also by W. B. Rogers, State Geologist. The talcose slates, which predominate in the gold districts, have a reddish color, and are finely laminated, their general strike being from 29° to 32° east of north. The laminæ stand nearly vertically, and contain intercalated masses of syenitic granite and protogine. The breadth of the auriferous belt is about fifteen miles. For some depth these rocks are entirely decomposed, so as to be easily worked with the pick and shovel. The matrix of the gold is invariably quartz, which, near the surface, usually has a cellular structure, resulting from the decomposition and removal of the iron pyrites with which it was filled.

It was the opinion of W. B. Rogers, that many of these quartz veins might be worked with profit. In 1836, according to his authority, numerous quartz veins had been wrought for some time in Spottsylvania, Orange, Louisa, Fluvanna, and Buckingham Counties, from many of which rich returns had been procured, and, under improved modes of operation, a still larger profit might be expected.

At the present time, the following companies are known to be working on a somewhat enlarged scale; the results of their operations, however, are not usually made public; and, indeed, but few of them have yet been got fully under way. Several English companies have purchased some of the most promising gold mines in the State, and have recently commenced working them, and it is to be presumed that they will thoroughly test the question of their permanent productiveness.

*Culpeper Mine*, on the Rapidan River, seventeen miles from Fredericks-

---

* Trans. Geol. Soc. Pa. i. 298.

† Report of the Geol. Reconnoissance of the State of Virginia, 1836, p. 67.

burg. In 1850, working twelve stamp-heads and two Chilian mills, with twenty-four men. Weekly expenses estimated at $120. Produce in seven weeks, 3,400 dwts.

*Freehold Gold Mining Company,* Orange County. This is an English company, formed in 1853, after examination of the property by Mr. Henwood. One vein, of twenty feet in width, is said to have been opened, and traced for a mile and a half.

*Liberty Mining Company.* (100,000 shares at £1.) This is an English Company, which has purchased the Vaucluse and Gryme's Mines, for which £50,000 was paid. The average yield of the ore is estimated at $8 per ton. Six shafts have been sunk, and preparations made for working a large quantity of ore. According to the Directors' report, made Sept. 30th, 1853, the amount of gold produced during the year, the mill running eighty days, was 556 oz. 6 dwts., of a fineness of 943½ thousandths. In December, 1853, the stamps were crushing 50 tons per day.

The vein or lode worked consists of five parallel bands of quartz, all bearing gold. Hydrated oxide of iron is the principal associated mineral, and the enclosing rocks are talco-micaceous slates. Previous to 1852, the mine had been worked in two open cuts, to the depth of 60 feet, 75 feet wide, and 120 feet long. From a description of these mines, published in 1848, the following information is extracted.

The Vaucluse Mine is situated in Orange County, a mile south of the Rapidan River, about seventeen miles from Fredericksburg. It was discovered in 1832, and, for a number of years, worked as a deposit mine, before any veins were discovered. There are numerous veins now known which are contained in talcose and other slates. The gold is contained in quartz, and in the adjoining slates, which are much mixed with decomposed iron pyrites. Sometimes the auriferous belt widens out to 30 or 40 feet. The present establishment was commenced in 1844 by an English Company, and extensive works were erected. A Cornish engine, of 120 horse power, was connected with six Chilian mills, and six batteries of stamps, of three stamp-heads each. The finer ores are ground in the Chilian mills, and the harder quartz rock is stamped. The amalgamation is effected by the "amalgamating bowls." The quantity of mercury annually consumed is stated at from 250 to 300 lbs. The amount of gold produced, however, is not given, nor the quantity of ore ground and stamped. The gold obtained is very fine, being from ·985 to ·990.

*Gardiner Gold Mining Company,* Spottsylvania County. Situated at the junction of the Rappahannock and Rapidan Rivers; but little work has ever been done here. A steam-engine is now erecting, intended to drive two of Gardiner's gold-crushing machines, which are estimated to crush and amalgamate 100 tons of rock per day, the expense being $3, and the yield of gold $12 50 per ton.

*Marshall Mine,* Spottsylvania County, on the Rappahannock River, twelve miles from Fredericksburg. This mine is said to be paying well. It is stated that $300,000 has been obtained up to this time, twenty hands being employed, and the depth of the workings being 100 feet.

*Whitehall Mine,* Spottsylvania County. According to Mr. Henwood, this mine is worked in a deep-blue clay slate, the veins bearing about southeast and northwest. Their principal veinstone is a hard, white quartz, sometimes marked with ferruginous stains. Occasionally they enclose large masses of slate. The gold is for the most part scantily, but pretty uniformly, scattered through the vein, perhaps more plentifully in the iron-tinged parts than elsewhere; and where the quartz is somewhat drusy, it has a tendency to a crystalline structure.

Native tellurium has been found here, according to Mr. Henwood,* disseminated through the quartz, like the gold, but in smaller quantity. Galena also occurs with these metals; and the association of ore in this part of the Virginia gold region is said by the same authority to resemble that of the Morro de San Vincente, in the province of Minas Geraes, Brazil. A telluret of bismuth, containing selenium, has also been described by Mr. C. Fisher, Jr.,† as occurring at this mine.

*Waller Gold Mining Company.* This mine is situated in Goochland County, nine miles from Columbia, and was taken up by an English company in 1853, under the advice of Professor Ansted, who recently examined and reported on it. According to that gentleman's statements, it appears that the property is traversed by the great auriferous belt of Virginia, which is made up of various schists and alternations of shale and quartz rock, more or less compact and crystalline. Some of the quartzite is hyaline and micaceous, and almost all the rocks are colored by oxide of iron. Distinct indications of gold have been found disseminated through the shales and quartz in almost every part of the property in which the experiment of panning has been tried, and especially near the veins, which have been proved to the depth of from five to thirty feet. The results are stated as having been very variable, and are not given; but they are reported as sufficiently encouraging to justify the recommendation that the mine should be opened and proved by working it to a considerable depth.

The estimated yield of the ore is £4 per ton. In October, 1853, the Company was erecting buildings and machinery preparatory to working the mine on an extensive scale.

*Garnett and Moseley Mines,* Buckingham County. These mines were purchased by an English company, some time since, and are now among the most extensively worked in Virginia. Three steam-engines, of from thirty to sixty horse power, have been erected, by which seventy-two stamps and other machinery are driven. At the time the present Company came into possession, several shafts had been sunk, the deepest of which was 110 feet, at which point the vein is said to be fifteen feet wide, and to be worth $20 per ton. These mines were taken up on the recommendation of Professor Ansted, who is said to have reported very favorably on their prospects. According to Mr. A. Partz,‡ there are five or six veins on the property, two of which are worked. The principal one has a course of north 35° east, and dips 40° to the southeast; and the other runs in nearly the same direction, and dips so as to unite with the first. About forty miners are employed, and sixty surface-hands.

---

* Eng. Mining Journal, No. 956.
† Sill. Am. Jour. (2) vii. 282.
‡ Mining Magazine, ii. 378.

*London and Virginia Gold and Copper Mining Company.* This Company was formed in London and incorporated in Virginia in 1853, for the purpose of purchasing and working the Eldridge Mine, in Buckingham County, near the Garnett and Moseley Mines. Operations have as yet hardly been commenced. The assays of the ore gave from 1½ to 1¾ oz. of gold to the 2000 lbs.

*Buckingham Gold Company.* This Company was also organized in 1853, to work a mine on the same vein as the one last-mentioned. According to Mr. Partz,* the vein has a course of north 25° east, and dips to the northwest 75°. A steam-engine of forty horse power is erecting, and a set of twenty-four stamps. The main shaft is to be sunk to the depth of two hundred feet.

The following description of the Buckingham Mines is given by W. J. Henwood, F.R.S., F.G.S., &c.† "They are situated near Maysville, in Buckingham County. In this neighborhood the lowest visible member of the series is a homogeneous, lead-colored clay slate, rather fissile, and somewhat contorted, which is succeeded by white quartzose mica slate, closely resembling the itacolumite of Brazil, but of no great thickness. The gold formation follows, and is overlaid by a thin-bedded, pale greenish-white talcose slate, which is the upper part of the deposit. The strike of all these beds is about (true) northeast and southwest, and their dip is 40° or 50° towards southeast. The auriferous deposit has been wrought to a depth of about twenty-six fathoms in the northeastern part of the mines; it varies in width from three to twenty feet, and the shallower portions consist for the most part of quartz, sometimes vesicular, sometimes granular, always very slightly coherent, generally much mixed with earthy, brown iron ore; irregularly dispersed through the silicious ingredients are masses of iron pyrites, copper pyrites, frequently invested with earthy, black copper ore, and vitreous copper ore; galena, and the phosphate of lead, occur in like manner, although in much smaller quantities; and in the cellular cavities of the quartz there are numerous little grains of gold, of which some present crystalline forms. In the southwestern extremity of the mines, the dimensions of the gold-bearing rock are both smaller and more regular than in the opposite, averaging from four to five feet in thickness; its composition is much more exclusively quartzose, and it includes masses (horses) of the adjoining micaceous and talcose rocks; its metallic contents are of much the same nature as those found at the other end of the mine, but their quantities are much smaller. Throughout the mines, in descending, the other ingredients of this formation, as well earthy as metallic, gradually give place to iron pyrites, which at length constitutes the entire body of the rock. The whole formation is very similar to that of Morro Velho, in Brazil, except that it is much more exclusively pyritous and softer. Gold and copper are not frequently found together, although copper pyrites is occasionally found among the auriferous pyrites of Morro Velho, and also near Dolgelly, in Merionethshire. Tellurium is associated with gold in many parts of Virginia, and there, as well as in Brazil, the gold thus associated is of excellent quality. The Garnett and Moseley, Waller and Whitehall Mines, are mentioned as localities of this rare substance."

---

* Mining Magazine, ii. 379. † E. M. J. Jan. 29, 1853.

In Fluvanna County, Commodore Stockton is well known to have been for some years engaged in working gold mines, and experimenting on gold-crushing and amalgamating machinery. No authentic account of these operations has been published.

At the so-called "Tellurium mine" an ore, considered by Dr. Genth a telluret of bismuth, has been found, and has given a name to the locality.

North Carolina.—Judging from the returns of gold hitherto furnished by this state, as well as from a personal examination of a portion of the gold district, I am inclined to consider the chances of successful mining as greater here than in any other of the Atlantic States. Several companies have been recently organized, nominally for the purpose of more energetically and systematically testing the value of some of the mines best known in the state, but with regard to some of them it is difficult to say how far they are *bona fide*, and how far mere speculative concerns.

*Gold Hill Mines.* These are the most extensive gold mines in the Atlantic States, and are supposed to have yielded more than any others. They are situated in Rowan County, thirty-eight miles northeast of Charlotte, and fourteen from Salisbury. There were, formerly, a great number of different parties owning and working the Gold Hill veins; but, during the year 1853, it is understood that the property was purchased by a New York Company, with a nominal capital of $1,000,000, divided into 200,000 shares. According to a report of Dr. Asbury, of Charlotte, the greatest depth to which the mines had been worked, at the time this Company took possession, was 340 feet; and the quartz is said to be wider at that depth, as well as richer in gold, than on the surface. According to the same authority, these mines have yielded $1,500,000 since 1843: the expense of raising it is not stated. The average yield is said to have been $1 50 per bushel, although a considerable portion of the gold remained in the tailings. These were purchased some time since, in order to be reworked by Berdan's machine; but the results of this operation have not as yet been made public. This mine is said to be paying regular dividends.

*McCullock Copper and Gold Mining Company.* This mine is situated about twelve miles from Greensboro, Guildford County. The vein, which crosses the tract diagonally, has a length on it of about half a mile and had been worked over a length of about 1600 feet, before the present Company entered into possession, in June 1853. The deepest shaft at that time was ninety feet perpendicular; but the principal workings were from the bottom of the sixty-feet level upwards. The inclination of the vein to the horizon varies from 24° to 30° to the eastward; but at the lowest level, the dip is somewhat more nearly vertical than it is above. Its average width is about six feet. The wall-rock is slate, talcose and micaceous. The principal portion of the vein is quartz, both compact and cellular; its metallic contents are gold, iron pyrites, copper pyrites, and carbonate and oxide of iron. The richest ore is a soft, ferruginous mass, resulting from the decomposition of spathic iron apparently. This mass lies on the foot-wall of the vein principally, and is variable in width and thickness, forming irregular bunches.

The auriferous portion varies from one to several feet in thickness. In the sixty-feet level, a thick belt of undecomposed iron pyrites rests upon it; and upon that is a mass of quartz, in which are disseminated bunches and masses of pure copper pyrites, in some places in considerable quantity. This ore is generally accumulated in the lower portion of the quartz. The workings were chiefly above the sixty-feet level, which had been driven about 1200 feet; the ninety-feet level had been extended about 300. Between these two levels there was a considerable amount of rich ore left standing. Eight shafts have been sunk; and the facility with which this kind of work can be executed in the soft decomposed rock, is evident from the fact that one of these was sunk to the depth of sixty feet, and timbered up in one week. There are peculiar facilities in working this mine, since the decomposition has extended to an unusual depth, and the decomposed portion of the vein lying on the foot-wall can be removed without blasting, and then the solid part above can be thrown down with little trouble or expense. About $31,000 of gold was said to have been taken out, in 1852, at an expense of $500 a month. The crushing and amalgamating machinery consisted of two Chilian mills, combined with the Sullivan bowls, and apparently much better and more economically arranged than is usual in the Southern gold mines; but, unfortunately, two miles distant from the mine.

The present Company is said to have twenty-five head of stamps and seven Chilian mills in operation, which number will be increased to ten. No reports of the quantity of gold obtained have been made public.

The stock is divided into 200,000 shares, at a nominal par value of $5, = $1,000,000 capital.

The mines, with machinery as purchased, cost $135,000.

*Conrad Hill Mining Company,* Davidson County, about six miles from Lexington. The property belonging to Governor Morehead is that properly known under the name of the Conrad Hill Mine; but a Company, formed on the adjacent lands, has taken the same name. According to Dr. Genth, there are six veins on the Morehead estate, which consist, as usual, of quartz, accompanied by a large amount of the hydrated oxide of iron, together with the specular and spathic ore. Near the water level considerable copper pyrites has been found with these ores. In some parts of the mine there are bodies of ore of four feet in width, worth $2 per hundred pounds.

On the property of the "Conrad Hill Mining Company," into which the same veins, in part, extend, I observed, in April, 1853, two shafts, one of which was said to be 115 and the other 100 feet deep, both well timbered up. The mine was at that time filled with water. In the rubbish about the mine, considerable spathic iron and copper pyrites was noticed.

*Vanderburg Mining Company,* Cabarrus County, twenty-two miles from Charlotte, adjoining the Phœnix Mine. Company organized in 1853. According to J. T. Hodge, Esq., the rock formation is greenstone, which is traversed by several veins, with a course of north 50° to 65° east. One of these has been worked to a depth of 100 feet, and found to have a thickness of from a few inches to 3½ feet. Besides rich bunches of ore, a large quantity has been mined yielding about $2 per bushel. Considerable copper pyrites is mixed with the veinstone. The property is considered, by Mr. Hodge, to be of great value, both for gold and copper.

*Phœnix Gold Mining Company,* Cabarrus County. From the report of the agent of the mine, it appears that there are several estates owned by the Company, of which the principal is the "Conner tract," so called, on which are three veins, on one of which, the "Sulphur Vein," workings have been carried down to the depth of 170 feet, and the vein has been traced for 3000 to 4000 feet, with a width of from one to three feet. The "Orchard Vein" is from one to six feet wide, and contains large quantities of heavy spar, with some copper pyrites. The average yield of the vein, as heretofore worked, is given at $1 per bushel.

A vein has been opened near Pioneer Mills, in Cabarrus County,* consisting of quartz, with brown hematite and pyrolusite, both in the form of carbonate of iron. These are accompanied by interesting minerals containing tungsten, such as wolfram, scheelite, tungstate of copper, and tungstic acid. The vein shows visible particles of gold, and the ore is considered to be worth $1 per bushel. The geological formation is granite.

*Long and Muse's Mine,* Cabarrus County. From a manuscript report on this mine, kindly furnished by Dr. Genth, the following notices are extracted. The geological formation is a diorite slate, passing into chloritic and talcose slates. There are six or eight veins on the property, of which four are considered as worth working. No. 1 has a course of about north 7° east, and dips to the west about 50°: its width is from two to five feet. It consists of white quartz, with iron pyrites, copper pyrites, brown hematite, and galena. It is reported to have yielded from 5 to 100 dwts. to the 2000 lbs. A sample of the refuse ore gave, on assay, 59 cents of gold to the 100 lbs. No. 2, a short distance east of No. 1, has nearly the same course, and is 18 inches in width. Its contents are nearly the same as those of No. 1. A rough experiment, in a drag-mill, showed a yield, in the ore from this vein, of 55 dwts. per 2000 lbs., of a fineness of 721·7. No. 3 intersects veins No. 1 and 2, and has a course of north 31° east. Near the surface it was 18 inches wide, and, at 35 feet in depth, about 3 feet. The ore of this vein yielded 65 dwts. to the 2000 lbs., of the same fineness as that obtained from No. 3. In order to show that by no means all the gold can be obtained by amalgamation, Dr. Genth operated on a portion of the ore, containing about 60 per cent. of iron pyrites, 10 per cent. of galena, and the remainder quartz, and obtained only $47 33 of gold and $1 12 of silver, by the aid of mercury; while the tailings contained $68 75 of gold, and $0 52 silver, per 2000 lbs. The pure galena of this vein yielded 23 oz. of silver per 2000 lbs., which contained from 3 to 4 per

---

* Communicated by Dr. Genth.

cent. of gold. No. 4 is from 18 inches to 2 feet wide, and shows visible particles of gold in connection with the other minerals named above. The ore yielded $1 12 per 100 lbs. by assay.

*Lemmond Mine.* From Dr. Genth I have also received some information with regard to this interesting mine, which is situated in Union County, eighteen miles from Concord. There are numerous quartz veins, contained in a very ferruginous talcose and clay-slate. On this property veins No. 3 and 8 belong to the fahlband class of deposits, consisting of belts of slate impregnated with sulphuret of iron, now decomposed into the hydrated oxide, and containing considerable quantities of gold. In No. 8 the rich part of the slate is ten inches wide. It produced by amalgamation 2·15 oz. of fine gold per 2000 lbs. and the tailings yielded by assay 1·85 oz. of an alloy of gold and silver in nearly equal quantities. In the space of about fifty yards are four veins, two of which are at present highly productive. Veins 11 and 12 are especially worthy of notice. No. 11 is a quartz vein of from six to eleven inches in width, and a course of north 43° east. It has been worked to a depth of twenty feet, and consists of laminated quartz, colored with oxide of iron, and showing, throughout, particles of gold. The ore is said to produce from 15 to 20 dwts. to the bushel. By amalgamation it gave, according to Dr. Genth, 25·6 oz. per 2000 lbs. of fine gold and 7·5 oz. of silver, and there still remained in the tailings 4·8 oz. of the former and 2·7 oz. of the latter metal. Vein No. 12 contains some rich pockets producing, by assay, 16·88 oz. of gold and 9·95 of silver to the 2000 lbs. One of the most interesting veins on this property is No. 15, which is a quartz vein having a course north 20° east. Its general width is about twelve to eighteen inches, but at twenty feet in depth it widened out to six feet. It contains, besides quartz, zinc-blende, galena and small quantities of iron and copper pyrites. Some very rich pockets have been met with, yielding ores worth $500 per bushel. Some of the refuse ore collected at this place gave by amalgamation 1·55 oz. of gold and 1·10 oz. silver per 2000 lbs.; but only a small part of the metallic contents were extracted by mercury, since the tailings gave, by assay, 11·76 oz. of gold and 17·90 oz. of silver to the 2000 lbs. An examination of the galena from this mine showed the interesting fact that it contained 29·4 oz. of gold and 86·5 oz. of silver, to the 2000 lbs. I know of no galena equalling this in richness.

There was in the spring of 1853 a great number of abandoned mines near Charlotte in Mecklenburg County, with regard to the value of which no opinion could be formed, since the workings had in almost every instance gone to decay and become filled with water.

At the *Capp's Mine* the openings were scattered about over a wide extent of ground as if a cluster of veins had been wrought. On panning a portion of the ore it was found to be rich in gold, although it was impossible, even with a magnifying glass, to discern a particle in the stuff taken out of the mine. A large amount of gold is reported to have been obtained here, the richest ore yielding as much as $8 per bushel. No copper pyrites was observed at this locality.

*Mecklenburg Gold and Copper Company,* Mecklenburg County, near Charlotte. This Company has been formed to work the Rhea and Cathay

Mines, which have been described by Mr. Partz.* According to him, the Rhea property is crossed by a cluster of veins which have formerly been worked superficially, the ore yielding from $1 to $5, per hundred pounds. This mine is about nine miles from Charlotte. The Cathay Mine, which is five miles from the same town, has also a promising vein, carrying ore rich in gold and considerable copper pyrites.

There are several other Companies which have been recently organized for the purpose of working some of the old mines of Cabarrus and Mecklenburg Counties, but no definite information with regard to them has yet reached me.

SOUTH CAROLINA.—The gold mines of this state had become almost entirely deserted, previous to 1852, when the discovery of the richness of the Dorn Mine, attracted the attention of capitalists in this direction.

*The Dorn Mine* is situated in the lower end of Abbeville District. It is the property of a single individual, and has a wide reputation as the richest gold mine of the Atlantic States. Some of the best ore is stated by Mr. Dorn's cashier to be worth $2000 per bushel, a yield of $5 per 100 lbs. being considered a poor one at this mine. After many years of persevering labor in exploring without success, Mr. Dorn struck the rich bunch of ore in February, 1852. Up to July, 1853, an amount equal to $300,000 is said to have been taken from an excavation a little over three hundred feet long, twelve feet deep, and fifteen wide. The expense is stated at $1200, $202,216 09 having been obtained in one year by a simple Chilian mill, driven by two mules, and working fifteen bushels per day. Pieces of gold weighing as much as sixty dwts. are said to have been found.

The vein is said to have a course of north 78° east, and to have widened out, from eighteen inches at the surface, to fourteen feet at a depth of ten feet below the water-level. Of course such rich bunches of ore cannot be expected to continue for any great distance; and the returns of the deposits at the Mint from the State of South Carolina, which rose from $19,000 in 1850 to $126,982 in 1852, seem to be again falling off, indicating that the best portion of the vein has been already excavated.

*The Dorn Mining Company* was formed in New York, to work a property adjacent to the above-described mine. They are reported as having erected a powerful steam-engine and two of Berdan's machines. Their lines run within a few hundred feet of the rich bunch of ore worked by Mr. Dorn. No report of their progress has been published.

Other mines, mentioned by M. Tuomey, which have been taken up within a recent period, mostly by New York capitalists, are Ezell's Mine, by the *Yorkville Mining Company*, Hale's Mine, and Lawson's Mine.

* Mining Magazine, ii. 380.

GEORGIA.—In this state the yield of gold, which once reached half a million of dollars per annum, has fallen off gradually, until, in 1853, the amount deposited at the Mints was only $58,896.

Efforts have been recently made to revive some of the abandoned mines. Among those taken up, are Moore's Mine, near Dahlonega, by a company called *The Georgia Gold Company*, and the Lawhorn Mine.

TENNESSEE AND ALABAMA.—These states have each produced a few thousand dollars annually of gold, during the last twenty years; but there is no reason to suppose that their future yield will ever be of any considerable importance.

NEW MEXICO.—In regard to New Mexico, it seems almost impossible to obtain any definite information, although it is known that gold is produced here in some quantity. The amount deposited at the Mints is very small, and varying from a few hundred to several thousand dollars. Dr. Wislizenus* describes two localities near Santa Fe, the Old and New Placer. The "Old Placer" is twenty-seven miles from that city, the predominating rocks being white and yellow quartzose sandstone, quartz, hornblende rock, sienite, and diorite. At the "New Placer," nine miles from the town, both washing and mining in the rock are carried on. Two mines were worked at that time. The vein of one of these is described by Dr. Wislizenus as being contained in sienite and greenstone, and having a quartzose and ferruginous gangue. The same description applies to the mines of the "Old Placer." Together, these washings and mines are said, by the same authority, to have yielded at various times from $30,000 to $250,000 per annum.

CALIFORNIA.—We come now to speak of a country whose golden wealth surpasses anything yet known to have been discovered, and which, in its influence on the march of the world, is to be ranked among the great events of modern times. Up to a very recent period, the Pacific side of our continent had remained an uninhabited and almost unvisited region, of which we knew almost nothing, save what might

* Memoir of a Tour to Northern Mexico, 1846-7, published by Congress, p. 24.

be gleaned from an occasional scientific expedition across the desert plains, or the returning ship of the whaler and dealer in hides; and it seemed to require some extraordinary event to direct the tide of emigration thither, to a land then inaccessible, except by a six months' voyage around Cape Horn, or a perilous journey, nearly as long, over unexplored and desert plains. This event came in the demonstration of the simple fact that, throughout a vast extent of that distant country, gold in unstinted quantity was to be had for the trouble of digging it, and that it was free to all who chose to go thither and take it. Then commenced a wandering of nations without a parallel in history. The perils and sufferings of the overland route were braved with indomitable energy; the Pacific was whitened with the sails of nations pressing to the land of gold; and, for the first time in the world's history, representatives of every people and climate hurried towards a common centre of attraction.

The existence of gold in California has been long known, although it was not until the region passed into the possession of the United States that this knowledge became a tangible and universally recognized thing. The same was the case with the Australian gold discoveries; it required the right moment and the right man, before they could acquire a national importance. There can be no doubt that the existence of gold was well known to the Jesuit fathers, who in reality governed the country before it was taken from the Mexicans. A vein of gold-bearing quartz was even worked, near the mission of San Fernando, by a Frenchman, M. Baric, in 1843, according to Duflot de Mofras.* An English naval officer brought a magnificent specimen of gold-bearing quartz rock from that region more than thirty years ago, and others had done the same.

The first effectual discovery of gold, however, in this state, was made in the spring of 1848, probably either in February or early in March. The earliest authentic information which reached this country must have been received in August or September: one of the first communications published was

* Duflot de Mofras, Ex. du Territoire de l'Oregon, i. p. 489.

a letter from Rev. C. S. Lyman, who happened to be in California at the time, dated San Jose, March 24, 1848, and published in Silliman's Journal for September. Mr. Lyman, in this communication, remarks as follows: "Gold has been recently found on the Sacramento, near Sutter's Fort. It occurs in small masses, in the sands of a new mill-race, and is said to promise well." The first gold discovery was entirely the result of accident. Col. Sutter, a retired Swiss officer of the guard of Charles X., had contracted with a Mr. Marshall for lumber which necessitated the erection of a saw-mill, on the South Fork of the American River, at a place now called Coloma. It was completed in the spring of 1848, and in letting the water flow with a strong current through the tail-race, a certain quantity of the glittering particles were washed out, so as to be quite conspicuous, and were immediately recognized by Mr. Marshall as gold. Naturally enough, it was desired to keep the discovery secret, but among inquisitive Yankees this was hardly possible, for any length of time. Col. Mason, who was then Governor of California, was led, by the rumors of gold discoveries, to visit the locality, and early in July he found the gold diggings a scene of the utmost excitement, the news having spread through California and completely emptied San Francisco of its inhabitants, then a few hundred in number. At the time of his arrival, the number of persons employed in washing, on the American River and its branches, was already as many as four thousand, who were supposed to be obtaining from $30,000 to $40,000 per day. In November, Mr. Lyman estimated the amount taken out, up to that time, at from four to five millions of dollars.* He asserts that from five to ten ounces was not an uncommon quantity for a single person to wash out in a day, while some had obtained as many pounds in the same time.

Information of these extraordinary discoveries spread far and wide, during the winter of 1848–9; and on the commencement of the dry season of 1849, the immigration may be said to have commenced on a large scale; and first from

* Sill. Am. Jour. (2), vii. 291.

Mexico, Peru, and Chili, and other mining states on the Pacific coast; afterwards from the Sandwich Islands, China, and New Holland; and last from the United States and Europe. The arrivals from the United States by sea did not begin to be numerous until July and August, 1849, and the overland emigration commenced pouring in early in September. Naturally, the first arrivals were chiefly foreigners, Mexicans and Chilians, and it is supposed that there were 15,000 of these at the mines in July, but the overwhelming tide of emigration from the United States swept them from the country in a short time. Not long after the receipt of Col. Mason's despatches, the Department of State, at Washington, selected Mr. T. B. King as bearer of despatches and agent to examine and report on the population, productions, and resources of California. This gentleman arrived at San Francisco, June 4, 1849, and his report, the first official document of any importance relating to California, was dated March 22, 1850. In the mean time, he had made a rapid tour through the gold region, and collected some interesting information in regard to it. Mr. King was not a scientific observer, and his opinions in regard to the geological structure of the country are very crude and incorrect, as was shown by Mr. P. T. Tyson, who accompanied him in a part of his tour of observation. His estimates, however, in regard to the future probable yield of gold, startling as they appeared at first, have proved to be rather under, than over, the mark. During the years 1848 and 1849, Mr. King supposes that about $40,000,000 were obtained from the washings, there being, before the close of the last-named year, between 40,000 and 50,000 Americans and 5,000 foreigners engaged in digging. This amount had been principally taken from the northern rivers, or those which empty into the Sacramento, the branches of the San Joaquin having been up to that time but little resorted to. Mr. King estimated the probable yield of gold for 1850 at $50,000,000, a sum of startling magnitude, which, however, the actual yield probably exceeded rather than fell below.

In regard to the geological features of California, as far

as detailed observations of scientific and unprejudiced observers are concerned, we are almost entirely destitute of them, so that great difficulty is experienced in giving anything like a complete idea of the position of the gold in all its relations to geology. It is to be regretted that a geological survey of this interesting region has not been instituted, since were such to be carried on by competent men, the results would possess a high degree of practical and scientific interest. At present it is impossible, from the conflicting accounts of the gold-region, to draw such general conclusions, with regard to the mode of occurrence of the gold, as would enable us to throw much light on the probable future of that country. The most reliable information is that conveyed by the figures stating the arrivals of the gold. These are sufficient to show that California has held, and still holds, the first rank as a gold-producing country, and that it is likely to continue to do so for a long period to come.

The great valley of California is drained by two principal rivers, the one flowing north and the other south, which unite midway, and make their way through a side-cut in the mountain ranges into the ocean. These streams are the Sacramento and the San Joaquin, and, with their numerous affluents, they drain the auriferous district of California. This immense basin or trough has a length of about 500 miles, and a breadth of from 50 to 100. On its eastern rim rise the grand masses of the Sierra Nevada, the continuation of the Cascade Range of Oregon, whose summits are elevated above the limits of perpetual snow. The central axis of the Sierra seems to be made up of granite rocks, through which volcanic fires have occasionally found vent, and piled up lofty masses of debris. This granitic axis is flanked by heavy accumulations of slaty rocks, in which the talcose varieties predominate: these alternate with trappean and serpentine masses, which extend to the valley of the Sacramento, where they are concealed by sedimentary deposits of recent origin. On the western edge of the great valley is the Coast Range, a series of elevations parallel with the coast, and but a few miles distant from it.

The slates of the Sierra Nevada, forming a belt forty or

fifty miles wide, and extending through the valley, are the gold-bearing rocks of the region.

They form precipitous and broken ranges cut through by deep gorges, through which, in the rainy season, impetuous torrents find their way to the valley. The average fall of the streams from the summit of the Sierra to the valley of the Sacramento is stated, by Mr. Tyson, at 180 feet to the mile, which would cause a current in the streams, when swollen by the melting snows, capable of wearing out those stupendous ravines, some of which exceed 3000 feet in depth.

The geological age of the gold-bearing slates remains undetermined; but they are similar in lithological character to those of other great gold regions of the world. There can be no doubt that they are very ancient, but whether of azoic or palæozoic age, it is impossible to say with absolute certainty. As similar formations, in some parts of the chain of the Andes, have been found to contain fossils of Silurian age, and as this is the period to which the Uralian gold rocks may, beyond a doubt, and those of Australia, with great probability, be referred, it is not improbable the auriferous rocks of California belong also to the palæozoic epoch. Their organic contents have been obliterated, however, and they have assumed a crystalline structure, being everywhere invaded and broken up by igneous masses.

Lower down on the flanks of the Sierra, the summits of the ridges are capped with sedimentary rocks, which rest nearly horizontally on the upturned edges of the slates, showing that they had been deposited after the latter had taken their present form and position. The conglomerates, which, according to Mr. Tyson, cover many of the hills between the Consumes and Calaveras, are found at the height of more than 2000 feet, and consist of pebbles, with a more or less ferruginous cement. They are interstratified with beds of sandstone, beneath which are heavy deposits of indurated clay. The whole thickness of this formation is estimated at 200 feet. This group of sandstone and conglomerates apparently extends under the valley of the Sacramento, and appears again in the ridges of the Coast Range, where its stratification is much broken and disturbed. They are considered to be identical, in geological age, with the rocks of similar character which occur in Oregon, and which belong to the Miocene division of the tertiary.

Thus far almost the whole of the gold of this region has been obtained from the superficial deposits, or the loose sand, gravel, and boulders lying upon the rocks in place, and they are found within the mountain districts of the western flank of the Sierra, in the river valleys, and far up in the mountain gorges. According to Prof. James Blake,* the extent of these deposits is commensurate, or nearly so, with the gold region itself. They are met with as we advance from the summit of the Sierra into the lower hills at its base, and extend some miles into the plain. Towards the base of the Sierra, the conglomerate and gravel are found in greater abundance, and the pebbles and boulders are larger; they often cover extensive plains surrounded by low ridges of porphyry and slate. As we ascend towards the axis of the chain, these deposits become more extensive; and, at a distance of twenty or thirty miles from the lower hills, they are found occupying the crests of almost all the highest ridges, when their extent frequently does not exceed a few yards in breadth. Their depth is extremely variable; sometimes a few rounded pebbles only remain, nearly the whole mass having been swept into the adjacent valleys; in other places, particularly on elevated plains, they are spread out in a pretty uniform thickness of a few feet. Frequently they attain a much greater extent, and are piled up in horizontal strata of different materials to a height of several hundred feet. These deposits are said by Professor Blake to be especially developed at Nevada and Mokelumne Hill. At the former place, they form the crest of a high mountain called the "Sugar-Loaf," 2000 feet above the level of the adjacent stream, of which the upper 600 feet are formed of stratified drift; the same material forms the upper 200 feet of the high mountain called Mokelumne Hill. In the lower valleys, among the less elevated ranges, the detrital materials consist of beds of gravel, mixed with boulders and fragments of the harder

* Sill. Am. Journ. (2), xiii. 386.

rocks. On the elevated flats, higher up in the mountains, the upper portion is made up of a reddish loam, mixed with small gravel, and, beneath, is a stratum of larger boulders, principally of quartz, scattered through the coarse drift. On the summits of the hills, the strata attain their greatest thickness, and are usually nearly horizontal. It appears that these accumulated masses of drift are highly auriferous, since the ravines descending from the ridges on which they are found are generally extremely rich; when they extend over a large surface on the elevated flats, gold is always met with diffused through the gravel immediately above the rock on which they rest, to the amount of from fifteen to forty cents to the bushel of dirt. Prof. Blake is of the opinion that these drift deposits are the secondary source from which the gold found in that region has been derived. He also remarks, that there is reason to believe that there is in these deposits a supply of gold which it will require centuries to exhaust.

It will be seen that the opinions with regard to the age of the recent stratified deposits resting on the flanks of the Sierra Nevada differ; by some they are referred to the tertiary, by others to the drift epoch. If Professor Blake's opinion is correct, and they are really of the period indicated by him, and the true depository of the gold, it is a fact of the greatest interest. In that case, California would present a striking analogy with the Uralian gold-fields, and the opinion of Murchison would be fully sustained, that the auriferous detritus is in no way to be confounded with the alluvial deposits of the present day, modern causes having done little more than concentrate the gold already abraded, by washing out the sedimentary matter with which it was imbedded.

From all the accounts published by numerous visitors to California, it would seem evident that the mode of occurrence of the gold in the rock is, in its main features, similar to that of the Southern Atlantic States. It is associated with quartz rock, which forms bands, or segregated masses, parallel with the stratification or lamination of the slates. But in California the quartz seems to be more solid and compact, less associated with gossan and iron pyrites, and less decomposed and softened in its superficial portions. On the other hand, the "veins" are much thicker, as is evident from the fact that they are almost universally called *ledges*, and probably richer; I say probably, for there are no certain data on which to build an assertion, nor could such be procured, even on the spot, for the majority of the gold-quartz mining companies have never gone on the principle of open and candid dealing with their stockholders or with the public. Such enormously rich specimens were shown in the Atlantic and English cities, and such stories of fabulous wealth visible in the quartz ledges were related, that it was enough to destroy all confidence in gold-quartz mining. Had one hundredth part of these stories been true, the companies ought ere this to have made the fortunes of their stockholders. At present, it is hardly possible to say that any of them have succeeded, but it is certainly true that the majority have proved utter failures. This does not necessarily demonstrate that there are not quartz-veins in California which might, under more favorable circumstances, be worked with profit. There are mines, like those of St. John del Rey, which, after long experience acquired in working them and with economical management, pay a handsome return for the capital invested, and that too when the yield is small in proportion to the quantity of ore stamped; but without judgment and economy a rock ten times as rich might be worked with loss.

In the beginning of 1853, there were at least twenty Anglo-Californian gold-quartz mining companies in the London market, representing nearly 2,000,000 shares, and an investment of about $10,000,000. How much of this nominal capital had been actually applied to mining it is impossible to say, but many of them had no existence except on paper; some, whose stock at that time stood considerably above par, had never obtained a foothold in California, and no one had paid anything like its expenses. There are many drawbacks to successful quartz-mining, aside from wilful dishonesty or simple incapacity. The first and greatest, undoubtedly, is the unequal distribution of the gold in the quartz; it cannot be depended on like the Brazilian gold-bearing rock, but is here rich and there poor, so that a large quantity of dead ground must be broken, and it is not always possible to separate the rich ore from the worthless. But the high price of labor, the cost of materials, expense of transportation, and scarcity of fuel and water, must be taken into consideration, and under all these disadvantages it will be evident that to make a paying mine would require a percentage of gold vastly higher than would be necessary in a country where labor was cheap, and other facilities abundant. As long as the placer-mines continue to yield as they are now yielding, so long it seems evident that the quartz mining companies will be laboring under a great disadvantage.

The most extensive operations thus far undertaken in California, belong rather to the department of hydraulic engineering than that of mining. They are chiefly of two classes, one comprehending the turning of the current of the larger streams, in order to lay bare their beds and wash out the gold; the other consists in digging canals and building sluices, so as to lead the water of some of the conveniently situated rivers on to the elevated flats, where there would otherwise be no means of washing. Very extensive works of the first-mentioned kind have been carried on in the valleys of the American River and its various branches, and over a million and a half of dollars are said to have been expended on this stream in this way; and it is reported that the yield of gold was not sufficient to repay the expenses. The companies engaged in these works labor under many serious difficulties, such as the scarcity of lumber, and the want of means of transportation, as well as the rapid evaporation and leakage from the long flumes they are forced to build. One of the most extensive of these undertakings, is the Bear River and Auburn Canal, which is said to be nearly forty miles long, and to have as much more in length of side-ditches, or sluices, for leading off the water to the various points where it is required.

The machinery used for washing the gold out of the earth is various in character. The common pan was the first implement, and one which still does good service, although much more complicated implements have been introduced. The "cradle," or rocker, of the southern mines, was the first improvement on the pan, and is still much used on account of its cheapness and portability. Various forms of the cradle and rocker have been introduced. That which seems most in vogue at present is the machine called a "tom," somewhat resembling the "Burke rocker" of the southern gold mines, but stationary, the gold being separated by the force of a current of water, and collected in the "riffle-box," or "riffle-board." The whole object of these various contrivances is to imitate nature, as nearly as possible, in the way in which she has separated the gold from the detritus in the channels of the mountain streams. "Sluicing," which is also much in use, and considered by many the best process, is the closest possible copy of what takes place where a stream of water flows through the auriferous sands of a rapidly inclining valley. The "sluice" is a sort of "washing-table," of extraordinary length, and comparatively narrow, some of them being one hundred feet in length, by two in width. One end is raised at a considerable angle, so that the current flows through, separating the dirt from the gold, which sinks and is retained, when it arrives near the bottom, by cleats or "riffles," placed across the sluice near its lower end. Many improvements have been attempted in the form of the riffles, but no one has been universally recognized as superior to the others.

The amount of gold produced by California is next to be considered; and the inquiry as to what that region has yielded and is still yielding, is one of great interest. And in the first place, in order to throw as much light as possible on this subject, the following table is appended, which shows the amount deposited at the United States Mint and all its branches by each state, from the period of the earliest discoveries up to the end of the first quarter of 1854. The amount for each state is specified separately, with the exception of the ten years from 1838 to 1847, during which time the accounts of the Mint were not kept in such a manner as to allow of the different states being separated, with the exception of Virginia.

TABULAR STATEMENT OF THE VALUE OF GOLD OF DOMESTIC PRODUCTION DEPOSITED AT THE UNITED STATES MINT AND ITS BRANCHES, FROM EACH STATE, FROM 1804 TO 1854.

| | Virginia. | N. Carolina. | S. Carolina. | Georgia. | Tenn. | Alabama. | New Mexico. | Oregon. | California. | Various. | Total for each Year. |
|---|---|---|---|---|---|---|---|---|---|---|---|
| 1804–23, . . . | | $47,000 | | | | | | | | | $47,000 |
| 1824, . . . | | 5,000 | | | | | | | | | 5,000 |
| 1825, . . . | | 17,000 | | | | | | | | | 17,000 |
| 1826, . . . | | 20,000 | | | | | | | | | 20,000 |
| 1827, . . . | | 21,000 | | | | | | | | | 21,000 |
| 1828, . . . | | 46,000 | | | | | | | | | 46,000 |
| 1829, . . . | $2,500 | 134,000 | $3,500 | | | | | | | | 140,000 |
| 1830, . . . | 24,000 | 204,000 | 26,000 | $212,000 | | | | | | | 466,000 |
| 1831, . . . | 26,000 | 294,000 | 22,000 | 176,000 | $1,000 | | | | | $1,000 | 520,000 |
| 1832, . . . | 34,000 | 458,000 | 45,000 | 140,000 | 1,000 | | | | | | 678,000 |
| 1833, . . . | 104,000 | 475,000 | 66,000 | 216,000 | 7,000 | | | | | | 868,000 |
| 1834, . . . | 62,000 | 380,000 | 38,000 | 415,000 | 3,000 | | | | | | 898,000 |
| 1835, . . . | 60,400 | 263,500 | 42,400 | 319,900 | 100 | | | | | 12,200 | 698,500 |
| 1836, . . . | 62,000 | 148,100 | 55,200 | 201,400 | 300 | | | | | | 467,000 |
| 1837, . . . | 52,100 | 116,900 | 29,400 | 83,600 | | | | | | | 282,000 |
| 1838, . . . | 55,000 | | | | | | | | | | 435,100 |
| 1839, . . . | 57,600 | | | | | | | | | | 404,208 |
| 1840, . . . | 38,995 | | | | | | | | | | 426,185 |
| 1841, . . . | 25,736 | | | | | | | | | | 542,117 |
| 1842, . . . | 42,163 | 2,898,505 | 406,040 | 3,582,033 | 50,446 | $155,107 | | | | 24,650 | 777,097 |
| 1843, . . . | 48,148 | | | | | | | | | | 1,045,445 |
| 1844, . . . | 40,595 | | | | | | | | | | 967,200 |
| 1845, . . . | 86,783 | | | | | | | | | | 1,019,281 |
| 1846, . . . | 55,538 | | | | | | | | | | 1,129,357 |
| 1847, . . . | 67,736 | | | | | | | | | | 889,085 |
| 1848, . . . | 57,886 | 473,543 | 40,577 | 257,063 | 7,161 | 14,462 | $682 | | $45,301 | | 896,675 |
| 1849, . . . | 129,382 | 485,793 | 24,564 | 236,349 | 5,180 | 10,700 | 32,889 | | 6,151,360 | 2,927 | 7,079,144 |
| 1850, . . . | 65,991 | 355,523 | 19,459 | 209,587 | 1,507 | 6,538 | 5,392 | | 36,273,097 | 1,220 | 36,938,314 |
| 1851, . . . | 69,052 | 326,883 | 41,052 | 157,213 | 2,377 | 3,962 | 890 | | 55,938,232 | 951 | 56,540,612 |
| 1852, . . . | 83,626 | 403,295 | 126,982 | 96,542 | 750 | 254 | 814 | | 53,794,700 | | 54,506,963 |
| 1853, . . . | 52,200 | 275,622 | 99,317 | 58,896 | 149 | | 3,632 | $13,535 | 55,113,487 | 5,213 | 55,622,051 |
| | $1,403,431 | $7,848,664 | $1,085,491 | $6,361,583 | $79,970 | $191,023 | $44,299 | $13,535 | $207,316,177 | $48,161 | $224,392,334 |
| 1854 (3 months), | 4,631 | 62,332 | 8,044 | 4,167 | | | | 445 | 11,031,479 | | 11,111,098 |

(The N. Carolina, S. Carolina, Georgia, Tenn., Alabama and Various values shown on the 1842 row are bracketed totals for 1838–1847.)

The striking superiority in richness of the California gold fields over those of the Atlantic slope of the continent is made very evident by an inspection of the above table. From it, it appears that up to the end of 1853 there had been deposited as follows:—

| | |
|---|---|
| From the Atlantic States, . . . . . . . . | $16,970,162 |
| New Mexico, Oregon, and various sources, . . . | 105,995 |
| California, . . . . . . . . . . . | 207,316,177 |
| | $224,392,334 |

In the above item of over two hundred and seven millions of dollars deposited from California, we have a groundwork on which to base a calculation as to the amount produced by that state. If we compare this sum with the actual shipments from San Francisco, as manifested, we find that it exceeds them by over 3,500,000 dollars.

The following sums are given as the value of the gold dust manifested and shipped from San Francisco for each year since the opening of the gold fields:—

| | |
|---|---|
| 1849 and 50, . . . . . . . . . . | $68,587,591 |
| 1851, . . . . . . . . . . . . | 34,492,634 |
| 1852, . . . . . . . . . . . . | 45,801,321 |
| 1853, . . . . . . . . . . . . | 54,907,005 |
| | $203,788,551 |

As a considerable portion of the amount manifested was for England, South America, and other foreign countries, it is seen at once that large sums must be taken from the country in the hands of passengers, and in other ways, without being manifested.

The highest yield seems to have been in 1851, when the auriferous fields were not yet in any degree exhausted, and the number of miners was perhaps as great nearly as now. There may have been a slight falling off in the latter part of 1853, but it is not apparent from any returns which could be obtained. It is probable that the year 1854 will show a falling off from 1853, and that the produce will continue slowly to decline from year to year.

The following statement is presented as an approximation to the grand total of the produce of California up to the end of 1853, most of the items being from actual returns.

| | |
|---|---|
| Deposited in United States Mint and branches up to Dec. 31, 1853, . . . . . . . . . . . | $207,316,177 |
| Shipments to foreign ports in 1848–49, and 50, est., . . | 10,000,000 |
| Taken out of the country by foreign miners, chiefly Mexicans, in 1848 and 49, est., . . . . . . . | 10,000,000 |
| Shipped to Europe in 1851, . . . . . . . | 3,392,760 |
| Shipped to South American ports in 1851, . . . | 2,372,000 |
| Shipped to Europe in 1852, . . . . . . . | 6,000,000 |
| Shipped to South American and Asiatic ports in 1852, . | 1,000,000 |
| Shipped to England in 1853, . . . . . . . | 5,000,000 |
| Shipped to other ports in 1853, . . . . . . | 1,600,000 |
| In circulation in California, in transitu, and otherwise absorbed, estimate, . . . . . . . . . . | 13,319,063 |
| | $260,000,000 |

After carefully examining all the statements made by the California bankers and others, and the amounts deposited each year at the Mints, I have divided this sum in the following manner, as the yield of each year:—

| | | | | lbs. troy. |
|---|---|---|---|---|
| 1848, . . . | $5,000,000 | representing | of pure gold | 20,150 |
| 1849, . . . | 20,000,000 | " | " | 80,600 |
| 1850, . . . | 45,000,000 | " | " | 181,400 |
| 1851, . . . | 65,000,000 | " | " | 262,000 |
| 1852, . . . | 62,500,000 | " | " | 252,000 |
| 1853, . . . | 62,500,000 | " | " | 252,000 |
| | $260,000,000 | | | 1,048,150 |

In the preceding pages of this chapter various statistics of the principal auriferous districts of the world have been placed before the reader. In order to facilitate a comparison of the results, they are herewith presented in a tabular form. In the first table the data for the Eastern Hemisphere are given, so far as they have been obtained, from the year 1800, the amounts, which are in lbs. troy, being specified at first for each tenth year, then for each fifth year, and, from 1845 up to the present time, for every year. The amounts given for Africa and the East Indies are mere estimates;

most of the other figures are from reliable authorities, and the spaces are left blank when such could not be obtained.

| | Russia. | Sweden. | Harz. | Austria. | Italy. | Spain. | Africa. | S. Asia and E. Indies. | Australia. |
|---|---|---|---|---|---|---|---|---|---|
| 1800, . . . . . | 1,440 | | | | | | 660 | | |
| 1810, . . . . . | | | | | | | | | |
| 1820, . . . . . | 1,930 | | | 2,682 | | | | | |
| 1825, . . . . . | 11,300 | | | 3,034 | | | | | |
| 1830, . . . . . | 16,600 | | 6 | 3.397 | 250 | | | | |
| 1835, . . . . . | 18,100 | 2 | 6 | 4,301 | | | | | |
| 1840, . . . . . | 25,650 | 2 | | 5,114 | | | | | |
| 1845, . . . . . | 60 800 | | | 5,406 | | ('44) 28 | | | |
| 1846, . . . . . | 75,600 | | | 5,720 | | | | | |
| 1847, . . . . . | 80,100 | | | 5,662 | | 49 | | | |
| 1848, . . . . . | 75,950 | | | 5,645 | | | | | |
| 1849, . . . . . | 71,700 | | | | | 28 | | | |
| 1850, . . . . . | 65,600 | 1 | 6 | | | | | | 0 |
| 1851, . . . . . | 68,550 | | | | | | | | 30,000 |
| 1852, . . . . . | 63,950 | | | | | | | | 330,000 |
| 1853, . . . . . | | | | | | | 4,000 | 25,000 | 210,000 |

In the subjoined table the same system is pursued with regard to the Western Hemisphere. Here, however, the reliable data are few and far between, and in hardly any instances are they much better than mere estimates. The zero (0) is introduced in both tables, to indicate the period previous to which no gold had been obtained in the country the name of which stands at the head of the column. California is separated from the other United States, as being geographically, if not politically, distinct.

| | Chili. | Bolivia. | Peru. | N. Grenada. | Brazil. | Cent. Am. and Mexico. | Mexico. | California. | U. States. |
|---|---|---|---|---|---|---|---|---|---|
| 1800, . . . . . | 7,500 | 1,600 | 2,400 | 12,600 | 10,000 | 4,300 | | | |
| 1810, . . . . . | | | | | | | | | |
| 1820, . . . . . | | | | | | | | | |
| 1825, . . . . . | | | | | 1,565 | | | | 80 |
| 1830, . . . . . | | | | | | | | | 2,000 |
| 1835, . . . . . | 2,500 | | | | | | | | 3,100 |
| 1840, . . . . . | 2,800 | | | | 6,700 | | | | 1,900 |
| 1845, . . . . . | 2,850 | 1,200 | 1,900 | 13,300 | 5,096 | | 7,925 | | 4,500 |
| 1846, . . . . . | | | | | | | 9,900 | | 5.000 |
| 1847, . . . . . | | | | | | | | 0 | 3.900 |
| 1848, . . . . . | | | | | | | | 20.150 | 3.750 |
| 1849, . . . . . | | | | | | | | 80.600 | 4,100 |
| 1850, . . . . . | | | | | 5,668 | | | 181,400 | 2,950 |
| 1851, . . . . . | | | | | | | | 262,000 | 2,700 |
| 1852, . . . . . | | | | | | | | 252.000 | 3,150 |
| 1853, . . . . . | | | | | | | | 252,000 | 2,200 |

In order still farther to elucidate the remarkable fluctuations which have taken place in the production of gold since the beginning of the present century, a third table is appended, in which is given the yield of the principal gold-producing countries at the commencement of the present century, in 1845, and for each year from 1850 to 1853, the blanks in the two preceding tables being filled, where necessary, with the amount produced in the nearest year for which returns could be obtained, or with the best estimate which could be formed of the probable quantity. After each absolute sum, its relative weight, in comparison with the grand total produced throughout the world, is given, so that the varying importance of each country at different epochs, in regard to its yield of gold, will be seen at a glance.

| | 1800. | | 1845. | | 1850. | | 1851. | | 1852. | | 1853. | |
|---|---|---|---|---|---|---|---|---|---|---|---|---|
| Russian Empire, . | 1,440 | 2·7 | 60,800 | 47 0 | 65,600 | 20·6 | 68,500 | 15·8 | 64,000 | 8·9 | 64,000 | 10·8 |
| Austrian Empire, | 3,500 | 6·5 | 5,400 | 4·2 | 5,600 | 1·7 | 5,650 | 1·3 | 5,700 | 0·8 | 5,700 | 0·9 |
| Rest of Europe, . | | | 300 | 0·3 | 100 | . . | 100 | . . | 100 | . . | 100 | |
| Southern Asia, . | 10,000 | 18·5 | 20,000 | 15·5 | 25,000 | 7·8 | 25,000 | 5·8 | 25,000 | 3·5 | 25,000 | 4·2 |
| Africa, . . . . | 660 | 1·2 | 4,000 | 3·1 | 4,000 | 1·2 | 4,000 | 1·0 | 4.000 | 0·6 | 4,000 | 0·7 |
| Australia, . . . | . . | . . | . . . | . . | . . . | . . | 30,000 | 7·0 | 330,000 | 45·9 | 210,000 | 35·2 |
| Chili, . . . . . | 7,500 | 13·8 | 2,850 | 2·2 | | | | | | | | |
| Bolivia, . . . . | 1,600 | 3·0 | 1,200 | 1·0 | | | | | | | | |
| Peru, . . . . . | 2.400 | 4·4 | 1.900 | 1·5 | | | | | | | | |
| New Grenada, . | 12,600 | 23·4 | 13.300 | 10·3 | 34,000 | 10·8 | 34,000 | 7·8 | 34,000 | 4·7 | 34,000 | 5·7 |
| Brazil, . . . . | 10,000 | 18·5 | 5.100 | 4·0 | | | | | | | | |
| Mexico, . . . . | 4,300 | 8·0 | 9,900 | 7·6 | | | | | | | | |
| California, . . . | . . | . . | . . . | . . | 181.400 | 57 0 | 262 000 | 60·7 | 252.000 | 35·1 | 252.000 | 42·2 |
| United States, . | . . | . . | 4.500 | 3·3 | 2,950 | 0·9 | 2,700 | 0·6 | 3,150 | 0·5 | 2,200 | 0·3 |
| | 54.000 | | 129.250 | | 318.650 | | 431.950 | | 717.950 | | 597.000 | |

## SECTION IV.

### PLATINA, AND ITS ASSOCIATED METALS.

The occurrence of platina resembles very much that of gold, with which it is generally found associated. It is one of a family or group of metals, which have a striking likeness to each other, and are only found in company. These are iridium, rhodium, osmium, ruthenium, and palladium. The first four of these metals are almost inseparable from platina, and closely resemble it in many of their properties: palladium differs from it in some important respects. They are distinguished by their infusibility at any but the very highest temperatures, their high specific gravity, and their insolubility and capability of resisting the action of air, moisture, and nearly all chemical reagents.

Platina, as it occurs in nature, is usually in small grains, although it is sometimes found in pieces weighing several pounds. It is always alloyed with other metals, and generally contains from 5 to 10 per cent. of iron and a trace of copper, as well as a small percentage of iridium, rhodium, osmium, and palladium; the platina itself usually constitutes from 75 to 85 per cent. of the alloy. This metal was first brought to Europe from the river Pinto, in South America, where it was discovered by Ulloa, about 1736. It was then known as *platina del Pinto*, platina being the diminutive of the Spanish, *plata*, silver. For many years it remained almost useless, on account of the very property which now makes it invaluable to the chemist and manufacturer, its infusibility. It was not until Wollaston introduced the method now in use for working this metal, that it could be applied in the arts to any extent. This process consists of a precipitation of the platina, in a solution effected by nitro-chlorohydric acid, by means of chloride of ammonium, which throws down a double chloride of platina and ammonium, in which the platina can be reduced to the metallic state by simple ignition. It is, as thus obtained, in the form of a very fine black powder, which is strongly heated and compressed in steel moulds, and thus welded together, and afterwards hammered into the proper shape. Its chief use is in apparatus for purifying sulphuric acid; the retorts which are constructed from it for this purpose being of large size and very costly. Besides, it is invaluable for the purposes of the analytical chemist, being infusible, and not liable to be acted on by the simple acids; were it a more common metal, it might be used to great advantage in many other ways.

The metal, when chemically pure, is harder than copper, but softer than iron; it is extremely ductile and tenacious, and is the heaviest substance known, being about 21½ times the weight of water.

*Iridosmine.*—This is an alloy of the metals iridium and osmium in varying proportions, which is usually associated with native platina. It is of a tin-white color, and excessively hard and heavy. It is found in the form of minute grains, the largest of which sell for a very high price, for the purpose of making the tips of gold pens.

*Platin-iridium.*—An alloy of platina and iridium, in which the iridium largely predominates. Found with the native platina of the East Indies.

Rhodium and Ruthenium have not been found, except in minute quantities, alloyed with native platina.

*Palladium.*—This metal occurs in nearly pure grains in Brazil, associated with native platina, and also in minute proportion in the latter substance itself. It was first discovered by Wollaston, in 1803. Quite large masses were formerly brought from Brazil. Its color is a light steel-gray, and it has a specific gravity of a little over 11. It has most of the valuable properties of platina, but in an inferior degree. It is, however, soluble in nitric acid.

The uses to which it has been put are very limited; it has been employed, in a few instances, in the construction of instruments for chemical and physical research; but the advantages which it presents over other metals do not seem to compensate for its costliness.

### GEOLOGICAL POSITION OF PLATINA AND THE ASSOCIATED METALS.

This family of metals is of rare occurrence, and has a marked geological position. For a long time they were only found in the gold-bearing alluvia, having been accidentally discovered in washing for that metal; but they were discovered in place in the province of Choco, in New Grenada, by Boussingault, in 1825, and, since that time, they have been detected in their native bed in the Ural, and other localities. The platiniferous rocks all have a similar character, and consist of greenstone, diorite, or a similar igneous eruptive rock, through which the platina is disseminated in grains and small masses. In the Ural, the rock in which it occurs is apparently a serpentine. A very large quantity of chromic iron is found in the washings, from which the platina is obtained, this being a mineral which ordinarily occurs in serpentine. At Nijny Tagilsk the platiniferous detritus is composed mainly of pebbles of serpentine, intermixed with chromic iron; and Le Play has proved that the platiniferous alluvia are only found near masses of serpentine, from which rock he even succeeded in obtaining a small quantity of the metal by careful washings. It is evident, however, that this metal is very little concentrated; but, on the contrary, disseminated through the rock in such a manner that it never could have been collected in any perceptible quantity had not natural causes, as is the case with gold, performed the washing process on an immense scale, removing a comparatively great amount of rock, and leaving the heavy, unoxidizable metal, which it contained, so concentrated near its original position, as to be quite easily gathered.

### GEOGRAPHICAL DISTRIBUTION OF PLATINA.

RUSSIAN EMPIRE.—Although platina was not discovered in Russia until nearly a hundred years after it had been found in South America, the quantity obtained there has much exceeded that from all other sources. The mode of its occurrence has already been spoken of. The gold washers first began to notice it about 1819, at the mines of Neiwin, and they called the platina grains *white gold.** They were not recognized, however, by any one as platina, until 1823, when Prof. Lubarski, at St. Petersburg, determined their true character.

According to G. Rose,† platina is widely distributed through the Ural Mountains. It occurs in the northern extremity of the chain, at Bogoslowsk and Kuschwinsk; towards the middle, especially at Newjansk and Werch-Issetsk; and in the southern districts, near Kyschtimsk and Miask. It has been remarked, that it is chiefly found in the western slope of the Ural, while the gold exists on the eastern side in the largest quantities. By far the largest portion of this metal has been obtained at Nijny Tagilsk; and next to that locality, at Goro Blagodat, near Kuschwinsk. The largest nugget ever found was from the former locality; it weighed 22·33 lbs. troy.

* Humboldt in Pogg. Annal. vii. 517. † Reise nach dem Ural, ii. 389.

The whole amount of native platina obtained, from its first discovery in 1824 up to 1851, was 2061 poods (90,451 lbs. troy), of which 1990 poods were from the washings of Nijny Tagilsk. The washings, at this locality, yielded, in 1828, 91 poods (3993 lbs. troy), averaging 40 zolotniks to 100 poods of sand, or ·01 per cent. The yield per annum of the Russian washings, during the years from 1827 to 1834, which seem to have been the most productive in this metal, was from 4000 to 5000 lbs. troy. Up to 1845, platina was coined in the Russian mint, but the use of it for that purpose was dropped at that time, and the produce has greatly fallen off, the locality of Nijny Tagilsk, formerly so rich, having ceased to be worked altogether. The whole produce of Russia is said to have amounted, in 1847, to only 2 poods, or 87·7 lbs; and in the following year to very little more. It is mostly exported to France.

France.—The existence of platina in France is but a matter of scientific interest thus far, as it has been found only in minute quantity. It was first discovered in the departments of La Charente and Deux-Sèvres, in the smallest possible quantity, in brown hematite ore.* Since that time, in 1847, M. Gueymard has found it in the Alps, on the mountain called the Chapeau, in the department of the Hautes Alpes.† Here it occurs, associated with gray copper ore, in a metamorphic limestone. The ores contain considerable silver, as well as antimony, lead, zinc, and other metals. It has since been discovered at four different points in the French Alps, at considerable distances from each other. The quantity thus far obtained has only been very minute, and not by any means sufficient to make the existence of this metal in France of any commercial importance.

Germany.—In 1848, Prof. Pettenkofer in Munich, made the interesting discovery that the old Brabant coins, called "Kronenthaler," which were worked up in large quantities in the mint at Munich in order to separate the gold which they contained, were platiniferous.‡ It had been frequently observed that, in separating gold from silver in the moist

* L'Institut. No. 46, p. 102.

† Ann. des Mines (4), xvi. 495.

‡ Pogg. Annal. lxxiv. 316.

way, certain silvery-looking particles could not be dissolved either in boiling sulphuric or nitric acids. This was found, on examination, to be owing to the presence of platina; 100 parts of the gold which remained undissolved containing on the average gold 97, silver 2·8, platina 0·2. These dollars were coined before platina had come into use at all, and it is evident that this metal must have formed a portion of the argentiferous ores from which they were derived.

It was the opinion of Pettenkofer, that platina was present in quite perceptible quantity in almost all the German silver coinage; and he estimated its average amount in the coins melted at the Munich mint at ·01 per cent. He also suggested a process by which he considered it possible to separate this very small quantity with profit; and actually produced, from the slags of the year 1847, fifteen ounces of pure platina.

SPAIN.—Vauquelin is said to have obtained platina from the argentiferous lead and copper ores of Guadalcanal. No farther confirmation of his results have yet been had; but, since the discoveries of Pettenkofer, it is not unreasonable to suppose that this metal may be present in small quantity in many silver ores.

EAST INDIA.—Although it is quite difficult to procure any reliable statistics of the metallic produce of the southeastern portion of Asia and the adjacent islands, it seems evident that platina occurs throughout that part of the world in comparatively large quantities. It was discovered in Ava (Burman Empire) in 1830,* in an artificial alloy, and afterwards in the native state. It occurs in the gold-bearing alluvia, associated with spinel, emerald, quartz, and magnetic iron sand.

BORNEO.—This island, so rich in metals, seems also able to supply a great quantity of platina. According to Dr. Ludwig Horner it occurs on the southeastern end of the island, where the rocks are chiefly serpentine, diorite, and gabbro. In the district of Pulo-Ari† these rocks are covered, in the valley and along the river, with a deposit of clay from ten to twenty feet thick, under which is an auriferous bed, not sharply defined, but varying from one to four feet in thick-

* Asiatic Researches, xviii. pt. 2, p. 279.

† Zeitschrift der Deutsch. Geol. Gesellschaft, ii. 408.

ness. This stratum contains, together with magnetic iron sand, grains of platina and iridosmine. Diamonds are also obtained here, and, according to Dr. Horner, in the washings carried on for the purpose of finding these precious gems, a large quantity of platina is obtained and allowed to remain useless, not being exported. It is said that the amount of this metal washed out is equal to one-tenth of the weight of the gold obtained.

More recently the value of this metal has begun to be better understood, and a considerable quantity is furnished to commerce by this island, but there do not seem to be any reliable statistics of its amount.

South America.—The first discovery of platina was made in South America, and for a long time this continent furnished all of this metal that was known. The gold washings of Choco and Barbacoas, on the western slopes of the Andes, were the first to yield platina, and it was afterwards found at numerous points along the flank of the Cordilleras, from the second to the sixth degree of north latitude. The districts which furnished the largest quantity were those of Condoto, in the province of Novita, Santa Rita, Santa Lucia, Iro and Apoto. For a long time that which was washed out was thrown away, under the mistaken idea that it might be used to debase the gold, its chemical properties and consequent commercial value being then entirely unknown. For many years the occurrence of this metal in the rock and its geological associations were not understood; but, in 1825–6, Boussingault succeeded in detecting it in place at the gold mines of Santa de Osos, in the province of Antioquia. At that point the gold was worked in quartz veins in a decomposed sienite, the quartz being associated with hydrated oxide of iron in large quantities. The gold and platina occur under precisely similar circumstances at this locality. There is very little platina coming from this region at the present time; but it seems quite impossible to procure any accurate statistics of this metal. Brazil has also furnished a large quantity of platina from its gold washings, especially those of Matto Grosso. In that country it is associated with palladium as well as with gold and diamonds.

The whole amount of the metal obtained from the western continent was estimated by Boussingault, some twenty years since, at between 800 and 900 lbs. per annum. Brazil still furnishes a small quantity, but it seems to be decreasing, and it is probable that we shall have to look to the East Indies for farther supplies of this essential metal.

The island of Hayti has also yielded a small quantity of platina, which was obtained from the sands of the river Jacky.

CANADA.—Platina has been detected, according to T. S. Hunt,* in the gold-washings of the Rivière du Loup, where it is found sparingly mixed with the gold, in minute scales and grains. Associated with it were found small plates of iridosmine, the native alloy of the rare metals iridium and osmium. Specimens of these metals are also said to have been obtained on the Rivière des Plantes.

UNITED STATES.—According to Dr. Genth, traces of platina have been found in the lead and copper ores of Lancaster County, Pennsylvania; but no grains of the native metal are known to have been discovered north of North Carolina. In that state a single grain is stated by Prof. C. U. Shepard† to have been obtained in Rutherford County, in gold-washings belonging to Mr. T. T. Erwin. The grain weighed 2·541 grains, and had a specific gravity of 18.

The occurrence of platina in California, associated with the gold, has been repeatedly noticed; but it has not been supposed to be present in sufficient quantity to become an object of commercial importance. M. Dillon, Consul-General of France in California, in an official communication,‡ mentions that platina is found almost everywhere in connection with the gold, and he thinks that a considerable quantity might be obtained, were it not rejected from ignorance of its nature. The latest official report of the United States Mint speaks of the existence of an appreciable percentage of this metal in the deposits of gold received from Oregon during the year 1853. I was informed by Prof. Booth, Melter and Refiner

* Geol. Survey of Canada: Report of Progress for 1851–52, p. 120.

† Sill. Am. Jour. (2), iv. 280.

‡ Ann. des Mines (5), i. 598.

at the Philadelphia Mint, in 1852, that the native gold received from California, did contain an appreciable trace of platina, but not a sufficient quantity to be worthy of being separated.

No statistics of the production or consumption of platina can be given beyond the value of the amount imported into this country, which was as follows:—

| | |
|---|---|
| 1848, | $12,778 |
| 1849, | 10,285 |
| 1850, | 11,283 |
| 1851, | 26,836 |

# CHAPTER III.

## SILVER.

### SECTION I.

#### THE MINERALOGICAL OCCURRENCE AND GEOLOGICAL POSITION OF SILVER.

MINERALOGICAL OCCURRENCE.—Silver occurs in nature in the following forms:—

NATIVE METALS AND ALLOYS.

*Native Silver;* sometimes almost chemically pure; but usually it contains copper, gold, platina, bismuth, antimony, or other metals. Native silver occurs very frequently in connection with the usual argentiferous ores, and sometimes in large masses. The most remarkable have been obtained at the mines of Kongsberg, in Norway, and at the Freiberg Mine. One is cited from the former locality exceeding 500 lbs. in weight. It is found generally, however, in arborescent and filiform shapes, sometimes looking almost like a bunch of dark-colored wool. It is frequently crystallized, the cube and octohedron, and intermediate forms, being most usual. It has been shown that native gold is never found except in association with silver; and sometimes the latter metal so predominates over the gold, that the alloy might be called native silver, rather than native gold. Almost all the silver obtained from the ores contains gold, but generally only in minute quantity.

*Bismuth Silver.*—An alloy of silver and bismuth, with a little copper and arsenic; a rare mineral from Chili.

### AMALGAM.

*Native Amalgam.*—Not a very unfrequent combination. It consists of one atom of silver and two or three of mercury.

There is an amalgam which is of economical importance; it is found in the mines of Arqueros, in Chili, and contains, according to Domeyko, six atoms of silver to one of mercury. Its percentage yield is, silver, 86·49, mercury, 13·51. In general appearance this latter variety resembles native silver.

### ORES.

The ores of silver which are of the greatest interest are the following:—

*Silver Glance.*—Vitreous Sulphuret of Silver; the ore of this metal which is of the greatest economical importance. When pure, it contains silver, 87·04, sulphur, 12·96.

*Stephanite.*—Brittle Sulphuret of Silver. The next ore in importance; a sulphuret of silver and antimony, containing silver, 70·4, antimony, 14·0, and sulphur, 15·6.

*Pyrargyrite.*—Ruby Silver. An important ore in the Mexican mines. Its composition, when pure, is silver, 58·98, antimony, 23·46, sulphur, 17·56. It contains the same substances as the Stephanite, but in different proportions.

*Horn Silver.*—Chloride of Silver. This ore contains silver, 75·33, and chlorine, 24·67. It is somewhat common in the Chilian mines.

Besides these, there is a great variety of other combinations in which silver occurs, and which are of greater or less importance, some of them being only mineralogical curiosities. A complete list of the known combinations is appended.

*Sulphurets, Arseniurets, Seleniurets, and Tellurets.*

1. Silver Glance.
2. *Hessite;* telluret of silver; found only in Siberia.
3. *Naumannite;* seleniuret of silver; occurs in the Harz.
4. *Eucairite;* seleniuret of copper and silver; found only in Sweden, in minute quantity.
5. *Stromeyerite;* sulphuret of copper and silver; rare.
6. *Antimonial Silver;* combination of silver and antimony; more common than 5, but not abundant.
7. *Flexible silver ore;* ferro-sulphuret of silver; very rare.
8. *Sternbergite;* sulphuret of silver and iron.

9. *Miargyrite;* sulphuret of silver and antimony; very rare.

10. PYRARGYRITE.

11. *Proustite;* light-red silver ore; sulphuret and arseniuret of silver; occurs in Saxony and Bohemia.

12. *Freieslebenite;* antimonial sulphuret of lead and silver; not uncommon. It contains about twenty-two per cent. of the latter metal.

13. *Polybasite;* complex combination of silver, copper, antimony, arsenic, and sulphur; not unfrequent in Mexico.

14. STEPHANITE.

15. *Xanthokon;* sulphuret and arseniuret of silver.

*Chlorides, Iodides, and Bromides.*

1. HORN SILVER; chloride of silver.
2. *Iodic Silver;* rare; found in Mexico and Spain.
3. *Bromic Silver;* occurs in Mexico and Chili.
4. *Embolite;* bromide and chloride of silver; very rare.

*Carbonate.*

*Selbite;* carbonate of silver; rare.

GEOLOGICAL POSITION OF THE ORES OF SILVER.—As already shown, hardly any one of the metals presents itself under a greater variety of combinations, and its mode of occurrence, in a geological point of view, is equally complex. It is a metal of almost universal diffusion. The researches of Malaguti and Durocher have shown that, out of 219 specimens of ores taken at random, only one in seventeen was found destitute of silver.* It is contained, in minute traces, in the water of the ocean and in the organic kingdom. With gold and lead it is most intimately associated; the fact that it is never absent from native gold has already been mentioned, and it is almost equally difficult to find a specimen of sulphuret of lead, in which at least a trace of it may not be detected.

The silver of commerce is drawn from three sources, each one of which requires to be separately considered. One of these, the native alloy of gold and silver, native gold, has already been sufficiently illustrated in the preceding chapter —it is of quite subordinate importance, in comparison with the entire quantity furnished by the world. But, in a region where the yield of gold is very large, the silver, combined

* Ann. des Mines (4), xvii. 3.

with it may amount to a considerable sum in value. In the United States, it is almost the only source from which the supply of this metal which we furnish is drawn; and, without a silver mine within our territory, we produce as much of this metal as Sweden, Norway, and France together, from their mines, some of which are of classic celebrity.

The second source of supply for silver is found in the mines of argentiferous lead ores, which are so numerous in every part of the world. Before the discovery of the Mexican and Peruvian mines, the silver obtained in this way was of great importance, and it forms still a notable portion of the whole amount produced. Of the silver furnished by Europe, that of Russia, Great Britain, the Harz, and Spain is derived almost exclusively from the working of silver-lead ores. This branch of the subject will be considered in the chapter devoted to Lead, in order to present, at one view, all the important facts with regard to the occurrence of the two metals.

In the present chapter, only those mining districts will be taken up where the ores are mainly of the third class, namely, the silver ores proper, where this metal predominates in importance over those associated with it. On the American Continent, Mexico and the Cordilleras of South America furnish examples of this mode of occurrence of pre-eminent magnitude; in Europe, on the other hand, the only mining districts of importance in which the silver ores are worked by themselves to any considerable extent, are those of the Erzgebirge in Saxony and Bohemia, of Norway, and of Hungary and Transylvania, which will be considered in the present connection.

The ores of silver have a wide range in the column of geological formations. They occur in true fissure-veins in the oldest crystalline rocks, and, in such positions, have been worked to very great depths without appearing to diminish essentially in richness. Of this form of occurrence, the mines of Freiberg and Kongsberg present the most striking examples. The great bulk of the silver ores of South America have, however, a different position, being in the calcareous rocks, and forming segregated veins and masses parallel with the stratification. The age of the rocks in which those of Bolivia and Peru are included seems to be that of the carboniferous. In Chili the stratified rocks, in which the ores of silver are chiefly developed, belong to the cretaceous formation. The particulars of their position will be given in the following section.

---

## SECTION II.

### GEOGRAPHICAL DISTRIBUTION OF SILVER ORES.

THE great source of the silver of the world is the American continent, and its discovery brought about a complete revolution in the relative production of the precious metals, and changed the whole aspect of commerce and prices. Previous, however, to the discovery of America, there were silver mines worked in Europe, which are still furnishing a quantity of the metal not entirely insignificant, and which have formerly been of great importance. Among the most interesting are those of Kongsberg, which will be first noticed.

NORWAY.—The celebrated mine of Kongsberg was discovered in 1623, and has been worked, with some interruptions, and with varying success, up to this time. In the early part of the eighteenth century, it was peculiarly prosperous. From the middle of that century to 1830, it languished along, under the direction of the government, which vainly attempted to dispose of it to a private company. From that period, however, a great improvement took place, and the works became very profitable, and continued so up to the present time, in a considerable degree. The whole production of the Kongsberg mines, in silver, was:—

| | | | |
|---|---|---|---|
| From | 1624 to 1805, . . . . | 2,360,140 marks, = | 1,580,800 lbs. troy. |
| " | 1805 to 1815, . . . . | 38,012 " | 25,460 " |
| " | 1815 to 1834, . . . . | 114,374 " | 76,600 " |
| | | 2,512,526 | |
| And in | 1835, . . . . . . . . . | 21,700 marks, = | 14,535 lbs. troy. |
| | 1836, . . . . . . . . . | 31,242 " | 20,924 " |
| | 1837, . . . . . . . . . | 23,240 " | 15,560 " |

For the five years from 1849 to 1853, the average annual profits have been about $160,000, the average production in that period being 16,971 lbs. per annum, and the total production 126,622 marks, or 84,857 lbs.

These mines are situated in the crystalline slates, which probably belong to the azoic period. Their direction is nearly north and south. The position of the silver ore in these slates is exceedingly curious, and has already been alluded to in a previous chapter.*

The peculiar mode of occurrence, in veins which are only productive when they cross the comparatively narrow fahlbands, renders mining very expensive, since the veins are narrow, and the productive portion of them very limited in extent; thus the workings are out of all proportion of length to their depth.

SAXONY AND BOHEMIA.—The chain of the Erzgebirge, which separates Saxony from Bohemia, extends for a distance of a hundred miles, in a direction from north 55° west to south 55° east. On the Saxon side, it slopes gradually to the north; but its southern declivities are steep and deeply ravined. The elevation above the sea of the highest points of this chain does not surpass 5000 feet. On both sides of the dividing line mines have been wrought for many hundred years, which are hardly surpassed in interest by any in the world; since, even if their produce is but small when compared with that of the Mexican and South American silver mines, yet the great number and complexity of the systems of vein-fissures, the vast extent and depth of the workings upon them, and the skill with which they have been developed, make them of peculiar importance to the mining engineer.

The commencement of mining operations in this district is said to date back to early in the tenth century; but the authentic records go no farther back than to the end of the twelfth century, when the still rich mines of the Freiberg district began to be worked. The Thirty Years' War had a ruinous effect on the Saxon mines, and many of the mining towns were entirely destroyed; some of which, up to this time, have never recovered their former prosperity.

* See page 42.

The rock in which these mines occur is almost exclusively gneiss, with a few intercalated beds of quartz-rock, and masses of porphyry, subordinate to those forming the principal crest of the mountains.

The principal mining centres are Freiberg, Johann-Georgenstadt, Marienberg, Annaberg and Schneeberg, and Schwarzenberg in Saxony, and Joachimsthal and Bleistadt in Bohemia. Of these, Freiberg is of by far the most importance; since it produces nine-tenths of the silver of Saxony, and about three-fourths of the whole yield of the Erzgebirge in that metal. The number of veins ascertained to exist in this district is over 900. They were divided by Von Weissenbach into four classes, according to the nature of their gangues, as follows:

1. Quartz veins with iron pyrites, blende, galena, and mispickel; ores of silver of moderate percentage.

2. Brown-spar veins, with the same ores as the preceding, but richer in silver.

3. Veins with gangues of oxide and carbonate of iron, fluor-spar and heavy spar; not so metalliferous as 1 and 2. These veins sometimes pass into the Zechstein formation.

4. Veins with calcareous gangues, sometimes carrying rich ores.

The characters of the different systems of veins do not, however, seem to be so strongly marked that any division of them into systems as above can be considered as absolute.

The Freiberg mines furnish the most interesting example which can be given of the persistence of the veins in richness at a considerable depth. Their mean depth is from 1000 to 1300 feet, and it is constantly increasing, although the machinery at present in use, according to Burat,* is not capable of allowing the works to be extended much farther. A few years since, it was proposed to drain this district by an adit-level, of the extraordinary length of twenty-four miles, driven from the river Elbe, which would cut the veins at a mean depth of nearly 2000 feet. This plan was vigorously supported by Von Beust and other distinguished mining

* Ann. des Mines (4), xi. 27.

engineers, and received the sanction of the Saxon government, thus indicating, on the part of those best qualified to judge, an entire confidence in the future permanence of the richness of the veins at great depths. In connection with this fact it should be noticed that in no mining district in the world do the metalliferous deposits bear more strongly the character of true fissure-veins than here. This gigantic work has not yet been commenced, on account of the expense and the length of time required for its completion; but a deep adit is now driving to drain the southern part of the Freiberg district, at a depth of 400 feet below the lowest point reached by the present system of drainage. This very extensive work, which may be regarded as the forerunner of the great Elbe adit, commences on a small stream called the Triebsche, near the village of Rothschönberg, and is to extend a little over eight miles, so as to communicate with all the mines of the upper part of the district. The adit is a little over eight feet wide, nearly ten feet high, and rises in the whole distance 12·6 feet.

The ores of the Freiberg mines are the various sulphurets, especially vitreous silver, pyrargyrite or dark-red silver ore, light-red silver ore (rothgültigerz), Freieslebenite, Stephanite, or brittle sulphuret of silver, and native silver; the latter in some instances in large masses. Besides these, there is argentiferous galena in some quantity.

The entire yield of silver from the mines of the Freiberg district, from 1524, back to which period accurate registers of the amount produced extend, up to the end of 1850, was 8,961,251 marks, or 5,611,900 lbs. troy. As a general rule, the amount produced by these mines has always been larger as their depth increased, although there have been many fluctuations in the yield of the different veins. Thus, at the time of the publication of Heron de Villefosse's great work, "Sur la Richesse Minerale," the mine of Himmelfürst, the workings of which are on several veins, was the most productive; so regularly so, that it divided thirty-two dollars quarterly on each share, for sixty-three years in succession; since 1831, however, it has fallen off considerably. The annexed table shows the ratio of the profits to the amount of silver produced by this mine for the periods specified:*—

| | Profit. | | Value of silver produced. |
|---|---|---|---|
| 1624 to 1850, | 1 | : | 7·82 |
| 1748 to 1850, | 1 | : | 8·03 |
| 1769 to 1831, | 1 | : | 6·63 |

Since 1830, the mine of Himmelfahrt has occupied the first rank, producing from that time up to 1850, 234,469 marks, or 146,824 lbs. troy, of silver, and paying a profit which was to that quantity of silver in the ratio of 1 : 22·355.

It is most interesting to observe that although the amount of silver produced in the Freiberg mines is on the whole increasing, yet the profits are rather falling off, with the exception of some favorable years, of such as 1847 and 1850. This is shown by the following statement:—

| Periods. | Ratio of profits to amount of silver produced. |
|---|---|
| 1530 to 1629, | 1 : 5·512 |
| 1710 to 1763, | 1 : 10·06 |
| 1764 to 1850, | 1 : 20·10 |
| 1530 to 1850, | 1 : 10·606 |

This decreased profit seems to be principally due to the increased expense of working the mines at so great depths, rather than to any falling off in the average yield of the ores.

The districts of Marienberg and Schneeberg were formerly of much more importance for silver than they now are. In the Annaberg district the production amounted in one hundred and ninety-one years, from 1654 to 1845, to 5,842,046 thalers in value; but the veins are not by any means as productive in depth as those of Freiberg. The whole amount of silver furnished by the Saxon mines was, in 1851, 59,798 lbs.; and it has not varied much from that for the last fifty years.

AUSTRIAN EMPIRE.—This country produces more silver

* Gätzschmann, Vergleichende Uebersicht der Ausbeute und des wiedererstatteten Verlages, welche vom Jahre 1530 an bis zum Jahre 1850, im Freiberger Revier vertheilt wurden.

than any other in Europe, with the exception of Spain. The mines of the metal are chiefly in Bohemia, as already noticed in describing the Erzgebirge, under the head of Saxony, and in Hungary and Transylvania. The average annual yield of the different provinces of the Empire, and their relative production in percentage, for the period between 1843 and 1847, were as follows:—

| | | | |
|---|---|---|---|
| Hungary, . . . . | 70,379 marks, or | 64·9 | per cent. |
| Bohemia, . . . . | 29,804 " | 27·5 | " |
| Transylvania, . . . | 5,794 " | 5·2 | " |
| Tyrol, Styria, &c., . . | 2,279 " | 2·1 | " |
| | 108,256 | 100·0 | |

The district of Joachimsthal, on the Bohemian side of the Erzgebirge, is of great interest, from the regularity and the great number of the veins which are there concentrated into a small space. There are two very distinct directions of fracture; one running east and west, and the others north and south. Those of the latter direction are of two ages, one set of them being heaved by the east and west veins, and the other set traversing them. They are accompanied by dykes of trappean rock, which run parallel with and are intimately connected with them. The ores of silver are associated here, as on the other slope, with those of nickel, uranium, bismuth, and cobalt.

*Hungary.*—The interesting mining region of Hungary is divided into four districts, namely, Upper Hungary, Lower Hungary, Nagybánya, and the Banat.* The mines of Lower Hungary, in the vicinity of the cities of Schemnitz, Kremnitz, and Neusohl, are still of very considerable importance, although much fallen off from what they once were.

Schmöllnitz, in Upper Hungary, and Orawicza and Szaska, in the Banat, offer examples of the treatment of copper ores of a very low percentage, and of a metallurgic process not elsewhere employed, the separation of silver from the black copper by amalgamation.

There are also mines in the vicinity of Nagybánya, on the

* Audibert, Ann. des Mines (4), vii. 85, and Rivot and Duchanoy (5), iii. 68.

western limits of Transylvania, which produce a considerable amount of silver.

The veins in the neighborhood of Schemnitz are in a rock, called by the Germans greenstone-porphyry; it is a mixture of hornblende and feldspar. The greenstone is surrounded by trachyte, in which the veins have not been found productive. There are seven principal lodes in one group, which pursue a nearly parallel course, and are distant from 1000 to 2000 feet from each other. The principal one, the Spitalergang, is known to extend for about 3 miles; it is from 12 to 22 feet in width, and has this peculiarity, in common with the other adjacent veins, that its western end yields only ores of silver, while in the eastern portion the predominating ore is galena; which seems, in fact, to take the place of the silver throughout the whole extent of the vein, where the workings have been extended to a considerable depth.

At the Windschacht, the depth attained is about 160 fathoms, and at this depth the ores are still chiefly those of silver, and are disseminated in bunches through the feldspathic veinstone.

The Biebergang is now pretty much exhausted, but has produced immensely. It has been worked to a depth which exceeds 1300 feet, and over a linear extent of more than three miles. The width is never less than sixty feet. The other principal veins, namely, the Wolfsgang, Theresiengang, Ochsenkopfgang, Johanngang, Stephangang, and Grünergang seem all to have lost a large part of their richness in silver as they have been worked down upon, the galena becoming less and less argentiferous, as the depth increases.

As an instance of the gigantic scale on which the mining works are laid out, the adit-level commenced in 1782 may be mentioned. Its object is to drain the Schemnitz mines, and it is intended to be nearly nine miles long. In August, 1850, it was about two-thirds completed, at an expense of 2,112,016 florins, about $1,000,000, having cost about $200 per fathom. It is expected to be completed in about ten years, and will, with the help of the powerful hydraulic machines used in that region, enable the workings to be carried to a depth of over 230 fathoms.

South America.—The argentiferous veins of that portion of South America which now constitutes the republics of Peru and Bolivia are next in interest and extent to those of Mexico; they are perhaps even more capable of producing silver in immense quantities than those of the latter country, as far as the quantity and quality of the ores is concerned, but their elevated position prevents their development, since no one, not goaded by an irresistible desire for riches, would live in such a desolate and frozen region as that of the argentiferous districts of the South American Andes. The mines of Potosi are worked at an elevation greater than that of Mont Blanc itself.

Peru.—The mines of Cerro de Pasco are the most remarkable of this country, and may be taken as a type of the others. The principal ores are the *pacos* so called, analogous to the *colorados* of the Mexican miners: they are ferruginous earths, mingled with argentiferous ores, and evidently resulting from the decomposition of the sulphurets. M. de Rivero considers the pacos of Santa Rosa, one of the richest mines in the Pasco district, not to be in a true vein, but a bed, since it is parallel with the formation, the hanging and the foot walls being of different character, and the gangues having no crystalline or comby structure.* All these deposits have grown sensibly poorer in descending upon them. The tenor of the ores, which at the surface sometimes amounted to 0·3 and averaged 0·0015, now hardly surpasses 0·0004. These deposits appear to be less to be depended on in depth than the true veins of Mexico, but their development superficially is enormous. Their exact geological position is not to be made out from the descriptions yet given, but they seem to be included in strata of the carboniferous epoch.

The mines of Yauricocha, or Pasco, were discovered, by accident, in 1630, and were among the worst-worked, as well as the richest, in the world. At one time, 300 miners were killed by the falling in of a mine, since called the Matagente (Kill-people). The town of Cerro de Pasco, which sometimes, when the mines are prosperous, contains 18,000 inhab-

* Ann. des Mines (3), ii. 169.

itants, is 13,673 feet above the sea. According to Tschudi,* there are two very remarkable argentiferous veins. One of them, the Veta de Colquirirca, runs nearly in a straight line from north to south, and has already been traced to the length of 9600 and the breadth of 412 feet; the other is the Veta de Pariarirca, which takes a direction from east-southeast to west-northwest, and intersects the Colquirirca vein, as is supposed, under the market-place of the city. Its known extent is 6400 feet in length and 380 feet in breadth. In 1814, for the first time, steam-engines were introduced for drainage,† but the acid water of the mine soon ruined the iron-work, so that one by one they gave out, and in 1832 only one was still at work. Notwithstanding this, within the last few years, some progress has been made in improving the system of mining, and the yield of silver is on the increase, the produce of the Pasco mines for 1851 being about $2,000,000, a slight increase over that of 1850.

The total amount of silver smelted at the principal works of Pasco, from 1784 to 1827, was 8,051,394 marks, or 4,967,710 lbs. troy.

Besides the Pasco mines, there are numerous other mining districts in Peru, especially in the provinces of Pataz, Huamanchuco, Caxamarca, and Hualgayoc. In the Cerro de San Fernando, belonging to the latter district, there were, in 1840, some 1700 "bocaminas" (mine-openings). There are also numerous silver mines in the southern provinces, but the amount produced is small, in comparison with what it might be expected to be, when the richness of the veins is considered.

Chevalier estimated the yield of Peru in silver, in 1845, at a little over 300,000 lbs. troy. Since that time, there has

* Travels in Peru, Eng. Trans. p. 327.

† By Uvillé, a Spanish mining engineer, under the direction of Richard Trevithick, the distinguished Cornish engineer, nine of whose engines were shipped to Peru at that time. Trevithick himself visited that country in 1817, where he was received with the highest honors, and it was even proposed to erect a statue of him in solid silver. Afterwards, in consequence of political dissensions and civil war, he was forced to escape, taking with him, as the only remnant of his former wealth, a pair of silver spurs.

probably been some increase, but it is doubtful whether the amount now produced much exceeds the above-named sum.

Bolivia.—The mines of Potosi, which were formerly within the viceroyalty of Buenos Ayres, are now included in the republic of Bolivia. They are so well known for their almost fabulous richness, as to have become proverbial throughout the world. The quantity of silver which they have poured out, since their discovery in 1545 up to the present time, is almost incredible: it may be estimated, according to the calculations of Humboldt and Chevalier, at $1,200,000,000. Thirty-two principal veins have been worked, besides numerous smaller ones, which intersect the isolated mountain, called the "Great Potosi," or Potocchi, whose summit rises to the height of 16,000 feet above the level of the sea. The establishment of a town, which at one time is said to have numbered 160,000 inhabitants, at an elevation far exceeding that of Mont Blanc, is one of the wonders to which this unheard of yield of silver gave rise. The period of the greatest productiveness of the Potosi mines was the century immediately following their discovery, ores of the most astonishing richness and abundance being first obtained near the surface, and smelted by the aid of the fuel of the region, until the introduction of the amalgamation process by Medina, in 1571. The average annual yield, from 1545 to 1556, was about $11,590,000, and that at a time when silver was over six times as valuable as now. After the first quarter of the seventeenth century, the production began to decline; but, even at the end of that century, it still amounted to between three and four millions of dollars; and during the eighteenth, it maintained a rank only second to that of the Mexican mines, yielding nearly twice as much as all the European mines together. The present yield of the Potosi mines is estimated, by Chevalier, at from 48,000 to 60,000 lbs. troy, equal in value to from $770,000 to $960,000.

Although the yield of the ores has become much less in descending than it was at the surface, still these mines are by no means exhausted, in the opinion of those best likely to know. The methods followed are of the most wasteful kind, and cruelly destructive to the lives of the workmen engaged in carrying them on. There were, in 1852, in the province of Potosi, 1800 abandoned silver mines, and only 26 at work; and in the remaining mining districts of Bolivia 2365 abandoned, and 40 working mines. From the present appearance of things, it is difficult to say when, if ever, this mining region will assume the importance which it once had. Under an enlightened management, and with a government which could be trusted, a vast change might be made, if only by the introduction of the simplest machinery and methods in use in Mexico.

The whole amount of silver produced by the mines of Peru and Bolivia, from the earliest period up to the year 1845, is estimated, by Chevalier, at 12,925,000,000 francs, or 155,839,180 lbs. troy.

Chili.—We are indebted to Domeyko for a very good description of the geological structure and a sketch of the mining districts of Chili.* He divides the rock formations into three groups: I. Secondary stratified, prior to the uplift of the Andes. II. Igneous eruptive masses of the period of the elevation of that chain. III. Tertiary beds posterior to that uplift. In general, the veins bearing copper and gold belong to Group II.; those of silver, argentiferous copper, and sulph-arseniurets and sulph-antimoniurets of silver, to Group I. The gold veins are principally in the granite, and the copper, uncombined with silver, arsenic, or antimony, is found in the diorites, porphyries, eurites, and other igneous eruptive rocks. The chlorides of silver and native amalgams are near the principal line of contact of I. and II.

The mining districts of this country are divided as follows:

*a.* Mountains to the north of the valley of Huasco. This part of Chili is richest in silver, but it also contains valuable mines of copper and gold. Among those of copper the mines of Carrisal are the most important.

*b.* District between Huasco and Coquimbo. In this region are the groups of veins of pyritous copper, and oxide of copper, of the mines of San Juan and La Higuera, which, in

* Ann. des Mines (4), ix. 22.

1845, were producing more than 5800 tons of 25 per cent. ore per annum for exportation, besides supplying a number of furnaces in Chili. These mines are in the diorites.

On a line between Arqueros and Agua Amarga, which line represents the line of contact of I. and II., there are numerous veins of chloride of silver, native silver, and amalgam.

*c.* District between the valleys of Coquimbo and Aconcagua. Here the granite rocks extend farther inland from the coast, and the gold-bearing veins acquire a greater development. The whole of the granite is more or less auriferous. On its borders are cupriferous veins, among which are those of Tamaya.

*d.* District south of the Valley of Aconcagua. Here, as in the region alluded to above, the granite is filled with auriferous veins, and mines of silver and argentiferous copper are worked above the level of the escarpments of the stratified rocks forming the elevated chain of the Andes.

The usual gangues of the copper ores are quartz and hornblende; of the gold, quartz and sulphuret of iron; while sulphate of baryta and carbonate of lime accompany the silver ores, which are very various and interesting. The most abundant is the chloride, which is associated with bromide of silver and the native metal. The chloride occurs in the usual gray, earthy, ferruginous deposits, called *pacos* and *colorados* by the South American miners. Besides these, there are a great variety of sulphurets and arseniurets. Their yield is from 0·003 to 0·008; the richest contain 0·002 of silver. It is very curious that the galena and blende so common in Chili are not by any means rich in silver, as would be expected in a region abounding in this metal. On the contrary, they hardly contain a trace of it.

The flourishing political state of Chili, and the internal quiet which it has so long enjoyed, in comparison with other South American republics, have enabled the mines to be opened on an extensive scale; and there has been a great increase in the production of silver and copper within the last few years.

The most interesting and important argentiferous district

is that of Copiapo, and from accounts published by Mr. Dillon,* and Col. J. A. Lloyd, English Chargé d'Affaires in Bolivia,† the following notices of that region are extracted:—

"All of that part of Chili which lies above the parallel of Valparaiso is a rocky and sterile country, excepting three tongues of land, from one-fourth of a mile to one mile wide, extending inwards from Coquimbo, Huasco, and Caldera. All the rest is a desert; and in it, at distances of from thirty to forty leagues apart, are several mining districts. Those of Coquimbo and Huasco produce chiefly copper, while that of Copiapo is remarkable for its richness in silver."‡

The principal mining centres are those of Los Tres Puntos and Chañarcillo, the first being thirty leagues north-northeast of Copiapo, the other sixteen leagues south. The Chañarcillo mines were discovered in 1832, by a poor muleteer, and were worked with large but gradually diminishing yield until 1836, when the ores were found to be cut off by a stratum of "tough and horny limestone," called, in that country, a "mesa" (table), and into which the veins did not penetrate. At first the miners gave up in discouragement; but one of them, more persevering than the others, had the courage to sink 266 feet through the unproductive rock, when, on the other side, a rich mass of silver was found; and, according to Col. Lloyd, seven of these intercepting belts have been pierced through, and it has been universally found that the veins are rich between them, and that there are large accumulations of silver near the plane of contact of the "mesa" with the adjoining rock.

The Descubridora was the first mine discovered in the district, and still continues productive. The district of the Tres Puntos, according to Mr. Dillon, has three principal and rich mines, La Buena Esperanza, La Salvadora, and the Al Fin Hallada. He remarks that "there may be in all Copiapo 20 good mines producing £1,300,000 annually, and 200 producing nothing or less."

* Eng. Mining Journal, No. 957.

† Jour. Geographical Society of London, xxiii. 196.

‡ Caldera is the port of Copiapo, between which places is a railroad fifty miles in length, extending through a desert country, where rain never falls. It is to be continued to Chañarcillo.

The Copiapo Mining Company is an English association with 10,000 shares, on which £15 per share has been paid in; present value 12½ to 12¼. They are working several silver as well as a number of copper mines.

The development of the mining interest of Chili has been more recent than that of the other South American states. Under the Spanish dominion, they produced little. In 1800 the amount of silver extracted was estimated by Humboldt at 18,300 lbs., that of gold at 7,500 lbs. Since that time the quantity of the latter metal has fallen off very much, the washings being nearly exhausted; but, on the other hand, the production of silver has been rapidly increasing since 1832, when the rich mines of the Copiapo district were first discovered. The total amount of silver extracted up to 1810 is estimated by Chevalier at 804,000 lbs.; from 1804 to 1845 at 1,803,636 lbs., we may add from 1846 to 1853 probably about 1,750,000 lbs. more, making as the entire amount of this metal produced from the Chili mines up to this time, 4,357,656 lbs.

Mexico.—It is well known that the Aztecs, before the arrival of Cortez, possessed the art of working gold, silver, copper, and tin, but there is no reason to believe that they had ever carried their knowledge of mining to any degree of perfection. On the other hand, everything leads to the supposition that the gold was procured entirely from washings, and the silver from the outcrop of the veins, where the metal exists in an exceedingly pure state. The art of reducing the metals copper and tin from their ores was well understood by them; but that does not imply any very extended knowledge of metallurgy, since the ores found in Mexico are very easy of reduction.

The data which we have on the subject of the Mexican mines are derived wholly from the works of European scientific men and travellers; and though their works are numerous and voluminous they necessarily leave much to be desired, especially touching the subject of the geology of the country. Until a thorough geological survey, based on a good map of Mexico, shall have been carried out, we shall remain very much in the dark as to many of the most interesting questions connected with the argentiferous veins, their geological position, and the mode of occurrence of their ores. Certain it is, that no country offers a more interesting field for the geologist than Mexico. What Humboldt said, in the early part of the present century, remains still quite applicable. He remarks,* "We are still far from understanding the structure of the metalliferous mountains of Mexico, and in spite of the great number of observations which I have been able to collect in traversing the country in different directions, over a distance of more than 400 leagues, I shall not hazard a general sketch of the Mexican mines considered in their geological relations."

The principal workings are on true veins.† Deposits and segregated masses, parallel with the stratification, are rare. The veins are mostly included in the so-called primitive and transition rocks, or, in the language of modern geology, the azoic (?), metamorphic palæozoic, and hypogene rocks. The limestone of the districts of Tasco and Catorce is called, by Humboldt, "Alpine limestone," and, by Burkart, "mountain limestone," from which it would appear to be of the carboniferous age, or of the age of the argentiferous strata of the Bolivian and Peruvian Cordilleras.

Granite, gneiss, and mica slate, form the crests of the highest mountains; but these rocks rarely appear on their flanks, which are covered by an immense thickness of beds of greenstone, amygdaloid, basalt, and other trappean rocks, under which the granite is effectually concealed. The porphyries in Mexico are pre-eminently metalliferous; but their relative geological age is pronounced, by Humboldt, one of the most difficult geological problems to solve. They are characterized by the presence of hornblende, and the absence of quartz.

The metalliferous veins have a direction approaching northwest and southeast; they dip at a high angle, more frequently to the south than the north, and they generally cut the enclosing rocks at a considerable angle. The great vein of Guanaxuato, the Veta Madre, the mother vein, or

* La Nouvelle Espagne, ii. 487.

† Duport, De la Production des Métaux précieux au Mexique, &c., p. 25.

champion lode, as the Cornish miners would call it, has, however, a direction so nearly coincident with that of the adjoining strata, that it might almost be considered a bed, which some mining engineers have supposed it to be; but it is pronounced by Humboldt to be a true vein, as it contains fragments of the hanging wall, and passes from one formation into another of a different geological character. This vein, the richest and most developed in the world probably, widens out sometimes to nearly 200 feet, and has been opened at various points for a length of more than three leagues; but the main workings are comprised within a distance of 2000 varas or yards. The deepest shafts had, in 1845, penetrated about 2000 feet beneath the surface. The Veta Grande of Zacatecas is next in magnitude to the Veta Madre; its average width is fully twenty-five or thirty feet, and in some places it widens out to seventy-five, although the whole of this extent is not metalliferous. The Mexican veins do not, however, in general, possess these extraordinary dimensions, but vary from a few inches to six feet in width, the narrow ones being often very profitable to work on account of their extreme richness.

The gangues are principally quartz, and their *crestones*, or outcrops, may generally be traced with great ease, owing to the imperishable nature of this material. The ores are the various simple and complex sulphurets of silver, which, near the surface, have been decomposed and converted into metallic silver, chloride of silver, oxide, carbonate, &c. This portion of the vein is generally much intermixed with ferruginous matter, and hence the ores are called *colorados* (*colorado* meaning red). Below the point of decomposition, where the sulphurets remain in their original character, the ores are called *negros* (*negro*, black); the latter kind, according to Duport, furnishes seven-eighths of the whole produce of Mexico.

About the year 1821, it having become evident to the Mexican Government that there was not capital enough in the country to work their mines properly, they offered to allow foreigners to become joint proprietors with natives, thus throwing open the door to foreign capital on advantageous terms. A lively movement, on the part of the English mining adventurers, immediately ensued; and in the expectation of enormous gains from the application of English skill to the great veins of Mexico, they entered with more zeal than discretion into the business of mining. The consequence was, that immense losses were incurred. In 1829, there were seven great English companies at work there, besides one German and two American, and about $15,000,000 of British capital was actually invested in their works. These seven companies were but a small portion of those which were started during the Mexican mining mania.* Their names were the Real del Monte, the Bolaños, the Tlalpujahua, the Anglo-Mexican, the United Mexican, the Mexican, and the Catorce Companies. Of all these there seems now to be only one in existence, the United Mexican, whose shares, on which £28¼ have been paid, are quoted at £4¾. The number of mines worked by this Company is two, out of some fifty to a hundred included in their leases and purchases. This frittering away of the means of a company over such an extent of ground as was embraced in the original plan, would alone be sufficient to account for its failure. Some of the mines which had been carried by the English almost to the paying point and then abandoned, are said to be now worked by the Mexicans with success. It is a curious fact, that all the skill of the English miners has failed to make any improvement in the Mexican methods of preparing the ores and extracting the silver from them, and the loss of mercury remains about the same that it was 200 years ago. In 1844, it was estimated by Duflot de Mofras, that the English companies produced only one-tenth of the whole quantity of silver raised in Mexico.†

The fact that the Mexican mines have produced, and are still producing, such large amounts of silver, taken in connection with the opinion universally given by scientific travellers, from Humboldt to Duport, as to the number and richness of the veins existing in that country, renders it an interesting subject of inquiry why the English capitalists were

* Ward's Mexico, i. 405.

† Duflot de Mofras, Exploration du Territoire de l'Oregon, i. 47.

so unsuccessful in their operations there. The most probable reasons may be found: 1st, in the want of concentration of their energy upon a few, and these the most promising, localities. Some companies had hundreds of concessions or leases, and instead of selecting one or two of the best, and expending their capital upon those, they tried to keep a great number of them in operation at once, and thus expended all their funds before arriving at a positive result anywhere; 2d. The cost of transportation and want of fuel, and consequent difficulty of using the powerful steam-engines intended to be employed in draining the mines; 3d. The attempts to introduce costly machinery and methods, which, after trial, were found unsuited to Mexican circumstances, and had to be abandoned; 4th. And chiefly, because, instead of opening new mines on fresh discoveries, the operations were principally directed to the working of those which had been previously, if not exhausted, at least carried down to a point where they ceased to be sufficiently productive to pay for working. Following the statements of Humboldt, those mines which were cited by him as having produced the largest amounts were most eagerly sought after, while, in reality, it by no means followed that, because they had already yielded immensely, they must continue to do so in the future. The fact is evident, indeed, from a candid examination of the Mexican mines, that the yield of silver does not continue to hold at a considerable depth. This is apparent, from the fact that the English companies were unsuccessful, and that the whole of the deep mines from which the water was removed and ore extracted, were found to present no encouragement for farther working. Duport himself,* who is most strongly of the opinion that immense wealth is still to be obtained from Mexico, admits that a mine which has been worked to a depth of 200 to 350 fathoms and then abandoned, however productive it may have been, offers little inducement for resuming the workings upon it, or, in other words, that the mines are no longer productive below that depth. At the same time, he considers the prospect of a

* Op. cit. p. 35.

future production of silver greatly surpassing that of the present day as highly encouraging. He remarks, that the principal causes which operate against the almost unlimited development of the metallic wealth of Mexico are, the unsettled political condition of that country, the want of scientific skill, and the high price of mercury. But, in view of the vast extent of metalliferous formations yet to be explored, and the number of veins known to exist, he considers the resources of that country as almost boundless, since he closes his work with the remark, "That the time will come, sooner or later, when the production of silver will have no other limits than such as are imposed upon it by the constantly increasing decline in the value of this metal."

The whole amount produced by the Mexican mines, from the earliest period up to 1845, was estimated by Chevalier at 13,507,000,000 francs in value, or 162,858,700 lbs. troy.

The mines of Zacatecas began to be worked in 1548, and those of Guanaxuato in 1558; while the Mexican process of amalgamation was invented at about the same time by Medina. At this time, the annual yield of the Mexican mines is estimated by Humboldt at from $2,000,000 to $3,000,000; during the 18th century, it gradually rose to $23,000,000. The production reached its highest point during the period of prosperity from 1800 to 1810, just before the war of Independence, when the average quantity of gold and silver coined at the various mints in Mexico, was $23,664,622, the ratio of the former metal to the latter being, by weight, 0·0029 : 1; and in value, 0·05 : 1. During the war, there was a very great falling off, from which the country has been gradually recovering; the average yield of silver from 1810 to 1845 being about $12,000,000, and of gold a little over $1,000,000. In 1850, and since that time, the mines seem to be producing more largely than during their most flourishing period; the various estimates indicating a yield of over $25,000,000 of silver, and $300,000 of gold, the Province of Guanaxuato furnishing about one-half of this sum.

United States.—As before remarked, the silver furnished by this country comes almost wholly from the native gold of California. There is no proper silver mine within our territory, although there are several localities where a small amount of this metal is obtained in connection with lead ores. The Washington Mine, Davidson County, North Carolina, formerly the most important silver producing mine in the country, is now suspended. This, as well as other localities where argentiferous galena is or has been worked, will be described in the chapter devoted to lead.

The statistics of silver show no such extraordinary fluctuations within the last half century, as are indicated by the tables of the yield of gold given in the preceding chapter. This will be apparent from an examination of the following tables, in which the amount of silver produced by the different countries throughout the world is given, so far as reliable statistics could be obtained. In the table immediately following are comprised the European states, and the years for which the results are given are each tenth year from 1800 to 1820; from 1820 to 1845, every fifth year; and from that period up to 1853, for every year. Blanks are left when data of approximate exactness could not be obtained. For Norway, Prussia, Saxony, the Harz, Austria, and France, the returns are taken from official documents, and may be relied on as very nearly correct. Great difficulty has been experienced in obtaining accurate returns from Spain, where the yield of all the metals has always been very fluctuating; this is the more to be regretted, as that country stands among the first in Europe in the production of silver, as well as of lead and mercury. The same may be said of Russia, whose statistics indicate for most of the years for which they have been obtained, a yield of the metal which would place it about on a par with Saxony. In England, it is only quite recently that any attention has been paid to the statistics of any other metal than copper. The exertions of Mr. R. Hunt, Keeper of the Mining Records, an office recently erected in connection with the Museum of Practical Geology, have made it possible to give the yield of both lead and silver for a few of the last years. Besides the production of Prussia, Saxony, the Harz, and the Austrian Empire, there has been within the last few years a small amount furnished by some of the minor states, estimated, in 1850, as equal to 2500 lbs. only. This has not been included in the table. The figures given represent the weight of pure silver in pounds troy.

| Year. | Russia. | Sweden. | Norway. | Great Britain. | Prussia. |
|---|---|---|---|---|---|
| 1800, . . . | 58,150 | | | | |
| 1810, | | | | | |
| 1820, | | | | | |
| 1825, | | | | | |
| 1830, . . . | 56,189 | | | | 12,776 av. |
| 1835, . . . | | 2,179 | 14,500 | | 13,336 " |
| 1840, . . . | | 2,473 | | | 15,368 " |
| 1845, . . . | | | ('44) 17,730 | | 15,240 " |
| 1846, . . . | 53,000 | | | 32,776 | 16,178 |
| 1847, . . . | | | | | 17,429 |
| 1848, . . . | 52,307 | | | | 18,120 |
| 1849, . . . | | | | | 14,108 |
| 1850, . . . | 2,984 | 3,418 | | 48,484 | 21,184 |
| 1851, . . . | 60,170 | | | | 26,493 |
| 1852, . . . | 59,985 | | | 68,194 | |
| 1853, . . . | | | 16,971 av. | | |

| Year. | Harz. | Saxony. | Austria. | Spain. | France. |
|---|---|---|---|---|---|
| 1800, . . . | | 31,700 | | | |
| 1810, | | | | | |
| 1820, . . . | | | 47,506 | | |
| 1825, . . . | | 37,093 | 60,924 | | 3,114 |
| 1830, . . . | ('31) 35,533 | 40,816 | 63,578 | | 4,841 |
| 1835, . . . | 32,115 | 39,830 | 63,578 | | 4,822 |
| 1840, . . . | | | 75,978 | | 5,131 |
| 1845, . . . | | | 81,510 | 108,236 | 7,566 |
| 1846, . . . | | | 83,093 | | |
| 1847, . . . | | | 86,992 | 68,825 | |
| 1848, . . . | | 51,150 | 93,545 | | |
| 1849, . . . | | 52,639 | | 61,132 | |
| 1850, . . . | 31,500 | 63,640 | | | |
| 1851, . . . | | 59,798 | | | |
| 1852, | | | | | |
| 1853, | | | | | |

In regard to the production of silver in those parts of Asia not included in the Russian Empire, and also of the East India Islands, and the continent of Africa, there is no satisfactory information. The amount obtained must be very trifling, and it does not in any way enter into or affect the commerce of the rest of the world; they have, therefore, been omitted from the list.

The next table presents such statistics of the western hemisphere as have been obtained. They will be found much more defective than the preceding one. So large a portion of the precious metals is smuggled out of those countries, without passing through the hands of government officers, that the most reliable statements must necessarily be based partly on estimates. There have been no investigations made into this subject by any competent persons on the spot, since the date of Chevalier's and Duport's writings, which were published nearly ten years ago. On that account the table has not been continued beyond 1850, except for the United States, since the fluctuations in the produce of Mexico and the South American states could not be ascertained. In general, the produce of all of them seems to be somewhat on the increase. Chili especially has exhibited within the last few years a rapid development of her silver mines.

| | Chili. | Bolivia. | Peru. | N. Grenada. | Brazil. | Mexico. | U. States. |
|---|---|---|---|---|---|---|---|
| 1800, . . . . . | 18,300 | 271,300 | 401,850 | | 1,200 | 1,440,500 | |
| 1810, . . . . . | | | | | | | |
| 1820, . . . . . | | | 230,296 | 6,550 av. | | | |
| 1825, . . . . . | | | | | | | |
| 1830, . . . . . | | | | | | | |
| 1835, . . . . . | 143,000 | | | | | | |
| 1840, . . . . . | 87,000 | | | | | 1,050,000 | ('41) 277 |
| 1845, . . . . . | 90,000 | 139,400 | 303,150 | 13,100 | 607 | 1,235,000 | 307 |
| 1846, . . . . . | 108,530 | | | | | | 197 |
| 1847, . . . . . | 108,637 | | | | | | 412 |
| 1848, . . . . . | 132,490 | | | | | | 400 |
| 1849, . . . . . | 190,653 | 130,000 | | | | | 2,521 |
| 1850, . . . . . | 238,502 | | | 13,000 | 675 | 1,650,000 | 17,354 |
| 1851, . . . . . | | | | | | | 25,103 |
| 1852, . . . . . | | | | | | | 26,072 |
| 1853, . . . . . | | | | | | | 26,895 |

Finally, the result of all these investigations is presented in a third table, which follows next in order. In this, the produce of silver throughout the world is summed up for the years 1800, 1845, and 1850, and the amount given in pounds troy, and to each is appended in the adjoining columns the relative quantity furnished by each country in percentage.

| | 1800. | | 1845. | | 1850. | |
|---|---|---|---|---|---|---|
| Russian Empire, . . . . . . | 58,150 | 2·5 | 53,000 | 2·4 | 60,000 | 2·1 |
| Scandinavia, . . . . . . . . | } | | 20,000 | 1· | 20,400 | ·7 |
| Great Britain, . . . . . . . | } | | 32,500 | 1·5 | 48,500 | 1·7 |
| Harz, . . . . . . . . . . | } | | 32,000 | 1·5 | 31,500 | 1·1 |
| Prussia, . . . . . . . . . | } | | 15,250 | ·7 | 21,200 | ·7 |
| Saxony, . . . . . . . . . | } 141,000 | 6· | 50,000 | 2·3 | 63,600 | 2·3 |
| Other German States, . . . | } | | 2,000 | ·1 | 2,500 | ·1 |
| Austria, . . . . . . . . . | } | | 81,500 | 3·7 | 87,000 | 3·1 |
| Spain, . . . . . . . . . . | } | | 108,200 | 4·9 | 125,000 | 4·4 |
| France, . . . . . . . . . | } | | 7,500 | ·3 | 5,000 | ·2 |
| Australia, . . . . . . . . . | | | | | 10,000 | ·4 |
| Chili, . . . . . . . . . . | 18,300 | ·8 | 90,000 | 4·1 | 238,500 | 8·3 |
| Bolivia, . . . . . . . . . | 271,300 | 11·6 | 139,400 | 6·4 | 130,000 | 4·6 |
| Peru, . . . . . . . . . . | 401,850 | 17·2 | 303,150 | 13·9 | 303,150 | 10·7 |
| New Granada, . . . . . . . | 5,000 | ·2 | 13,100 | ·6 | 13,000 | ·5 |
| Brazil, . . . . . . . . . . | 1,200 | | 600 | | 675 | |
| Mexico, . . . . . . . . . . | 1,440,500 | 61·7 | 1,235,000 | 56·6 | 1,650,000 | 58·4 |
| California and United States, . | | | 300 | | 17,400 | ·7 |
| | 2,337,300 | | 2,183,500 | | 2,817,425 | |

The great preponderance of the American continent will

be noticed at a glance, and Mexico will be seen to have furnished, alone, considerably over half of the whole produce of the world.

By comparing these tables with those given at the end of the last chapter, it will be seen, that while the production of gold has risen suddenly and enormously, at various periods within the last few years, that of silver, on the other hand, has been, with slight fluctuations, slowly increasing, having risen from about two and a quarter millions of pounds in 1800, to a little over two million eight hundred thousand in 1850. Gold, on the other hand, has vibrated between 54,000, and over 717,000 lbs., since the commencement of the century, the great increase being, of course, principally due to California and Australia.

If we compare the produce of the two metals we shall have the following result:—

| | Relative weight of | |
|---|---|---|
| Year. | Gold. | Silver. |
| 1800, | 1 | 43 |
| 1845, | 1 | 17 |
| 1850, | 1 | 8·8 |
| 1852, | 1 | 4 |

The effect of this extraordinary change in the relative production of the two metals has been felt everywhere. Already gold has driven silver almost out of the field, in those countries where it was before the principal medium of circulation. That which has thus been withdrawn as coin, has been partly brought into use again in the form of plate; but a large part has been shipped to Asia. Up to this time there has been no very considerable change in the relative value of the two metals; but should California and Australia continue to supply gold at anything like the present rates for some years to come, it seems impossible that such a change should not take place. As this metal rose, while the Mexican and South American mines were pouring out their treasures, from a value only ten times as great as that of silver to that which it now has, of between 15 and 16 to 1 of the latter metal, so now there is no reason why it should not gradually return to something like the former ratio.

Silver is in a geological point of view the metal best adapted for a standard of value, since, possessing all the valuable qualities which make gold suitable for that purpose, it is not liable to those fluctuations in its production to which this latter is exposed. There is no discovery of a new continent to be looked forward to, whose mines shall deluge the world with silver, and any increase in the amount of this metal produced must come chiefly from the working of mining regions already known by the application of increased skill and capital. As it is obtained mostly from mines wrought in the solid rock, any additional development they may acquire must be gradual, while gold from the very nature of its occurrence can never be produced with that steadiness which characterizes the metals wrought chiefly in deep and permanent mines.

# CHAPTER IV.

## MERCURY.

## SECTION I.

### MINERALOGICAL OCCURRENCE AND GEOLOGICAL POSITION OF THE ORES OF MERCURY.

MINERALOGICAL OCCURRENCE.—Native mercury is not of uncommon occurrence, but it is from the sulphuret of this metal, or native cinnabar, that nearly the whole of the mercury of commerce is obtained. The combinations in which this metal occurs in nature are the following:—

#### NATIVE METAL.

*Native Mercury, or Quicksilver*, occurs in small quantity in almost all the mines of this metal, as a product of the decomposition of its ores, and especially of its sulphuret.

#### AMALGAM.

*Native Amalgam*, amalgam of mercury with silver, containing two or three atoms of mercury to one of silver, equal to 65·2 and 73·75 per cent. of mercury. The *Arquerite* is another variety of amalgam, containing, according to Domeyko, one atom of mercury to six of silver. This would give silver 86·49, mercury 13·50 per cent. This is a valuable silver ore in Chili.

#### ORES.

*Cinnabar*, native sulphuret of mercury. This is the ore of mercury, from which nearly all the mercury of commerce is obtained. It consists of one atom of mercury, and one of sulphur, or, in percentage, 86·2 of the former to 13·8 of the latter.

*Native Calomel*, horn quicksilver; a chloride of mercury, containing two atoms of the metal to one of chlorine; or chlorine, 14·88; mercury, 85·12. This is a rare ore, as are also the following:—

*Coccinite.* Iodide of mercury; found in Mexico.

*Onofrite.* Seleniuret of mercury; a sulphuret and seleniuret of mercury, from Mexico.

*Ammiolite.* An antimoniuret of mercury; found accompanying ores of antimony, copper, and mercury in Chili. These last four are rather objects of mineralogical curiosity, at present, than of commercial importance.

GEOLOGICAL POSITION.—The ores of mercury are very unequally distributed over the world, being confined to a comparatively small number of localities, in most of which, however, they occur in very considerable quantity, so that a few mines supply almost the whole of this metal furnished to commerce. The geological formations in which they are found range from the lowest to the highest in the scale, but the principal mines seem to be worked in the Silurian and Carboniferous. When the mercurial ores occur in the older metamorphic rocks, they appear to exist either in true veins or in connection with them; but the great masses of ore belong rather to the class of contact deposits. From the volatility of the metal, it would naturally be expected that it should be found diffused through the strata adjacent to the source in which it originated, and not confined within very narrow limits, like other metals whose ores are of a less volatile character. And such is the case, for where it occurs in considerable quantity it is associated with the rocks in such a way as to make it evident that it has been absorbed into their mass while in a state of vapor, impregnating them with metallic matter in every direction.

There are instances where it appears evident that this metal must have been sublimed from below during the most recent geological epochs, since it is found in considerable quantities, in the metallic state, associated with the superficial deposits, even in the alluvium itself. These facts will be sufficiently set forth in the description of some of the principal localities of this metal and its ores.

## SECTION II.

### GEOGRAPHICAL DISTRIBUTION OF MERCURY AND ITS ORES.

ALTHOUGH the mines of Almaden in Spain, and Idria in the Austrian Empire, furnish nearly all the mercury supplied by Europe, there are a few localities where it is found, which have formerly been of importance, and which still possess a certain degree of interest. The principal deposits will be hastily passed in review.

FRANCE.—Mercury has been found in several places in this country, but only in small quantities. The localities mentioned are: at Montpellier, in the tertiary marls; Saint-Rome-de-Tarn, in marl of the age of the lias; at Ménildot near Mortain, department of the Manche, in the older metamorphic rocks; and a few others of no importance. None of them are worked.

BAVARIA.—The mercury mines in Rhenish Bavaria have now almost ceased to be worked, but previous to the discovery of America they were of considerable importance; up to the beginning of the seventeenth century they produced from 150,000 to 180,000 lbs. per annum, the working dating as far back as 1420. They belong to the carboniferous system, and according to Dechen, form regular veins; but their mode of occurrence seems to be a complicated and difficult one to decipher. Although an English company attempted to take up these mines about the year 1836, they do not seem to have accomplished anything. One of the principal mines, that of the Dreikönigszug, on the north side of the Polzberg, was opened on a vein which is rarely more than an inch wide. It consists of argillaceous matter (flucan), and the ore is chiefly found adjacent to this seam in the rocks rather than in the vein itself. The mine of Theodor's Erzlust, in the Königsberg, near Wolfstein, produced, from 1771 to 1794, 134,000 lbs. of mercury; it was still at work in 1848, and had produced about 8000 lbs. of mercury from 1843 to 1845.

AUSTRIA.—*Mines of Idria.*—At Idria, in Carinthia, mines of mercury have been worked for several hundred years, which are still of great importance, and second only to those of Almaden. The ore is chiefly the sulphuret, with some native mercury; it is contained in a black compact limestone, associated with shales, in which are fossils of the age of the Jura limestone. The ore is intimately combined with the shales, having evidently penetrated every portion of them in the form of a metallic vapor.

The yield of the Idria mines averaged, during the five years from 1843 to 1847, 358,281 lbs. per annum.

Bohemia formerly produced mercury in some quantity, but the workings for this metal have now entirely ceased.

Hungary furnishes also a very small quantity of this metal from the copper ores obtained near Schmöllnitz.

SPAIN.—The mines of mercury in Spain are the most important of this metal in the world, whether we take into consideration the length of time during which they have been worked, the quantity of ore they have furnished, or that which they will be able in the future to produce. The mines of Almaden are situated in the province of La Mancha, near the frontier of Estremadura; the chief workings are near Almaden, but mercurial ores are found over a wide belt, which runs in an easterly and westerly direction, through that place, and extends from the town of Chillon to that of Almadenejos; at the last-mentioned place considerable mining operations are also carried on. These mines have been worked longer and more uninterruptedly, perhaps, than any others in the world. Pliny asserts that the Greeks procured vermilion from them at least seven hundred years before the Christian era, and, according to the same authority, they yielded annually to the Romans 100,000 lbs. of cinnabar. They will be capable yet, for an indefinite period, of furnishing the amount at present obtained from them. These mines were formerly worked by condemned criminals; but paid miners are now employed in them, who, it need hardly be added, live but a short and wretched life.

The Almaden mines are opened in three parallel beds, which belong to the class of contact-deposits, as they are situated at the junction of the Silurian slates and sandstones with a metamorphic rock, called locally Fraylesca, which forms a zone between the stratified deposits and the eruptive rocks. The three metalliferous beds are from twenty to forty feet in thickness, and follow all the flexures of the stratification of the enclosing rocks, both in a vertical and horizontal direction. The rock which is called Fraylesca, is considered by Burat as a sandstone metamorphosed by contact with the dioritic masses on which it rests, and which appear at the surface at no great distance from the mine. The mercurial vapors seem to have been introduced by sublimation from below into the strata, and are accumulated in such masses that twenty-five centuries of exploitation have only excavated them to the depth of about 1000 feet. According to Le Play,* the ore, at the bottom of the principal workings in the main vein, is from forty to fifty feet in width, all of which is rich enough to pay well for taking out, being entirely unmixed with barren rock. The average yield for the ore is about 10 per cent., but a considerable quantity of the metal is lost by the imperfect processes adopted.

The present yield of the Spanish mercury mines appears to be about 2½ millions of lbs. per annum. The mines are the property of the government, and their produce is a considerable source of revenue, having for a long period been monopolized by the Rothschilds and other eminent European bankers. Until recently, in consequence of this state of things, the price of the metal has risen considerably; the Spanish government having raised the price, at successive lettings of the contract, from $51 25, in 1839, to $59; and afterwards, in 1843, to $82 50 per quintal (106 lbs.). Recently, the price has been lowered on account of the competition of the California mines. Under the head of Mexico, some particulars of the amount required to supply the silver mines of that country will be given.

Tuscany.—At Ripa, in Tuscany, mercury is obtained in small quantity. The mine is wrought in small veins of cinnabar disseminated through mica slate. Their aggregate thickness only amounts to about twenty-eight inches. No

* Observations sur l'Histoire Naturelle et sur la Richesse Minerale de l'Espagne, p. 31.

statistics of the amount produced here have been obtained, but it must be quite small.

South America.—*Peru.*—The mines of Peru have hitherto been the principal source of mercury on the American continent, although they now seem likely to be eclipsed by those of California. In addition to the information given by Humboldt, and other scientific travellers, in regard to these extensive deposits of the mercurial ores, we have an elaborate account of them, of a much more recent date, by M. Crosnier, who was sent by the Peruvian government in 1850 and '51, to explore and report upon some of the principal mining districts of that country.* According to his statement, it appears that the deposits of this metal which occur in South America are not confined to any particular formation, but range from the granite to the carboniferous; those of Chili are in the former rock, while the Peruvian mines appear to be confined to the sandstones of the latter age.

Mercury ores are widely scattered through Peru, and were well known to the inhabitants before the invasion of Europeans into the country, although it is not supposed that they knew how to procure the metal itself from them, and they probably only used the native cinnabar for painting their faces.

The deposits of the province of Huancavelica are the most important in the country, both in number and richness, as the presence of mercurial ores has been proved in forty-one different localities in that district. The most important one is that of Santa Barbara, which is still known by the inhabitants as "The Great Mine." It has been worked since 1566, although its product has very much fallen off, and now hardly exceeds 100,000 lbs. per annum. For a long time the mine was held by the government, and farmed out to be worked; of course those who leased it were only desirous of producing the greatest amount of mercury during the period of their possession, and the consequence was, that it was most negligently wrought, and the whole of the upper and richer part of the works was ruined by being allowed to fall in. The

* See Ann. des Mines (5), ii. 1.

excavations, which are of immense extent, having been going on for more than three centuries, are more like a series of quarries, arranged one above the other at different depths, than like a regular mine. The metalliferous strata are about 350 feet in thickness, and consist of a series of sandstones and shales, intercalated in other sandstones and conglomerates, the whole dipping to the west at an angle of about 64°. The excavations in these strata impregnated with mercurial ores, are a third of a mile in length, and extend to a depth of about 1200 feet. Two hundred workmen were killed at one time by the caving in of a part of the mine, and many such accidents have happened within the last two hundred years, so that a tunnel-shaped cavity has been formed on the surface, of 30 to 40 feet in depth, and 200 or 300 feet in circumference. The most important engineering work of this mine is the deep adit, called the Socabon de Belen, which is nearly 2000 feet long, 10 to 12 feet wide, and equally high.

It is a very curious fact that native mercury has been found, sometimes in considerable quantity, in the alluvial deposits near the city of Huancavelica: 600 lbs. were taken out from this position by digging a simple ditch only from three to six feet in depth. This seems to indicate that a constant slow sublimation of the metallic particles is still going on, and a condensation in the superficial deposits. Such facts have been repeatedly observed in different countries, and it seems impossible that the presence of the metallic mercury could in all these cases have been the result of accident, and not of natural causes.

In regard to the mode of occurrence of these ores of Huancavelica, M. Crosnier remarks, that they have nothing in common with true veins; on the contrary, everything seems to indicate that the metalliferous matter was introduced into the strata in the state of vapor, at the time when they were elevated into their present almost vertical position.

According to Humboldt, this mine yielded from 1570 to 1789, 1,040,452 quintals of metallic mercury, or about 47,285 tons; amounting in value, at the price paid by the government, $73 per quintal, to $75,954,257. The average was about 6000 quintals yearly, and the produce in the best years went up to 10,500 quintals. From 1790 to 1845 it has only produced about 66,000 quintals, or a little less than 3000 tons.

There are some other mines of this metal worked in Peru, but they are of less importance and extent than those of Huancavelica; the total yield of the country amounts to about 203,000 lbs. per annum, about one-half of which comes from the Santa Barbara mine.

Mercury is found in many other localities in different parts of South America, but none of them appear to be of much importance. Humboldt mentions cinnabar as occurring in New Grenada in three places; in the Province of Antioquia, in the Valle de Santa Rosa; in the Quindiu Mountain; and between the villages of Azogue and Cuença, in the Province of Quito. At this locality the ore is obtained in a quartzose sandstone, of great thickness, containing fossil wood and bitumen.

Mexico.—An immense amount of mercury is consumed in the process of separating the silver supplied by this country from its ores; but there seem to be no mines wrought within its territory, although there are numerous localities where the ores of this metal have been found. Humboldt and Duport mention the following as the most important: Gigante, near Guanaxuato; Rincon de Centeno, near Queretaro; Durasno; Sierra de Piños, and other places in the department of San Luis Potosi; Melilla, in that of Zacatecas; and El Doctor, in that of Queretaro.

According to Humboldt, the ores are found in two positions quite distinct from each other; they form beds in the secondary strata, or veins which traverse the porphyritic trappean rocks. At Durasno, cinnabar, mixed with many globules of native metal, forms a horizontal bed resting on the porphyry, and covered by beds of shaly clay containing fossil wood and coal. The excavations are only pits a few feet in depth. This seems to have been only a limited deposit, although several hundred quintals were at one time procured from it. The cinnabar vein of San Juan de la Chica is from seven to twenty feet in thickness; the ores are rich, but not abundant. The geological position is remarkable, since it is found in a pitchstone-porphyry, having a globular structure. It had been worked, at the time of Humboldt's visit, to a depth of over 150 feet. There were then only two mines of this metal wrought in Mexico; one called the Lomo del Toro, and the other the mine of Nuestra Señora de los Dolores, southeast of Gigante, yielding from 70 to 80 lbs. a week. About 1844, workings were carried on near Guadalajara, from which 400 to 500 quintals were produced.

At the beginning of the present century Mexico consumed annually 16,000 quintals of mercury in the separation of silver from its ores. This was furnished by the Spanish government, to which alone belonged the right of supplying this necessary metal, which was procured mostly from the mines of Almaden and Huancavelica. From 1770 to 1802 a small amount was purchased of the Austrian government at a price fixed by treaty, namely $52 per quintal; this was on account of the partial stoppage of the Almaden mines, owing to their being flooded with water. From 1762 to 1781, 191,405 quintals of mercury were thus absorbed by Mexico. The price at which it was furnished was, in 1590, $187 the quintal, but it gradually sunk, and was in 1777 only $41¼, its consumption increasing rapidly as the price fell. In 1844, the Spanish government having no longer any interest in supplying the mercury at a low price, but having made a monopoly of it, it cost in the harbors of Mexico about $120, and in the mining region of Zacatecas about $165 the quintal. At this latter period, according to Duport, the value of the mercury lost in separating the silver amounted to about one-tenth of the whole cost of its production. At the usual standard of the ores, the cost of the mercury would represent 0·0002 of silver, while at the price at which it was furnished by Spain, during the early part of the century, it would only amount to 0·000064. The difference in the cost of producing silver caused by the rise in the price of mercury is not considered by Duport and others who have investigated the subject to be a very serious obstacle to the development of the Mexican mines. The superintendent of the Pachuca and Real del Monte mines makes the following statement:* "It may safely be assumed that no known

* Stryker's Ann. Register.

mine with a remunerative content of silver in its ores is at present idle from the high price of mercury; and we must therefore only look for an increased produce of this metal (silver) from mines whose ores, although abundant, are too poor to pay the cost of reduction." A poor ore is considered by him to be one not yielding over 27 oz. to the ton. In his specification of the expenditure for one year at Real del Monte, the cost of the mercury absorbed is given as only ·046 of the entire expense of the establishment. The loss of mercury is estimated by him at ¾ lb. for each mark of silver produced.

UNITED STATES.—No mercury is known to have been found this side of the Mississippi River. The newspapers have recently brought reports of the discovery of important deposits of this metal in its native state, in New Mexico, about 40 miles north of Santa Fé. According to these statements, the mercury is found in the superficial formation, in small globules, which are easily separated from the dirt by washing. These accounts require confirmation.

The existence of this metal in California was known, and works were established there, prior to the late gold discoveries. In 1845, a company was formed to work a very important and extensive deposit of cinnabar, at New Almaden, in one of the side valleys of the San José, two or three miles from the main valley. From a description, by W. P. Blake, Esq., recently published,* the following notice of this interesting mine is chiefly extracted.

The ore is found in connection with sedimentary strata, composed of alternating beds of argillaceous shales and layers of flint, which are tilted up at a high angle, and much flexed. They are considered by Mr. Blake to be of Silurian age, but their position has not been determined with certainty. With these rocks the mercurial ores are mingled in a series of beds and laminations, of great number and extent, so that the whole of the workings are very irregular and contorted. The masses of ore are separated by intercalated strata of rock of variable thickness, which are them-

* Sill. Am. Jour. (2), xvii. 438.

selves often filled with seams and bunches of the sulphuret. Numerous veins of carbonate of lime traverse the rock in various directions, cutting through the ore and dislocating the small veins; and the same mineral lines cavities in the masses of cinnabar, being there finely crystallized, and sometimes containing bitumen in minute globules. The sulphurets of iron and copper and arsenical pyrites are associates of the ore, but they occur in very small quantities. An analysis of the ore by Prof. Hoffmann gave:—

| | |
|---|---|
| Mercury, | 67·25 |
| Sulphur, | 10·33 |
| Silica, alumina, &c., | 22·55 |
| | 100·13 |

The mine and works are now under the superintendence of Capt. H. W. Halleck, formerly of the U. S. Engineer Corps. An adit-level has been driven in for 900 feet, cutting the old works about 200 feet below the former entrance to the mine; this adit is about 10 feet by 10, and well timbered.

The furnaces for reducing the mercury from the ore are described by Mr. Blake as being well-arranged and effective, the reduction being effected without the use of lime, and without crushing the ore. About a hundred workmen are said to be employed at the mine and smelting-works: most of them are Mexicans and Yaqui Indians. A few Cornish miners have also been introduced.

The only statistics of the actual yield of this mine which I have seen are in a statement in the San Francisco Herald, to the effect that, during the first six months of 1853, 9047 flasks of mercury, of 100 lbs. each, were exported from San Francisco. This would give, at the same rate for the year, nearly 2,000,000 lbs. as the annual yield at present, which is probably too much, as the works are described by M. Dillon* as intended to produce only 10,000 cwts. It appears certain that the supply of ore is very large, and, if worked properly, capable of holding out for a long time.

* Ann. des Mines (5), i. 597.

The annexed table shows the amount (in lbs. avoirdupois) of mercury produced throughout the world, so far as it has been ascertained.

| Year. | Austria. | Spain. | Peru. | California. |
|---|---|---|---|---|
| 1800, | | | | |
| 1810, | | | | |
| 1820, | | | | |
| 1825, . . . . . | 391,406 | ('27) 2,264,000 | | |
| 1830, . . . . . | 299,746 | | | |
| 1835, . . . . . | 465,326 | | | |
| 1840, . . . . . | 337,814 | | | |
| 1845, . . . . . | 489,227 | ('44) 2,360,000 | | |
| 1846, . . . . . | 415.430 | | | |
| 1847, . . . . . | 448,571 | 2,240,000 | | |
| 1848, . . . . . | 460,398 | | | |
| 1849, . . . . . | | 2,500,000 | | |
| 1850, . . . . . | | | 203,000 | |
| 1851, | | | | |
| 1852, | | | | |
| 1853, . . . . . | | | | 1,000,000 |

The present production may be estimated as follows:—

| | lbs. | Per cent. |
|---|---|---|
| Austria, . . . . | 500,000 . . | 11·9 |
| Spain, . . . . | 2,500,000 . . | 59·6 |
| Peru, . . . . | 200,000 . . | 4·7 |
| California, . . . | 1,000,000(?) . . | 23·8 |
| | 4,200,000 | 100·0 |

The value of the mercury imported into this country, and retained for consumption, was, according to the official tables, as follows:—

| | |
|---|---|
| 1840, . . . . . . . | $43,513 |
| 1841, . . . . . . . | 59,587 |
| 1842, . . . . . . . | 30,321 |
| 1843, . . . . . . . | 35,114 |
| 1844, . . . . . . . | 77,464 |
| 1845, . . . . . . . | 54,993 |
| 1846, . . . . . . . | 155,813 |
| 1847, . . . . . . . | 143,078 |
| 1848, . . . . . . . | 2,092 |
| 1849, . . . . . . . | 21,979 |
| 1850, . . . . . . . | 79,350 |
| 1851, . . . . . . . | 62,767 |

# CHAPTER V.

## TIN.

### SECTION I.

#### MINERALOGICAL OCCURRENCE AND GEOLOGICAL POSITION OF TIN.

MINERALOGICAL OCCURRENCE.—It is doubtful whether this metal occurs in nature in its native state. A few grayish-white metallic grains were detected by Hermann in the gold-washings of the Ural, which proved, on examination, to be tin alloyed with a little lead; but there is reason to doubt whether these may not have been of artificial origin. Certainly, if native tin does occur, it must be an extremely rare substance. Its ores, and the combinations in which it is found, are very few in number. Two only are worthy of notice; these are—

*Cassiterite*, or Tin-stone; an oxide of tin, containing one atom of the metal and two of oxygen, or, in percentage, 78·62 of tin and 21·38 of oxygen. This is an ore which is destitute of a metallic appearance. Its color is usually a dark brown or black. It not unfrequently occurs finely crystallized in right square prisms; frequently in twin crystals, which sometimes weigh several pounds. The finest crystallizations are found in Cornwall and the Erzgebirge. *Wood tin* is a common form of this ore, and consists of botryoidal and reniform masses, having a radiated structure. This is the ore from which nearly the whole of the tin of commerce is derived.

*Tin pyrites*, Bell-metal ore. A sulphuret of tin and copper, with a little iron and zinc. It is, when pure, of a steel-gray color, but has often the appearance of bronze; hence the name, *bell-metal ore.* This is a species of rare occurrence: its principal locality is Wheal Rock in Cornwall, and it is found, in small quantity, in the Saxon and Bohemian tin mines.

A few traces of tin have also been found in some of the ores of titanium and uranium; but, compared with the other metals in common use, it is a rare substance. Notwithstanding this, it was one of those best known and most used by the ancients, from the earliest historic times. Long before the art of reducing iron from its ores had been acquired, tin, alloyed with copper, forming bronze, was generally applied, by those nations which were most advanced in civilization, to the fabrication of utensils of household and warlike use; but, in most cases, the sources from which their ores were derived are no longer known.

At present, although tin mines are worked in several countries, two stanniferous districts may be said to supply the world with this metal, since the amount obtained from other sources is but trifling in comparison with what they furnish. The great tin-producing regions are Cornwall in England, and the islands of the Malayan Archipelago, especially Banca.

GEOLOGICAL POSITION.—Tin, more than almost any other metal, has a peculiar and characteristic mode of occurrence. It is pre-eminently an old metal, since it is not found at all in the newer rocks. Neither does it occur disseminated through nature, like silver, copper, or iron, or even arsenic, which are present almost everywhere, if not in quantity, at least in minute traces. Tin ore is confined almost exclusively to the azoic, metamorphic palæozoic, and hypogene rocks. The latter is its characteristic position.

There are four forms in which the deposits of the ores of this metal present themselves: 1st. In flat sheets or beds lying between the laminæ of the slates and granites, and parallel with them and each other; each deposit is usually quite limited in its dimensions, although frequently accompanied by similar ones at no great distance. Such sheets of ore are called, in Cornwall, *floors*, and, when they consist of tin ore, *tin floors*, although this name is also given to deposits, to which the name of stockwerk would be more properly applied. They seem to be allied in character to contact deposits, or segregated masses, and pass into the next class, which is that most characteristic of the ores of this metal. 2d. The stockwerk, in which form of deposit the stanniferous mass is made up of an assemblage of veins of small size, in which the ore is mostly concentrated, and which ramify through the rock, which, itself, contains oxide of tin disseminated through it in fine particles in the neighborhood of the veins. These evidently do not originate in fissures, although frequently approximately parallel with each other. Quartz almost invariably forms the principal gangue of the stanniferous veins, and the rock itself, in their vicinity, is usually more quartzose than elsewhere. 3d. The ores of tin are frequently found in true fissure-veins; but they are generally believed, in such cases, not to continue to a great depth, being frequently replaced by copper and other metals. It is usually allowed, that where there are several sets of veins in one district, those which carry ores of tin are the oldest. 4th. Tin-stone is very extensively obtained from washings, or "stream-works," as they are called in Cornwall, the ore being scattered through the superficial detritus, and separated from it by the same methods which are applied to gold and platina, as already noticed. This is the character of the deposits of Banca and the Malayan Peninsula, which have been long worked, and have yielded extensively, no mining in the solid rock having been as yet practised in those regions.

The metalliferous substances which are chiefly obtained from washings, are necessarily such as are not liable to undergo decomposition when exposed to air and moisture. Gold, platina, and the associated metals are of this character, and would remain for ever unaltered, except from the action of mechanical causes. Oxide of tin possesses similar characters, being an ore which does not readily enter into new combinations with carbonic and other acids with which it is brought in contact in the superficial deposits. Almost all the other metallic ores under the same circumstances form various salts, some of which are soluble, and are washed away entirely, while others are earthy and pulverulent, and for this reason, and on account of their low specific gravity, could not be collected by washing, at least without great loss.

The vein-stones and minerals which are associated with the oxide of tin are remarkably constant in their nature, all over the world. They are, wolfram, or tungstate of iron and manganese, apatite, topaz, and mica; sulphuret of molybdena, native bismuth, and arsenical pyrites are also rarely wanting where tin is found. Tourmaline is another almost constant companion of this metal. Sometimes it forms a part of the veins themselves, but more usually it occurs disseminated through the rock adjacent to them. Where these minerals occur, veins of tin ore may be reasonably expected to be found.

---

## SECTION II.

### GEOGRAPHICAL DISTRIBUTION OF TIN.

Great Britain.—Cornwall is alike celebrated for its copper and tin, but the latter metal has been worked there for a very much longer time, and it was only at a comparatively recent period that copper became of importance.* The Phœnicians were the earliest traders in the metal obtained from that region, and it is supposed by Mr. Hawkins that Gades, on the western coast of Spain, was the entrepot of the trade between Phœnicia and Cornwall. Diodorus Siculus, who wrote in the time of Julius Cæsar, about 60 B. C., gives an account of the tin trade of Britain; and St. Michael's Mount, on the Cornish coast, is considered to have been identified as the market where it was carried on. When bells came to be generally used, the demand for tin increased rapidly, and still more on the introduction of bronze cannon, and for a long time the continent of Europe was supplied from Cornwall with the necessary metal for these purposes,† Bruges

* Trans. Geol. Soc. Cornwall, iii. 116. † De la Beche, Geology of Cornwall, p. 525.

being the emporium of the trade; and had not the East Indian deposits of this metal been discovered, England would still have almost a complete monopoly of it.

### GEOLOGICAL STRUCTURE OF THE CORNISH MINING REGION.

It will here be a proper place to introduce some account of the geological formation of Cornwall, a country so important in the development of the metallic wealth of the world, and especially interesting to us, from the fact that a large portion of the miners employed in our own mines come from thence, bringing with them peculiar ideas, based on their experience in a region which presents the most complicated and extensive network of metallic veins, which is thus far known to exist. We consider it important to give a succinct account of these phenomena, in order to show the peculiar character of the Cornish mining region, a district which has nothing analogous to it in this country, where the modes of occurrence of the metals are so different, as to make the experience obtained in Cornwall of little value, when applied to our own veins.

The best descriptions of Cornwall have been given by De la Beche, in his "Report on the Geology of Cornwall, Devon, and West Somerset," and by the French mining engineers, who have also investigated the metallurgic treatment of the ores with consummate ability.*

There are three well-marked groups of rocks, distinguished everywhere by the miners under the names of growan, killas, and elvan, or, in the usual geological language, granite, slate, and porphyry. De la Beche distinguishes between the mica and hornblende slates, and the grauwacke group. The former are developed in the Lizard district, and these rocks he seems disposed to refer to the azoic, while the grauwacke group, including sedimentary deposits, varying from the finest roofing slates to conglomerates, some of the component parts of which weigh more than half a ton, belongs to

* See Dufrenoy, Elie de Beaumont, Coste, and Perdonnet, Voyage Metallurgique en Angleterre, 2 vols. and atlas, Paris, 2d ed., 1839. Also, Annales des Mines, for various articles, especially an elaborate one by Le Play, Ann. des Mines (4), xiii. 1.

the palæozoic system. So complete is the metamorphosis which has been effected in these rocks, and so difficult the task of unravelling the complicated foldings which they have undergone, that it will be long before they will be classified with precision.

The miners call all the slaty rocks killas, but the type of the killas is a greenish argillaceous slate, in which are the principal copper mines. The name elvan is given to any rocky masses which occur in the slates and granites, and displace the veins, or derange the stratification of the rocks. In general, however, it is applied to the long lines of porphyritic rock, which differ from trap dykes only in the mineralogical composition of the material, being rather a porphyritic granite, and more feldspathic than hornblendic. These elvans, though narrow, varying from a few to 300 or 400 feet in width, may frequently be traced for a great distance; the longest one being at least 12 miles in length. With regard to their mineral components, De la Beche says: "These elvans have, for the most part, a common mineral character, being chiefly composed of a feldspathic, perhaps often a quartzo-feldspathic base, containing crystals of feldspar and quartz, either singly or together, in the same rock; schorl less frequently, and, still more rarely, mica."

There are six principal, isolated masses of granite, besides smaller patches, extending in a general linear direction from north 66° east, to south 66° west. This rock, as it is exhibited in Cornwall, is made up of the usual ingredients, quartz, feldspar, and mica, with the addition of schorl or black tourmaline, which is very common along the confines of the granitic masses, and in the vicinity of the stanniferous lodes; this is a general fact with regard to other accidental minerals; they are almost entirely concentrated near the junction of the two formations, the granite and the killas; and here also are the stockwerk deposits and the veins of tin. These two rocks do not pass into each other by insensible gradations, but the only change that has taken place in the killas seems to be a hardening, the granite penetrating it frequently in ramifying veins and interlacing with it.

The metalliferous region of Cornwall and Devon is divided

by De la Beche into six groups; that of Tavistock; of St. Austell; of St. Agnes; of Gwennap, Redruth, and Camborne; that of Breague, Marazion, and Gwinear; that of St. Just and St. Ives. Of these divisions, the first bears copper, tin, and silver-lead; the second, that of St. Austell, is stanniferous; the third, the great mining district of Gwennap, Redruth, and Camborne, is chiefly cupriferous; the fourth is of a mixed character, bearing tin and copper with some lead and silver; the district of St. Just and St. Ives is chiefly stanniferous. The ores of copper and tin are invariably found in the vicinity of the granite or elvans.

We come now to speak of the veins of the Cornish mining region and their relation to each other, both of age and position, and the mode of occurrence of the ores in them. It will be easily understood that these things have been thoroughly investigated and studied out in a district where there was such a large amount of capital invested in mining operations, and that the accumulated experience of hundreds of years has made many things very plain, which were at first undoubtedly difficult to decipher.

The first thing to be noticed in this relation is the fact of the occurrence of numerous faults, as they are termed, or fractures of the strata, which stretch along sometimes for great distances, entirely interrupting the continuity of the rocks on each side. This actual severing of the strata is also accompanied by a vertical displacement, so that they are no longer in their original position, portions of the region having been lifted up or sunk down below their former level. These faults have either a north and south or an east and west direction, though of course, with many minor variations. When such dislocations have taken place among the stratified and fossiliferous rocks, it is easy to ascertain the amount of vertical displacement which has taken place, since beds of dissimilar character, either in respect to the organic remains enclosed in them or in mineralogical composition, are thus brought in contact with each other. In the unstratified and older slaty rocks, on the other hand, where masses of the same general appearance are brought together, this is not always possible. If, however, there are fissures filled with mineral matter, and two such fissures cross each other, it will be easy to ascertain whether they were contemporaneous, and, if not, which was first formed, since the oldest will be cut through by the more recent one, and will have its continuity completely interrupted.

Mr. Carne divided the veins and fissures of this district into seven groups, as follows: 1, elvan courses; 2, tin lodes; 3, east and west copper lodes; 4, contra copper lodes; 5, cross-courses; 6, more recent copper lodes; 7, cross-flucans and slides. This classification, however, is in some respects fanciful, since the copper lodes cannot be separated into so many groups as he imagined. It is clear, however, that there were several distinct periods of disturbance, and consequently as many sets of fissures. The following are definitely ascertained: First, the *elvan courses*, which are subordinate to the main granitic masses of the district, and traverse them, but which are cut by all the metalliferous veins. Their direction is nearly that of the granite ranges themselves. The second epoch is that of the east and west veins, or the *great metalliferous lodes*, which have in general the same direction as the elvans; but which intersect, and are therefore posterior to them. De la Beche, who has certainly studied the subject with as much care as any one, and who is remarkably cautious in advancing mere theoretical opinions, does not divide the east and west lodes, or those having approximately that direction, into groups, or profess to be able to distinguish between their relative ages. It is generally asserted, however, that the tin veins are the oldest; but it would seem that this could hardly be considered as a general rule, since there is in reality no set of tin veins distinct from those bearing copper, the same lode often carrying the two metals at different points of its course. The third set of fissures are the *cross-courses*, or veins and fissures with a general north and south course. The cross-courses cut the east and west lodes at a considerable angle, which in the vicinity of Tavistock varies from 70° to 90°, and it is frequently very nearly a right angle. In that district, the cross-courses carry argentiferous galena; and at Beer Alston are extensively mined. The Dartmoor lodes run nearly east and west, and some of them are cut by an argentiferous north and south cross-course. In the St. Austell district, while the general direction of the fractures remains the same, there are probably three epochs of dislocation, a part of the region having been heaved or moved nearly a tenth of a mile. In the Polgooth mines, however, the fractures have a radiated character, diverging from a common centre. In the great Gwennap, Redruth, and Camborne district, the veins preserve a remarkable degree of parallelism, and have a west-southwest and east-northeast course. They are traversed by cross-courses running nearly at right angles to them. One of these, called the Great Cross-Course, can be traced for many miles, heaving the country from seventy to eighty fathoms horizontally. Many others traverse this district; one of which intersects the east end of the Dolcoath mine, separating it from the Cook's Kitchen mine.

These north and south fractures, or the great cross-courses, were evidently the result of the last great disturbance, as they cut through all the veins and elvans, and there are no later fissures except the so-called east and west slides. These slides and cross-flucans, as they are called, are barren of ore, being filled with clay; they are on this account of a good deal of practical importance, inasmuch as they act as an effectual barrier of the water. They are of a comparatively modern date, certainly later than the chalk in some cases, and probably in all.

### OCCURRENCE OF TIN IN CORNWALL.

Having thus given a general description of the vein-phenomena of the Cornish mining district, we come to speak more particularly of the modes of occurrence of the tin in that region. These are four in number: 1st. In beds between the strata, or so-called *tin floors;* these are thin masses of ore lying in the direction of the stratification. They are found in the killas, generally not far from the granite, and are sometimes accompanied by floors of schorl, called in Cornwall, *cockle.* Although these floors do not always seem to be connected with the stanniferous veins, yet they are in most cases subordinate to them. 2d. *Stockwerk deposits:*

these are also called tin floors by the miners. They occur in the granite and elvan. The Carclase mine near St. Austell is in the granite and was formerly extensively worked, and was open to the day like a quarry. It presented an aggregation of small veins traversing the decomposed granite in various directions, and made up of tourmaline and quartz with oxide of tin disseminated through it. 3d. *The true veins.* Much the largest part of the tin is obtained from this source. The greater number of tin veins are found in the granite, although the richest ones are in the killas, and the most productive portion of them is near the junction of these two rocks. 4th. *Stream works.* These are now nearly exhausted, after centuries of working. The most extensive and productive, in 1839, were those of Pentowan, near St. Austell. The tin-stone was found in the form of fine grains and pebbles at the bottom of the ancient alluvial deposits, covered by from twenty to seventy feet of alternating beds of sand and clay.

According to Mr. J. Y. Watson,* the tin mines of Cornwall, taken in the aggregate, have not been profitable for many years. From the reign of Charles I. to George I., the black tin produced averaged 1500 tons annually; the produce then increased, and in the year 1742 a proposal was made by the Mines Royal Company, in London, to raise £140,000 to encourage the tin trade, by farming that commodity for several years at a certain price. A committee of Cornish gentlemen was appointed to consider the proposals, and they reported, "That the quantity of tin raised yearly in Cornwall, at an average, for many years last past, hath been about 2100 tons," and resolved, that £3 9*s.* for grain tin, and £3 5*s.* for common tin, are the lowest prices for which such tin will be sold to the contractors, exclusive of all coinage duties and fees. From the year 1750 to 1837, the annual produce never exceeded 5000 tons, but generally averaged 2500 to 3500 tons. In the latter year, an application having been made to the government to abolish the duty, the state of the mines was officially ascertained, and the result showed that, upon 58 mines, the loss was £80,517,

* English Mining Journal, No. 692.

and upon 10 the profit was £20,358, showing a net loss of £60,159. Within the last year, however, the tin mines of Cornwall have been much more profitable, owing to the rise in the price of that metal.

The principal tin district is that of St. Just, which produces this metal chiefly; it is about three miles long and one and a half broad. In this district are several tin stream-works, the chief of which is Carnon. The mines which have produced most considerably in the past year are: Botallack, Balleswidden, Boscean, Levant, Spearne Consols, and Wheal Owles.

Other important mines now yielding a considerable amount of tin are: Polberro and Wheal Kitty, in the parish of St. Agnes; Lewis and West Providence, in St. Erth; Trelyon Consols and St. Ives Consols, in St. Ives; Great Polgooth, in St. Austell; Drake Walls, in Calstock; and, finally, Dolcoath, in Camborne.

Up to 1838, the number of blocks of tin produced in Cornwall and Devon is accurately known, although the exact weight of a block appears to be somewhat doubtful; taking it at the amount fixed by De la Beche, 3·34 cwts., we have the following table of the annual produce of this metal from 1750 up to 1838; the sums given for the years up to 1835 being the average of each period of five years:—

| | Tons. | | Tons. |
|---|---|---|---|
| 1750–54, | 2804 | 1805–9, | 2681 |
| 1755–59, | 2955 | 1810–14, | 2406 |
| 1760–64, | 2827 | 1815–19, | 3540 |
| 1765–69, | 3083 | 1820–24, | 3592 |
| 1770–74, | 3092 | 1825–29, | 4549 |
| 1775–79, | 2866 | 1830–34, | 4023 |
| 1780–84, | 2890 | 1835, | 4027 |
| 1785–89, | 3520 | 1836, | 3862 |
| 1790–94, | 3689 | 1837, | 4562 |
| 1795–99, | 3341 | 1838, | 4887 |
| 1800–4, | 2810 | | |

Since 1838 the production of tin is supposed to have increased considerably. The ore, as taken from the mines, is very variable in richness, being sometimes as low as 2 or 3 per cent., but it is raised, by washing and dressing in various ways, so as to produce from 66 to 70 per cent. In 1850, according to De la Beche,* 10,052 tons of ore were produced, which would give as the product of metallic tin about 7000 tons, a quantity which agrees very nearly with other estimates which have been made.

SAXONY.—The most important tin localities in Saxony are those of Geyer, Altenberg, Zinnwald, and Auersberg, in the Erzgebirge. At Zinnwald, the stanniferous district is partly in Bohemia.

At Geyer, the rock most directly connected with the tin veins is a granite, consisting chiefly of decomposed feldspar, containing apatite, tourmaline, and fluor-spar. The ore is in numerous small parallel veins, and disseminated in the adjacent rock. The veins are rarely over two inches wide, and they have no defined selvages, but shade off gradually into the granite.

The tin ore of Zinnwald is found in a granitic rock, composed mainly of quartz and mica, which forms a flattened dome-shaped mass rising through the porphyry. Frequently the whole mass of the rock is stanniferous, but the most productive deposits are in nearly horizontal layers, composed of quartz, mica, and oxide of tin, with a great variety of other minerals. About thirty of these deposits are known, of which nine are worked. These beds or veins are not strictly horizontal, but seem to form layers concentric with the granitic mass, which dip at a greater angle as they approach its limits in every direction.

At Altenberg, the stanniferous rock is a porphyritic mass, which is intersected by a great number of small veins running in every direction and interlacing with each other, constituting a true stockwerk deposit. Not only are these veins stanniferous, but the adjacent rock is impregnated with metallic particles, though not so richly as the veins themselves, the best portion being at their intersections. The whole mass is stamped and washed, the yield of metal not being more than one to two per cent.

The principal yield of tin, at present, is from the Altenberg mines, and amounts to about $50,000 in value. The

* Lectures on the Results of the Exhibition. Lect. II., p. 62.

whole product of the Saxon and Bohemian tin mines, in the most favorable years, is not more than one-fifteenth of that of Cornwall.

AUSTRIAN EMPIRE.—The only tin mines worked in this country are those of the Erzgebirge, which, as before remarked, are partly in Saxony and partly in Austria. They have already been sufficiently alluded to. The principal mines are at Zinnwald, Schlackenwald, and Abertham. The total production only amounts to about fifty tons a year.

FRANCE.—There are no workable mines of tin in France, although this metal is found in several localities. Its mode of occurrence seems to be in every respect similar to that of Cornwall. At Vaury, in the Department of Haute Vienne, the ore is found in narrow quartz veins in a species of granite, accompanied by the usual minerals. In the Morbihan, at Villeder, there is also a powerful quartz vein, near the junction of granite and slate; the tin ore is disseminated through the vein where mica and schorl occur in it, but the quantity appears to be too small to be worked. The same may be said of the ore occurring at Piriac, near Nantes.

SPAIN.—There are several localities in this country where tin ore has been found; but the produce of metal amounts to only a few tons annually. According to Schulz and Paillette,* tin is found in the districts of Penouta and Romilo, in the Province of Orense, in narrow veins traversing granite and mica slate; between Verin and Monterry, on the borders of Portugal, in a similar position; and finally, near the line between the provinces of Orense and Pondevedra, in veins from one to eight inches wide, contained in mica and hornblende slate, near its junction with granite. The usual minerals which accompany tin ores occur here also. All these localities are in Gallicia.

MALAYAN ARCHIPELAGO.—BANCA.—The mines of Banca are said to have been discovered in the year 1710–11. They are, in general, worked by the Chinese. Crawfurd says:† "The quantity of tin which the mines of Banca are capable of affording is immense, as the supply of ore is nearly indefi-

* Bull. de la Soc. Geol. de France, vii. 16.

† Hist. Indian Archipelago, iii. 462 (1820).

nite, and the facility for working great. About the year 1750, or forty years after their discovery, they yielded, as has been calculated, much above 120,000 slabs, or 66,000 piculs, = 3,870 tons. About the year 1780 the produce had fallen to 30,000 piculs, or to less than half its maximum, and, from 1799 until the British conquest, seldom exceeded one-third of this last amount, or 10,000 piculs. In 1817 the produce reached 2083 tons, higher wages having been paid to the workmen."

The tin is obtained exclusively from the recent (alluvial?) deposits, which are horizontally stratified, and consist of beds of variously-colored clays and sands, the tin being found at a depth of from ten to fifteen feet below the surface, resting on a bed of white clay of a peculiar character, which is considered by the Chinese miners as an unequivocal indication of the termination of the stanniferous stratum.* The process of mining is of the simplest kind, and the excavations are mere open pits, the whole of the surface being turned over. The fragments of rocks found accompanying the ore indicate that the formation from which the tin was derived was a granite, containing schorl in considerable quantity. The process of decomposition and denudation must, however, have been on the most extensive scale, since the rock seems hardly to be found in place anywhere in the neighborhood of the mines.

The Banca tin of commerce is very pure, as is shown by the following analysis, made by Mulder. According to him, it contained—

| | |
|---|---|
| Tin, | 99·961 |
| Iron, | ·019 |
| Lead, | ·014 |
| Copper, | ·006 |
| | 100·00 |

The tin of Banca finds its way into almost every part of the world. A portion of it goes to China and the continent of India; but how much, it is impossible to ascertain. That which is destined for Europe is sold at auction at Rotterdam and Amsterdam, annually or semi-annually. The following

* Thomas Horsfield, M.D., in Journal of Indian Archipelago, iii. No. 7 (1848).

are the amounts which have been thus sold for the last sixteen years:—

| Year. | Slabs. | Tons. | Year. | Slabs. | Tons. |
|---|---|---|---|---|---|
| 1837, | 28,041 | 940 | 1846, | 60,090 | 2000 |
| 1838–9, | 58,133 | 1940 | 1847, | 119,955 | 4000 |
| 1840, | 27,520 | 920 | 1848, | 84,943 | 2830 |
| 1841, | 54,241 | 1810 | 1849, | 249,937 | 8330 |
| 1842, | 78,299 | 2610 | 1850, | 117,766 | 3925 |
| 1843, | 95,439 | 3180 | 1851, | 111,181 | 3705 |
| 1844, | 63,160 | 2105 | 1852, | 156,702 | 5225 |
| 1845, | 75,102 | 2505 | 1853, | 112,305 | 3745 |

MALAYAN PENINSULA.—The first tin mine worked in this region was opened in 1793.* The principal mine, at Cassang, is worked by 2200 Chinese. The excavations, mines they can hardly be called, are in the swampy flats at the base of the hills, and are not more than from six to twenty feet deep: they follow certain "streams of ore," which extend for two or three miles. The Malayans themselves are too lazy to work the mines to any depth. The quantity annually exported from Malacca is estimated at from 900 to 1000 tons.

SOUTH AMERICA.—The only mines of tin known to be worked in South America, are in Bolivia, at Guanuni. They are considered, by D'Orbigny, to be of great richness,† although only worked, at present, to a very trifling extent, in order to provide return freight for mules coming from Peru with brandy. He estimates their produce at from 180 to 225 tons annually.

MEXICO.—At the time of Humboldt's visit, tin was obtained from washings at Gigante, San Felipe, Robledal, and San Miguel el Grande, in the province of Guanaxuato, and between the towns of Xeres and Villa Nueva in Zacatecas. The ore, according to Humboldt,‡ appeared to have originated in veins traversing the porphyritic trap, but the only workings were in the alluvial deposits.

The amount of this tin exported in 1803 was only 58½ quintals; and it appears that the washings must now be

* J. B. Westerhout, in E. Mining Jour., Aug. 26, 1848.

† D'Orbigny's Travels, vol. iii. of History, p. 316.

‡ La Nouvelle Espagne, ii. 581.

pretty much exhausted and abandoned, as there is no information of any production of that metal in Mexico at the present time.

United States.—A single crystal of oxide of tin, weighing fifty grains, was found by President Hitchcock, State Geologist, many years since, at Goshen, Massachusetts.* It was contained in granite. This was the first discovery of this metal within the territory of the United States. Since that it has been found at various places in small quantity. Prof. C. U. Shepard has noticed it, in minute crystals, at Chesterfield, in Massachusetts; and Prof. Rogers has detected it in the talco-micaceous slates of the gold mines in Virginia.

The only locality in this country where this ore has been found in any noticeable quantity, is at Jackson, New Hampshire, where it was discovered by Dr. C. T. Jackson in 1840, on land belonging to Mr. Wm. Eastman. It occurs in mica slate, in three or four small veins, which run in different directions and intersect each other within a space of 200 to 300 feet square. The widest part of the principal vein was eight inches, and there it yielded thirty per cent. of tin; the others are small strings, mostly less than an inch in width. Associated with this ore is an abundance of arsenical pyrites, and small quantities of copper pyrites, together with the usual minerals found with tin, namely, fluor-spar, tourmaline, sulphuret of molybdena, and others.

The ore was considered by Dr. Jackson to be sufficiently abundant to be profitably worked; but it does not appear to have become the object of mining enterprise.

The production of tin of the various parts of the world is summed up in the following table, so far as it can be ascertained with any certainty. In regard to the product of the Malayan Archipelago the estimates are very vague, since we do not know what portion of it goes to China and East India. But it will be seen at once how insignificant that portion is which is furnished by Europe, apart from England, and the American continent. The weights are given in tons.

* Final Rep. on the Geol. of Mass., i. 205.

| Year. | Great Britain. | Saxony. | Austria. | Spain. | Malayan Peninsula and Archipelago. | Bolivia. |
|---|---|---|---|---|---|---|
| 1800, . . | 2.800 | | | | | |
| 1810, . . | 2,400 | | | | | |
| 1820, . . | 3,600 | | | | | |
| 1825, . . | 4,550 | 142 | 35 | | | |
| 1830, . . | 4,050 | 149 | 61 | | | |
| 1835, . . | | 125 | 49 | | | |
| 1840, . . | | | 57 | | 4,000 | |
| 1845, . . | | | 55 | 14 | | 200 |
| 1846, . . | | | 55 | | | |
| 1847, . . | | | 54 | 10 | | |
| 1848, . . | | 81 | 49 | | | |
| 1849, . . | | 78 | | 4 | | |
| 1850, . . | 7,000 | | | | 5,000 | |
| 1851, . . | | 106 | | | | |
| 1852, | | | | | | |
| 1853, | | | | | | |

The whole produce of the world of this metal at present probably amounts to about 12,000 tons. Of this a very large portion is consumed in England, especially in the manufacture of tin plates. From elaborate tables given by Mr. Carne,* it appears that all the foreign tin brought to that country, amounting, from 1830 to 1840, to from 1000 to 2000 tons per annum, was exported, and in addition to this a large portion of the metal produced in Great Britain. The following table shows the amount produced, consumed at home, and exported from that country for the period from 1780 to 1837.

| Years. | Produced. | Consumed at home. Total. | An. av. | Exported. | | |
|---|---|---|---|---|---|---|
| 1783–1790, | 24,593 | 7,412 | 926 | 17,281, | = | 7-10 |
| 1791–1800, | 31,777 | 7,545 | 754 | 24,232, | = | 3-4 |
| 1801–1810, | 76,144 | 11,179 | 1,118 | 14,965, | = | 4-7 |
| 1811–1820, | 30,473 | 16,000 | 1,600 | 14,200, | = | 7-15 |
| 1821–1830, | 44,194 | 26,158 | 2,616 | 18,036, | = | 2-5 |
| 1831–1837, | 29,749 | 23,542 | 3,363 | 6,207, | = | 1-5 |

The United States are the largest consumers of tin plates. It is said that over 900,000 boxes are now manufactured in South Wales and Staffordshire, of which two-thirds are exported from Liverpool, nearly all of which comes to this country. In 1852, 512,400 boxes were shipped to all America, of which all but 69,502 boxes was for Boston, New York, and

* Jour. Stat. Soc. London, ii. 260.

Philadelphia. The consumption is rapidly increasing here, as is evidenced by the annexed table of shipments of tin plates from Liverpool to the three above-mentioned ports.

| | Boxes. | | Boxes. |
|---|---|---|---|
| 1846, . . . . . . . | 193,409 | 1850, . . . . . . . | 338,538 |
| 1847, . . . . . . . | 137,546 | 1851, . . . . . . . | 344,602 |
| 1848, . . . . . . . | 297,255 | 1852, . . . . . . . | 442,898 |
| 1849, . . . . . . . | 236,297 | | |

The following table, compiled from official documents, presents as complete a view as can be given of the amount of this metal consumed within our own borders, it being the value of that which was imported and retained for domestic consumption, from the year 1840 to 1851.

| Year. | Pigs, Bars, and Foil. | Plates and Sheets. | Manufactured. |
|---|---|---|---|
| 1840, . . . . . . . . . | $184,047 | $863,842 | $28,774 |
| 1841, . . . . . . . . . | 284,967 | 1,143,321 | 28,912 |
| 1842, . . . . . . . . . | 276,054 | 912,572 | 25,255 |
| 1843, . . . . . . . . . | 107,443 | 577,174 | 3,314 |
| 1844, . . . . . . . . . | | | 28,599 |
| 1845, . . . . . . . . . | | | 12,564 |
| 1846, . . . . . . . . . | | | 9,136 |
| 1847, . . . . . . . . . | 274,532 | 600.951 | 8.287 |
| 1848, . . . . . . . . . | 450,794 | 1,558,712 | 25,146 |
| 1849, . . . . . . . . . | 586.659 | 2,271,334 | 22,305 |
| 1850, . . . . . . . . . | 673.527 | 2,443,918 | 19.179 |
| 1851, . . . . . . . . . | 339,494 | 3,536,350 | 23,810 |

Since the above was in type, there has come to hand (in English Mining Journal of April 22d, 1854) an abstract of a statistical paper by R. Hunt, Esq., on the produce of copper, lead, and silver, in Great Britain, in which the production of tin in that kingdom, during the five years 1848–52, is given as 50,407 tons: an annual average of a little over 10,000 tons. There is no higher authority than Mr. Hunt on such a point; but as tin does not properly form one of the subjects of his paper, it is impossible to say precisely what value should be placed upon his estimate, until the paper itself can be seen.

# CHAPTER VI.

## COPPER.

### SECTION I.

#### MINERALOGICAL OCCURRENCE AND GEOLOGICAL POSITION OF THE ORES OF COPPER.

MINERALOGICAL OCCURRENCE.—The metal copper, in some one of its forms of combination, may be found almost everywhere in nature. It has been detected in numerous soils, in the ochrey deposits of mineral springs, in sea-water, and even in plants and animals: were our means of determining its presence as sensitive as they are for silver, it would probably be found, like that metal, almost universally diffused in infinitesimal traces; this is a natural consequence of the great variety and the solubility of its combinations. Copper has been known from the most remote periods. In the early history of nations its use has always preceded that of iron; and bronze has been the material for a great variety of tools before the art of working steel had been learned. As gold is the only yellow metal, so copper is the only red one; its softness and toughness make it very valuable for a great variety of purposes, but its great use is for the sheathing of ships, either by itself or alloyed with a small portion of zinc. In the various alloys of which it forms a part, especially in the form of brass and bronze, it is one of the most valuable of metals, and the consumption of it is rapidly increasing.

##### I. NATIVE METAL.

Metallic or native copper is not at all uncommon, being a frequent result of the decomposition of cupriferous ores; but until within a very few years, it had nowhere been found in sufficient quantity and with such a mode of occurrence as to make it a special object of exploitation. It frequently occurs crystallized, and usually in regular octohedra or some form intermediate between that and the cube. The specific gravity of the pure native metal is 8·838. It is often found in a state of absolute purity.

### II. ORES.

#### *a. Combinations with Sulphur, Selenium, Arsenic, and Antimony.*

*Copper Glance*, Vitreous Copper, Sulphuret of Copper. A sulphuret of copper with one atom of sulphur to two of copper, or 20·2 of the former to 79·8 of the latter. This is one of the important ores of this metal, although not as much so, by any means, as copper pyrites. The principal locality of it in this country is at the Bristol mine, Connecticut.

*Covelline*, Indigo Copper. A sulphuret of copper with one atom of each constituent, or 33·5 per cent. of sulphur to 66·5 per cent. of copper. This ore occurs in some quantity in South America.

*Phillipsite*, Variegated Copper, Peacock Ore, Horse-flesh Ore. A sulphuret of iron and copper, containing, according to the formula adopted by Rammelsberg, sulphur 28·1, copper 55·5, and iron 16·4: it is frequently mixed with copper glance, and usually contains from 60 to 70 per cent. of copper. This is an ore which is of considerable commercial importance, being often associated with copper pyrites and other cupriferous ores. The principal locality in this country is Bristol, Connecticut. The Tuscan mines abound in this ore.

*Barnhardtite.** This is a new ore of copper, recently discovered by Dr. Genth, which is intermediate in composition between Phillipsite and copper pyrites. It is of a pale yellow color, much resembling iron pyrites, but having a duller lustre, and tarnishing with bronze-colored and pavonine tints. Its composition is expressed by the formula, $2\ Cu_2\ S + Fe_2\ S_3$; and it contains, according to theory, sulphur 30·53, copper 48·14, and iron 21·33. It was first

* Communicated by Dr. Genth.

found on land of Daniel Barnhardt, and afterwards at Pioneer Mills, Cabarrus County, North Carolina, and since at the Phœnix and Vanderburg mines.

*Copper Pyrites*, Yellow Copper Ore. A combination of the sulphurets of copper and iron, with the following percentage: sulphur 35·05, copper 34·47, iron 30·48. This ore has a great resemblance to iron pyrites, from which it may, however, be easily distinguished by its softness, since it may be cut with the knife, while iron pyrites is so hard as to strike fire with steel. It is the great ore of copper, being that from which nearly all the Cornish copper is obtained; it furnishes probably two-thirds of the copper of the world, and is the source from which many of the other ores have been chiefly derived; since the true cupriferous veins, which generally bear other ores near the surface, on being wrought in depth, gradually pass into this variety, and bear it almost exclusively.

*Domeykite*, an arseniuret of copper, is a rare mineral from Chili.

*Gray Copper*, Fahlerz. A mineral of a complicated constitution, essentially a sulph-antimoniuret of copper, but generally containing also arsenic, iron, silver, and sometimes zinc, in small quantities. The amount of copper in it varies from 35 to 50 per cent. It is frequently of more value for the silver it contains than for its copper. It is one of the important ores in the Harz mines, but has not yet been discovered in this country.

*Tennantite*, another rare ore, which very much resembles the last; it is a combination of sulphur, arsenic, and copper, in different proportions from those of the gray copper.

*Wolfsbergite*, a sulphuret of copper and antimony; rare.

*Wölchite*, a sulphuret of lead, antimony, and copper.

*Bournonite*, another sulphuret of lead, antimony, and copper, with only from twelve to fifteen per cent. of the latter mineral. This is found in a good many localities, but is not of importance as an ore, either of lead or copper.

*b. Combinations with Oxygen and Chlorine.*

*Red Oxide of Copper*, Red Copper Ore. This contains two atoms of copper and one of oxygen, or 88·88 of metal, and

11·12 of oxygen. This is a valuable ore of copper, and is one of the frequent products of the decomposition of copper pyrites; the Australian, Cuban, and South American ores contain a large amount of the red oxide. The lodes which produce this ore on the surface generally bear the yellow sulphuret after they have been mined to any very great depth.

*Black Oxide of Copper*, Tenorite, Copper Smut. The pure black oxide of copper contains one atom each of copper and oxygen, or 20·14 of the latter to 79·86 of metal. This, however, is a rare substance, and had only been found in minute crystals in the lava of Vesuvius, and called Tenorite, previous to the discovery of a vein of this ore, mixed with silicate of copper, near Copper Harbor, on Lake Superior. Here the ore was found of an almost perfect purity, and in considerable quantity, but is no longer mined.

Copper smut is a black substance, found in many copper veins near the surface, and consists of a mechanical mixture of oxide of copper with the oxides of iron and manganese, together with earthy impurities. It is one of the products of the decomposition of the pyritiferous ores.

*Atacamite*, Chloride of Copper. A combination of the oxide and chloride of copper, and water. It occurs in some quantity in the form of sand in the Atacama desert; some of it comes intermixed with the Chili ores.

### c. *Silicates of Copper.*

*Dioptase*, Emerald Copper. A hydrated silicate of copper, very rare.

*Chrysocolla*, Copper Green. A hydrated silicate of copper, containing 34·82 of silica, 44·82 of copper, and 20·36 of water, when perfectly pure; it is, however, generally much mixed with earthy matter. This is an ore of very common occurrence, but it is not usually found in large quantity. It is a valuable ore, and easy of reduction. Like the carbonate and oxides, it is confined chiefly to the upper portion of the veins in which it occurs, or to beds and deposits which have not much depth.

### *d. Carbonates and Sulphates.*

*Azurite*, Blue Copper, Blue Carbonate. This beautiful ore differs from the green carbonate, in containing a hydrated oxide, as well as a carbonate. The principal localities are Chessy, in France, and Siberia, where splendid crystallizations have been obtained. Traces of it are common in many mines, but fine specimens are rare, and it is by no means of so much importance as an ore, as the next species.

*Malachite*, Green Carbonate of Copper. A hydrated carbonate of copper, with 71·82 oxide of copper, 20·00 carbonic acid, and 8·18 water. This is not only valuable as an ore, but is highly esteemed for ornamental purposes. It is most abundant in Siberia, and is sawn into thin plates, and used for veneering vases, tables, doors, &c. Some of the African ores contain a considerable quantity of malachite. It is much more common than azurite.

*Blue Vitriol*, Sulphate of Copper. This salt of copper, being soluble in water, is found in the waters issuing from mines of this metal, having been produced by oxidation of the sulphuret. In some localities it is an object of considerable commercial importance, the metallic copper being precipitated from it by immersing pieces of iron in the cupriferous solution.

*Brochantite*, *Lettsomite*, and *Connelite* are rare minerals, of which the principal ingredients are oxide of copper and sulphuric acid.

The phosphates and arseniates of copper form a numerous family of salts, but as they are none of them of any importance as ores, they will be passed over here without enumeration.

GEOLOGICAL POSITION.—Copper, occurring as it does in so many forms of combination, and so extensively distributed geographically, is, as might be expected, a metal which has a wide range in the list of geological formations. This will be sufficiently seen on examining the descriptions of the principal cupriferous districts of the world which are comprised in the next sections. There are, however, two characteristic positions in which these ores are found, and to one of which most of the great mining districts may be referred. These are:—

1st. Veins in the older crystalline rocks, especially the metamorphic palæozoic, and the igneous formations associated with them. Such is the character of the great mining districts of Cornwall and Australia, and also of the Lake Superior copper region, as well as of most of the localities in the Atlantic States. The veins belong either to the class of segregated, or to that of fissure-veins. Some of the segregated masses are of immense extent and have produced large amounts of ore; but the districts in which the workings are upon true veins are those in which the yield is of a more permanent character, and where the mining interest has attained the greatest development.

2d. The cupriferous ores are largely distributed through certain strata, which in the geological series belong between the carboniferous and the period of the Lias. In such cases the ore is not in veins, but is disseminated through the beds of rock, which are usually sandstone and slates; the metalliferous substance being usually in fine particles, although occasionally concentrated into bunches. The most characteristic localities of this form of deposit are those of the Kupferschiefer of Mansfeld and the Permian strata of the Ural, which are quite analogous to each other both in the mode of occurrence of the ores, and in the geological age of the rocks in which they occur, which is that of the "Magnesian Limestone," or "Zechstein" of the German geologists; they lie next beneath the Triassic group.

In this country the strongly marked group of rocks usually known by the name of the "New Red Sandstone," which has been usually supposed to belong to the Triassic group, but is now referred by some geologists to the period of the lower Oolite, is found to contain cupriferous ores, which occur under circumstances somewhat analogous to those just noticed.

Above the New Red Sandstone, there are few deposits of copper of any importance; the principal concentration of that metal out of the true metalliferous formation, the lower palæozoic, apparently having taken place near the limit between the older secondary and newer palæozoic periods.

Enough has been said to convey a general idea of the position of the cupriferous ores, and, in the following sections, the detailed description of the principal mining districts in which they are wrought will make their mode of occurrence, in its relation to the geological series, sufficiently clear.

---

## SECTION II.

### GEOGRAPHICAL DISTRIBUTION OF COPPER IN FOREIGN COUNTRIES.

THE ores of copper cannot be said to be limited to a few countries; they are scattered all over the world, and no one country can be considered to be far in advance of all others in their development. Nor is this a metal in whose yield there have been, or are likely to be, great fluctuations; new cupriferous regions are discovered from time to time, but the increased supply thus afforded does not more than keep pace with the growing demand of the arts. After a brief description of the principal copper-bearing districts, it will not be difficult to form a general idea of the resources of the civilized world in regard to this metal.

RUSSIAN EMPIRE.—The principal mines of this metal within the Russian territory, are in the Ural Mountains, the Altai, the Caucasus, and in Finland. Those of Finland are of minor importance. In the Caucasus, copper ores are said to be very abundant, and there is proof that they have been extensively worked at some very remote period. Their present produce is not considerable. The same may be said of the Altai, which yields about 400 or 500 tons yearly. The most extensive mines are in the vicinity of the Ural Mountains, on whose western flanks, in the governments of Perm and Orenburg, the beds of the Permian system of Murchison, corresponding to the Zechstein of Germany, are cupriferous, and possess a remarkable analogy with the copper schists of Mansfeld.* The strata in which the copper is found, consist

* Murchison's Russia, i. 144.

of thick, flag-like grits, of gray and dingy color, through which, at intervals, the ores of copper, chiefly the green carbonate, are disseminated.

The cupriferous beds contain only about 2½ per cent. of ore, but owing to its wide dissemination throughout vast masses of rock, its extraction is profitable, though by no means as much so as in the copper works of the Ural. The strata in which copper is found do not extend west beyond 400 or 500 versts from the Ural chain, and as the mineral matter in them decreases in quantity in receding from the mountains, it is evident that it must have originated there. The metal smelted from these ores is very pure and ductile, and much sought for making bronze. The imperial Zavods near Perm are said by Murchison to produce 16,000 poods per annum (257 tons), and to yield a profit to the government of about $40,000.

The copper deposits on the east side of the Ural are of great interest. They produce much the largest portion of the whole amount of this metal which is furnished by Russia. The principal mines are the Gumeschewskoi, the Bogoslowskoi, and those of Nijny Tagilsk. The Gumeschewskoi mine, which has been worked for more than a hundred years, is opened on bunches and nests of copper ore, mostly malachite and red oxide, contained in an argillaceous shale. There are no regular veins. A single mass of malachite, a cube of 3½ feet in diameter, was taken from this mine, and is now in the collection of the Russian School of Mines at St. Petersburg. The mines near Bogoslowsk, or the Turjinsk mines, as they are sometimes called, from the river Turja on which the smelting-works are situated, have little resemblance to veins; the rock in which they are found is a Silurian limestone, the strata of which alternate with beds or dykes of trap, and along whose lines of contact occur deposits of clay, in which the copper ores are found in bunches and nests. The crystals of native copper which are obtained here are unsurpassed in beauty and regularity. At Nijny Tagilsk, the general features are the same, namely, deposits of ore in connection with igneous rocks, associated with the upper Silurian limestones. "The cupriferous deposit," to use the language of

Murchison, "seemed to resemble a slightly consolidated heap of detritus which had been tumultuously aggregated in this hollow at a period of convulsion, when the subjacent rocks were invaded by some sort of igneous action, and all the strata were broken up and re-arranged." The ore lies in the hollows of the eruptive rocks, and is mixed with lumps of limestone and other rocks. There, at the depth of 280 feet, an enormous mass of malachite, estimated to weigh 1,320,000 pounds, or more than 580 tons, was found. This mass appears to have been deposited from a cupriferous solution, in a manner analogous to that of a stalagmite.

During the ten years previous to 1848, the mines of the Ural produced yearly an average of 3720 tons of copper;* but since that time the amount has increased considerably, and was, in 1850, over 5400 tons.

The amount of copper furnished yearly by Russia, according to an average of the last ten years, was 4540 tons; but it seems to be rapidly increasing, as the yield of 1849 was 6546, and that of 1850, 6449 tons. Owing to various reasons, and, probably, chiefly to the cost of transportation, the quantity of this metal which is exported is very small, and rapidly decreasing.

NORWAY AND SWEDEN.—The quantity of copper furnished by the Scandinavian peninsula is small, and increasing but slowly. The quality of the Norwegian metal is very good, the ores from which it is manufactured being quite free from arsenic, antimony, and lead; but that of the Falun mine is less valuable, on account of the occurrence of the sulphurets of lead and zinc in connection with the copper pyrites.†

The mines of Alten are in the most northern position of any in the world, being in the latitude 70°. They have been worked by an English company since 1826, and are now paying a respectable profit. The formation belongs to the metamorphic palæozoic or azoic, perhaps to the latter, with which are associated diorite and hornblende rock. The cupriferous veins of Kaafjord are only productive in the igneous rock, and they do not, generally, even extend into the

* Tschewkin, in Jour. des Mines, extracted in Ann. des Mines (5), iii. 801.

† Durocher, Ann. des Mines (4), xv. 272.

slates. Several mines are wrought here with greater or less success. A few miles east-southeast of Kaafjord is another group of veins on the Raipasvara Mountain. Here the lodes are contained in a compact subcrystalline limestone. The principal one is about six feet wide, and dips vertically. On the back of the lode, gossan, with carbonates and arseniate of copper, is found, but at some depth the ore is principally the variegated.* The veins impoverish rapidly on passing from the limestone into the slates.

The mines of Röraas are also of importance: here, however, there are no veins, but the ore is disseminated in chlorite slate, forming what are called "fahlbands," or metalliferous beds. There are three principal mines; in that of Storwartz the ore yields from five to six per cent. of copper after being picked over by hand.

There are numerous other localities in Southern Norway where copper ores are found, but they are worked only to a trifling extent.

The copper deposits of Sweden are quite analogous to those of Norway, except that they are principally in quartzose, micaceous, and calcareous beds, subordinate to gneiss, that formation being much more developed than the slates. Among these may be mentioned those of Åreskuttan, which are in fahlbands, formed by the dissemination of copper pyrites in the crystalline schists. At Gustavsberg the cupriferous fahlband is from thirteen to sixteen feet thick, and sometimes much more. The ore yields only from three to four per cent. of copper. There are eight groups of mines or mining districts in Sweden, principally in the province of Dalecarlia. The mines of Garpenberg, in that province, have been worked since the twelfth century, and are now 1000 feet deep; but diminishing in productiveness.

Falun has been long celebrated for its copper mines, but its importance is very much diminished, and it is now nearly exhausted. The ore is poor, not giving above 3½ to 4 per cent. after hand-sorting. The character of the deposit is peculiar, and hardly admits of classification. The rock is a

* Russegger in Karsten & Dechen's Arch. xv.

gray, quartzose mass, with little plates of mica scattered through it; but whether it is an igneous or stratified mass remains undecided. It is divided up into irregular ovoidal masses, by what are called "skölar," or curved and undulating belts of chlorite, along which the ore is principally concentrated. The "great mine," so called (Storgrufva), has been worked for centuries on the largest of these masses, which is found to be an immense cone, with a rounded apex, turned upside down. At the surface the dimensions of the mass, the base of the cone so to speak, are 800 feet in an east-northeast and west-southwest direction, and 500 feet in breadth. The interior of this great cone is principally iron pyrites, the copper ore lying on its exterior, forming a sort of envelope around it. The depth of the workings is about 1100 feet.

The mines of Åtvedaberg in East Gothland are next in importance to those of Falun. They produce, however, less than 200 tons per annum. The whole amount of this metal raised in Sweden was, in 1850, only a little over 1400 tons; that of Norway averaged, during the five years preceding 1850, 567 tons.

Great Britain.—We come now to speak of the copper region of Cornwall and Devonshire, where the cupriferous veins have been worked longer, and have produced more, than anywhere else in the world, and where the art of metallurgy has been carried to a high pitch of perfection, so that comparatively poor ores have been smelted with large profits.

In describing the stanniferous deposits of Cornwall, a general account has been given of the metalliferous veins of that region, their direction, the different systems of fracture, and the phenomena of a more general character exhibited by them; it now remains to consider the facts more closely connected with the occurrence of the copper, such as the nature of the ores, the varieties of the gangues, and the changes in the character of the lodes at various depths.

Among the first things to be noticed, is the decomposed state of the upper portion of the veins, giving rise to what the miners call "gossan," which usually forms the backs of the productive copper lodes. This substance is, all over the world, looked upon as a favorable indication of valuable ores at some depth below.* It is a mixture of quartzose matter with more or less oxide of iron, and usually contains traces of other metals in the form of oxides, carbonates, and sulphurets, as would naturally be expected, since it is in all cases the result of the decomposition of copper and iron pyrites and other sulphuretted ores which formed the original vein. The sulphuret of copper is oxidized into a sulphate, which is soluble in water, and is carried away in the streams and by the rains, while the iron remains as an insoluble hydrated oxide. If there was gold or oxide of tin in the veinstone, it remains in its original form, and is generally concentrated together. Silver is also present in some gossans in sufficient quantity to be of value. A great many of the English gossans contain gold from 1 to 2 ounces per ton, and in some cases much more. Whether auriferous gossan is sufficiently abundant to allow of its being worked for gold with profit is a question which will soon be settled, as the attention of the English mining adventurers seems to be most extensively turned to that subject. The depth to which this decomposition has extended is usually from twenty to thirty fathoms. By long experience, the miners are enabled to form an opinion, with some confidence, of the value of the lode, from the character of its outcrop, certain kinds of gossan being considered more "kindly," or favorable to the existence of ore, than others. In some cases the backs of the copper lodes have been extensively worked in ancient times for the tin ore which was disseminated through the gossan.

The metalliferous part of the lodes below the line of decomposition, consists, in the copper-bearing veins, chiefly of copper pyrites, with some vitreous and variegated ore and other sulphurets. Black-jack, or sulphuret of zinc, is frequently associated with the copper, and occurs in the greatest abundance throughout Cornwall. The ores are not uni-

* The German miners have a proverb:

> Es ist nie ein Gang so gut,
> Der trägt nicht einen eisernen Hut.

The "iron hat" or *chapeau de fer* of the French miner, is equivalent to the gossan of the Cornish, being the ferruginous covering of the lode.

formly scattered through the lodes, but lie in "bunches," or gathered together into masses, which are frequently of very great size, and connected together by mere threads and strings of metalliferous substance, in quantities too small to be profitably worked, yet sufficiently evident to be easily followed by the miners. Many facts have been developed, in the extended workings on the Cornish lodes, in regard to the position of these bunches of ore with reference to the nature of the "country," or rock enclosing the vein, and the displacements to which the lodes have been subjected. If the lode ramifies, and divides into branches, as it descends, the quantity of ore becomes less; on the other hand, where the branches reunite, it is frequently accumulated in a large bunch. So, too, where different veins unite, there is generally an increase in the quantity of ore, and the best part of the veins is in the vicinity of their junction. Thus the great cross, or counter-lode, of the Gwennap district, cuts through a great number of east and west veins, enriching them all at its intersection with them, and giving rise to a number of productive mines along its course. In general, however, it is considered that the smaller the angle at which two lodes cross each other, the greater the likelihood of there being a rich bunch of ore at the point of meeting. The intersection of the elvan dykes with the lodes has almost always an important influence on the quantity of ore. Some lodes have only been productive in their passage through an elvan; and it has been remarked by a good observer, that most of the ore in some of the most productive districts in Cornwall has been found in or near elvan courses. The miners have not failed to recognize the fact, that the nature of the "country" has an important influence on the character of the lode. It is well established, that the most productive portions of the lodes are in the vicinity of the junction of the granite and the killas: sometimes the lode is richest in the one rock, and sometimes in the other; thus, in the case of two parallel tin lodes, Wheal Vor and Great Work, near Breague, which are only a mile apart, one of them, Wheal Vor, was rich in the slates, but when it entered the granite it was no longer worth working, while exactly the opposite was true of the other. It was formerly considered that the killas was the natural mining ground for copper, but very rich mines have been worked in the granite, although not at a very great distance from the junction of the slates. In Wheal Tresavean and Wheal Jewel, the ore lies in the granite in rich bunches, close under the intersection of that rock with the killas.

The nature of the rock is examined with great care by the miner, some kinds being considered much more favorable to the development of ore than others. In the Gwennap district, for instance, it is said that the reddish killas is unproductive, and that those rocks only are rich which are of a bluish-gray color. These differences, although considered by the miners as of importance, can hardly be conveyed by words; and it may be doubted whether their opinions are not often based as much on imagination as on facts. It is certain, however, that the metalliferous contents of the lodes vary with the character of the enclosing rocks, although there are no positive rules by which it can be decided beforehand what particular kind of stratum will be the most productive.

One of the most important subjects is that of the variations in the character of the lodes in depth. It is very frequently said that true veins enrich in descending, but this opinion is not supported by facts; the records of mining, with few exceptions, show that although this statement may be true when applied only to a very inconsiderable depth, yet, in general, a point is soon reached, beyond which there is, on the whole, no farther increase in the valuable metallic contents of the lode. Indeed, it is maintained by some that there is an actual diminution in the value of the Cornish ores, as the mines have increased in depth; and Burr has made the following statements, which he asserts to be borne out by the results of working in Cornwall: 1st. Where the vein is poor at the surface, it increases in richness to a certain, not very great, depth, where it reaches its maximum; from that point downwards it becomes poorer, as far down at least as any mines have yet been worked. 2d. In those cases where the lode is rich at the surface, it continues good for a certain depth, which is much less than in the first case, and then decreases in value. Others deny that the veins have decreased in richness, and point to some of the deepest mines in Cornwall which are still yielding largely, as proof of an undiminished yield at the greatest depth yet attained. An examination of the statistical tables will show that there has been a gradual and constant falling off in the percentage of the ores sold for the last hundred years; this may be due, however, to improvements in the metallurgic processes, which permit the working of a poorer ore with profit than could formerly be smelted; and so allow the old ores, formerly rejected as halvans, to be dressed and smelted with profit.

The amount of dip is an element of importance in calculating the value of a lode; it very rarely happens that a vein with an inclination varying very considerably from the perpendicular is of much value. In a lode which varies in its dip, those portions which approach nearest to a perpendicular are richest.

The mines of Cornwall are worked with great skill. The steam-engine was created almost by the necessity of powerful means of raising the water from the Cornish mines; and there are no more perfect or beautifully constructed machines than here. The deepest mines are down about 350 fathoms.

The great Cornish adit commences near the village of Ferney Splat, in the Carnon Valley, and extends to the Cardrew Downs mine, which is nearly five and a half miles from its mouth. Its total length, with all its ramifications, is about thirty-five miles. At the deepest point it is seventy fathoms from the surface.

There have been great fluctuations in the yield of the different mines, but the fact that the lodes do not increase in richness in descending sufficiently to counterbalance the increased expense of working, seems to be abundantly proved by the successive decline, one after the other, of most of the great mines, each of which had its period of great prosperity when at a moderate depth. Thus, in 1815, Dolcoath was the mine from which the copper ores produced the greatest amount of money (£66,839). In 1817, the

United Mines took the first place, their ores yielding £63,116; in 1822, the Consolidated Mines were at the head of the Cornish mines, producing in that year £80,311; while, since 1845, the Devon Great Consols have maintained the front rank. The following notices of these mines are appended, as worthy of especial attention.

*Dolcoath Mine.* This is celebrated as the oldest mine in Cornwall, having been worked with little interruption for a century past. It is 300 fathoms deep, and has paid dividends to the amount of £300,000. It is now hardly paying expenses. From 1814 to June 1848, the returns of ore from this mine were 238,059 tons, yielding £1,361,681. From June 1849, to June 1850, it furnished 1218 tons, worth £6083.

*Great Consolidated Mines.* These are among the most celebrated mines in the world for their extent and former productiveness. In 1848, the workings on the Carharrack, West Wheal Virgin, Wheal Virgin, and Wheal Fortune mines, which belonged to this Consolidated Company, were sixty-three miles in length, and had made a profit, under different proprietors, of £700,000. From 1819 to 1840, these mines were under the management of the distinguished mining engineer, John Taylor, and during that time a profit was made of £500,000, besides expending £100,000 in opening the United Mines.* In the last twelve months of Mr. Taylor's management, the mines yielded 17,823 tons of ore, which produced £100,279. In 1840, a new lease was obtained by another Company, by whom £100,000 was paid for the mine and plant, there being 100 shares, with £1000 paid in on each share. Up to June 1848, the ores sold by this Company amounted to 83,660 tons, in value £490,543, of which only £32,000 was divided among the stockholders, owing to the heavy expense of working the mine. No dividend has been paid since January 1851, and in March 1854, the price of shares is quoted at £100 for £1000 paid in.

Eight large and thirty smaller steam-engines form a part of the plant of this mine, which employed in 1848 about 1100 persons.

*Devon Great Consolidated Copper Mining Company.* As a specimen of what Cornish mining is, when the highest prizes are drawn, a short notice is given of the progress of the above-named Company, whose stock now stands higher than that of any mining Company in the world.

This Company commenced operations in August 1844, having paid £1024 on the same number of shares. The ground was leased of the Duke of Bedford for twenty-one years, at $\frac{1}{15}$ dues, to be raised to $\frac{1}{12}$ as soon as £20,000 profit had been made. The mine is about five miles from Tavistock. The lode, at fourteen fathoms from the surface, was eighteen feet wide, "carrying an immense gossan." At the depth of 17½ fathoms, the great deposit of ore was struck, and since that time the yield has been truly wonderful.

---

* J. Y. Watson in Eng. Mining Jour., No. 811.

The first three months' operations showed a profit of from £15,000 to £16,000: the next year, 1846, the clear profit realized was £73,622, the mining expenses having been £30,590, against £116,068 received for ore sold. At this time the shares, on which £1 had been paid, went up to £800. The average cost of sinking the shafts, during that year, was £8 14*s.* 9*d.* per fathom; that of driving levels, £3 8*s.* 4*d.* per fathom.

Up to 1852, the produce of the mine had been as follows:—

| Year. | Tons Ore. | Value. | Av. price per ton. |
|---|---|---|---|
| 1845, | 13,293 | £116,068 | £8 14*s.* 7*d.* |
| 1846, | 14,398 | 93,610 | 6 10 0 |
| 1847, | 14,413 | 101,916 | 7 1 5 |
| 1848, | 16,584 | 100,058 | 6 0 8 |
| 1849, | 15,432 | 104,622 | 6 15 7 |
| 1850, | 17,290 | 117,361 | 6 14 11 |
| 1851, | 18,946 | 110,379 | 5 16 6 |
| 1852, | 20,886 | 138,728 | 6 12 10 |

In 1850 there were ten shafts sunk on the lode, the deepest being 100 fathoms. There were at that time 653 fathoms of shafting, averaging £12 per fathom in cost; 440 fathoms of winzes, costing from £5 to £6 per fathom: 5039 fathoms of drivages, averaging about £4 per fathom. There were five water-wheels, two of which were used for drawing, two for crushing, and one in pumping. For this latter, the power is carried 390 fathoms in horizontal and 64 fathoms in vertical distance to the shaft, by means of flat rods carried over rollers erected on 130 substantial supports, the whole arrangement being considered a master-piece of mining work. At the same time about 1000 persons were employed in and about the mine.

Up to January 1853, £358 had been paid in dividends on each share, and the shares of £1 stood at that time at £430.

Out of the £329,014 18*s.* 6*d.* paid in dividends by sixty English mines during the year 1853, two mines alone, Devon Consols and Wheal Buller, paid £110,464, of which Devon Consols paid £65,024, leaving large reserves of ore in the mines to meet future contingencies. In January 1854, shares in the Wheal Buller stood at £1,025 for £5 paid in.

As to the date at which copper was first raised in Cornwall there is much uncertainty. Two hundred and fifty years ago (1602) Carew writes that "copper is found in sundrie places, but with what gaine to the searchers I have not been anxious to enquire or they hastie to reveal." Borlase gives an account of the Cornish mining region, as it was a hundred years ago. At that time there were steam, or rather, fire engines for raising the water, one of which had a cylinder of 70 inches in diameter. At the time Borlase wrote, copper mines were in active operation and had been profitable since the beginning of the eighteenth century. It was computed that the average annual value of copper raised for the fourteen years previous to 1758 was about £14,000. The copper ores are still sold, exactly as they used to be 120 years ago, by public ticketings, at which agents of the various smelting companies attend, and make their bids for lots of ore which have been previously notified to be sold at that time, and which they have had assayed. The first sale of ores on record was in 1729, when 2216 tons, being the produce of twelve months, were sold. In 1764 the quantity had increased to 16,437 tons; in 1800, to 55,981 tons, yielding 5187 tons of copper and £550,925 in money. For some years after this the quantity varied from 60,000 to 78,000 tons of ore per annum. In 1830 it was 141,263 tons; in 1840, 147,266 tons; in 1848, 155,616 tons, and in 1853, 180,095 tons, yielding 11,839 tons of fine copper. Thus it will be seen that there has been a constant increase from the earliest period up to the present time, both in the quantity of ore raised and the yield of copper. The produce of the mines of Cornwall and Devon will be found presented in the following tables of the ores raised and sold, together with the amount received in money, and the percentage yield of the ores.

The following is an account, by Pryce, of the copper ores sold in Cornwall from 1726 to 1775 inclusive.*

| | Ten years' tonnage. | Av. price per ton. £ | s. | d. | Amount. £ | Av. annual tonnage. | Av. annual amount. £ |
|---|---|---|---|---|---|---|---|
| 1726–1735 | 64,800 | 7 | 15 | 10 | 473,500 | 6,480 | 47,350 |
| 1736–1745 | 75,520 | 7 | 8 | 6 | 560,106 | 7,552 | 56,010 |
| 1746–1755 | 98,790 | 7 | 8 | 0 | 731,457 | 9,879 | 73,145 |
| 1756–1765 | 169,699 | 7 | 6 | 6 | 1,243,045 | 16,970 | 124,304 |
| 1766–1775 | 264,273 | 6 | 14 | 6 | 1,778,337 | 26,427 | 177,833 |

The subjoined table shows the amount of copper ores

* De la Beche, Geol. of Cornwall, p. 606.

raised in Cornwall from 1771 to 1853, for each year ending June 30.*

| Years. | Tons of Ore. | Tons of Copper. | Value in pounds sterling. | Standard. | Percentage yield. | Years. | Tons of Ore. | Tons of Copper. | Value in pounds sterling. | Standard. | Percentage yield. |
|---|---|---|---|---|---|---|---|---|---|---|---|
| 1771, | 27,896 | 3,347 | 189,609 | £81 0s. | | 1813. | 86,713 | 8,166 | 685,572 | £113 0s. | |
| 1772, | 27,965 | 3.356 | 189,505 | 81 | | 1814, | 87,482 | 7,936 | 766.825 | 128 | |
| 1773, | 27,663 | 3,320 | 148,431 | 70 | 12 | 1815, | 79,984 | 6,607 | 582,108 | 121 | 8·5 |
| 1774, | 30,254 | 3,630 | 162,000 | 68 | | 1816, | 83,058 | 7,045 | 541,737 | 109 | |
| 1775, | 29,966 | 3,596 | 192.000 | 78 | | 1817, | 75,816 | 6,608 | 422,426 | 96 | |
| 1776, | 29.433 | 3,532 | 191,590 | 79 | | 1818, | 80,525 | 6,714 | 587,977 | 121 | |
| 1777, | 28.216 | 3,386 | 177,000 | 77 | | 1819, | 93,234 | 7,214 | 728,032 | 136 | |
| 1778. | 24,706 | 2,965 | 140,536 | 72 | 12 | 1820, | 92.672 | 7,364 | 620,347 | 119 | |
| 1779, | 31,115 | 3,734 | 180,906 | 73 | | 1821, | 98,803 | 8,163 | 628,832 | 111 | 8·1 |
| 1780, | 24,433 | 2,932 | 171,231 | 83 | | 1822, | 106,723 | 9.331 | 676,285 | 104 | |
| 1781, | 28,749 | 3,450 | 178,789 | 77 | | 1823, | 97.470 | 8,070 | 618,933 | 110 | |
| 1782, | 28.122 | 3,375 | 152.434 | 70 | | 1824, | 102,200 | 8,022 | 603.878 | 110 | |
| 1783, | 35,799 | 4,296 | 219.937 | 76 | 12 | 1825, | 110.000 | 8,417 | 743.253 | 124 | |
| 1784, | 36.601 | 4.392 | 209,132 | 72 | | 1826, | 118,768 | 9,140 | 798,790 | 123 | |
| 1785, | 36.959 | 4,434 | 205,451 | 71 | | 1827, | 128,459 | 10,450 | 755,358 | 106 | |
| 1786, | 39,895 | 4,787 | 237,237 | 75 | | 1828, | 130,866 | 9,961 | 759,175 | 112 7 | 7·9 |
| 1787, | 38,047 | . . | 190,738 | . . | | 1829, | 125,902 | 9,763 | 725.834 | 109 14 | |
| 1788, | 31.541 | . . | 150,303 | . . | | 1830, | 135,665 | 10,890 | 784,000 | 106 15 | |
| 1789. | 33,281 | . . | 184,382 | . . | | 1831, | 146,502 | 12,218 | 817,740 | 99 18 | |
| 1790, | . . | . . | . . | . . | | 1832, | 139,057 | 12,099 | 835.812 | 104 14 | |
| 1791, | . . | . . | . . | . . | | 1833, | 138,300 | 11,185 | 858,708 | 110 | 8·1 |
| 1792, | . . | . . | . . | . . | | 1834, | 143,296 | 11,224 | 887,902 | 114 4 | |
| 1793, | . . | . . | . . | . . | | 1835, | 153,607 | 12,271 | 896,401 | 106 11 | |
| 1794, | 42.816 | . . | 320,875 | . . | | 1836, | 140,981 | 11,639 | 957,752 | 115 12 | |
| 1795, | 43,589 | . . | 336,189 | . . | | 1837, | 140,753 | 10.823 | 908,613 | 119 5 | |
| 1796, | 43,313 | 4,950 | 356,564 | . . | | 1838, | 145,688 | 11,527 | 857,779 | 109 3 | |
| 1797, | 47,909 | 5,210 | 377,838 | . . | | 1839. | 159 551 | 12,451 | 932,297 | 110 2 | 7·8 |
| 1798, | 51,358 | 5,600 | 422,633 | . . | | 1840, | 147,266 | 11,038 | 792.758 | 108 10 | 7·5 |
| 1799, | 51,273 | 4.923 | 469,664 | 121 | | 1841, | 135,090 | 9.987 | 819,949 | 119 6 | 7·4 |
| 1800, | 55,981 | 5,187 | 550,925 | 133 3 | | 1842, | 135,581 | 9,896 | 822,870 | 120 16 | 7·3 |
| 1801, | 56,611 | 5,267 | 476,313 | 117 5 | | 1843. | 144.806 | 10,926 | 804.445 | 110 1 | 7·5 |
| 1802, | 53,937 | 5,228 | 445,094 | 110 18 | | 1844, | 152,667 | 11.247 | 815,246 | 109 17 | 7·4 |
| 1803, | 60,566 | 5,616 | 533.910 | 122 | | 1845, | 157,000 | 12.293 | 835.350 | 103 10 | 7·8 |
| 1804, | 64,637 | 5.374 | 570,840 | 136 5 | 8·8 | 1846, | 158,913 | 12,448 | 886.785 | 106 8 | 7·8 |
| 1805, | 78,452 | 6,234 | 862,410 | 169 16 | | 1847, | 148,674 | 11,966 | 830,739 | 103 12 | 8 |
| 1806, | 79,269 | 6,863 | 730.845 | 138 5 | | 1848, | 155,616 | 12,870 | 825.080 | 97 7 | 8·3 |
| 1807, | 71,694 | 6.716 | 609,002 | 120 | | 1849, | 144,933 | 12,052 | 716.917 | 92 11 | 8·3 |
| 1808, | 67,867 | 6.795 | 495,303 | 100 7 | | 1850, | 150,890 | 11,824 | 814.037 | 103 19 | 7·8 |
| 1809, | 76,245 | 6,821 | 770.028 | 143 12 | 9·1 | 1851, | 154,299 | 12.199 | 808,244 | 101 | 7·9 |
| 1810, | 66,048 | 5.682 | 569,981 | 132 5 | | 1852, | 152,802 | 11,706 | 828,057 | 106 12 | 7·6 |
| 1811, | 66.499 | 5.948 | 563,742 | 126 | | 1853, | 180,095 | 11,839 | 1,124,561 | 136 15 | 6·6 |
| 1812. | 75.510 | 7,248 | 608.065 | 113 | | | | | | | |

In addition to the copper raised in Cornwall, there is a considerable amount of this metal obtained from mines in Wales and Ireland, which is sold at the Swansea ticketings. The quantity thus disposed of from these portions of the British Empire will be seen on examining the table of the sales of

* This table is taken from De la Beche's Geology of Cornwall, with a few additions, up to 1838, and has been completed up to 1853 from various authentic sources, especially "Gryll's Mining Sheet," an annual record of the transactions of the Cornish mines.

ores at Swansea, accompanying the statistics of copper at the close of this chapter.

The total yield of copper of the United Kingdom can only be given with exactness down to 1834, from which time the amount smelted from English ores cannot be accurately distinguished from that of foreign origin.*

The following table is given by Mr. Porter as approximate only from 1835 to 1848.

| Years. | Tons of copper. | Years. | Tons of copper. |
|---|---|---|---|
| 1820, | 8,127 | 1835, | 14,470 |
| 1821, | 10,288 | 1836, | 14,770 |
| 1822, | 11,018 | 1837, | 10,150 |
| 1823, | 9,679 | 1838, | 12,570 |
| 1824, | 9,705 | 1839, | 14,670 |
| 1825, | 10,358 | 1840, | 13,020 |
| 1826, | 11,093 | 1841, | 12,850 |
| 1827, | 12,326 | 1842, | |
| 1828, | 12,188 | 1843, | |
| 1829, | 12,057 | 1844, | 14,840 |
| 1830, | 13,232 | 1845, | 14,900 |
| 1831, | 14,685 | 1846, | 14,950 |
| 1832, | 14,450 | 1847, | 13,780 |
| 1833, | 13,260 | 1848, | 14,720 |
| 1834, | 14,042 | | |

Prussia.—The most important cupriferous district in Prussia is that of Mansfeld, where mining has been carried on for centuries, in a formation known as the Kupferschiefer, a part of the Zechstein. The copper-bearing stratum is not over two or three feet thick, but it stretches for miles in length with wonderful regularity. The ore is gray copper, somewhat argentiferous, and it is scattered in fine particles through the metalliferous stratum, a bituminous marly slate. The works are carried on with consummate skill; and as the thickness and richness of the rock remain very constant, the produce of metal does not vary much, and is not likely to do so for many years to come.

Near Stadtberg, in the district of Siegen, a silicious slate is found which has fine particles of carbonate of copper disseminated through it in sufficient quantity to be worthy of

* G. R. Porter, Progress of the Nation, London, 1851.

being worked. The metal is dissolved out by sulphuric acid, and precipitated by metallic iron.

Copper ore occurs in regular veins near Kupferberg, in Lower Silesia; at Camsdorf, in the district of Henneberg-Neustadt; and at Rheinbreitenbach, in the district of Siegen. The ores are mostly pyritous; the last-named locality is the only one of importance.

The entire yield of the kingdom, which has not varied materially for several years, amounted in 1850 to 1450 tons; and the smallest quantity produced in any of the preceding ten years was 796 tons.

AUSTRIAN EMPIRE.—On the European continent, the mines of Austria are those which produce the largest amount of copper; the greater part of the Russian cupriferous deposits being in Asia. The production of this metal has been slowly increasing for several years; a result due rather to the perfecting of the metallurgic processes than to the opening of new mines. The poorer ores, which were formerly rejected, in the Schmöllnitz district, are now worked with profit. The following table shows the relative production of the different provinces of the Austrian Empire on an average of five years, from 1843 to 1847:—

| | | |
|---|---|---|
| Hungary, | 79·2 | per cent. |
| Venice, | 6·0 | " |
| Tyrol, | 5·6 | " |
| Gallicia, | 4·1 | " |
| Transylvania, | 2·4 | " |
| Styria, Salzburg, and Bohemia, | 2·7 | " |
| | 100·0 | |

The districts of Schmöllnitz in Upper Hungary, and Tsiklova in the Banat, are exceedingly interesting on account of the peculiar processes employed for the treatment of argentiferous copper ores.* In the Banat, the ores are in irregular deposits, near the junction of the sienite and metamorphic strata of Jurassic age; and consist of pyritous copper and argentiferous gray copper, with blende, iron pyrites, and

* Rivot and Duchanoy, Ann. des Mines (5), iii. 63.

sometimes a little gold. They do not yield over 4 per cent. of copper after being picked over by hand.

The principal mine at Tsiklova furnishes argentiferous mispickel, mixed with copper pyrites. The black copper obtained by smelting is treated by the amalgamation process, for separating the silver, which is present in small quantity; the minimum amount which can be advantageously separated, is 0·00125. It is heated to the highest possible point which it will bear without fusing, and then stamped. The pulverized substance is then mixed with about 10 per cent. of common salt, and a little sulphuret of iron, if it is not present in the ore in sufficient quantity, and heated in the reverberatory furnace. The resulting mass, which contains the silver in the form of chloride, is then amalgamated with mercury in barrels, the reduction of the silver being effected by metallic copper, which at the same time converts the small quantity of chloride of copper which had been formed in the process of chloruration, into a bichloride. The amalgam is then distilled, so that no mercury is lost in the operation, and the silver remains behind. The copper in the fine mud of the amalgamation barrels is smelted and refined.

Besides the ore proper of the Schmöllnitz district, a large quantity of copper of cementation is obtained in the old workings, and a considerable amount of poor ore, formerly rejected, is now found to be worth smelting.

The copper mines of Lower Hungary, or in the Schemnitz region, are no longer of much importance.

The annual production of this metal in the Austrian Empire amounted, in 1847 and 1848, to a little over 3300 tons, having increased nearly 1000 tons in the preceding ten years.

The produce of the other German States is very small, amounting altogether, probably, to about 200 tons.

France.—There were formerly mines in this country of some interest, more from the beautiful crystallizations of their ores than from the amount of their product, at Chessy and Sain Bel, a few miles northwest of Lyons.* These were deposits at the junction of the mica and talcose slates

* A. d. M. (3), iv. 393.

with strata of the age of the Triassic and Jura limestone. The ores were principally the blue carbonate and the red oxide, with pyritous copper, black oxide, and green carbonate.

The pyritous ores were in the ancient slates, but were too poor to be worked with advantage. The carbonates and oxides were obtained from the overlying strata of sandstone, belonging to the "grès bigarré," or Variegated Sandstone. This, being a mere deposit, was soon exhausted, after having furnished the European cabinets with fine specimens of azurite and crystallized red oxide of copper.

These mines have been abandoned for several years, and it does not appear that there is at present a single copper mine worked in France, at least with profit. In 1845, the produce amounted to 142 tons.

Spain.—The amount of copper produced in Spain, though small, is on the increase. The most remarkable mine is that of Rio Tinto, a few leagues north of Seville, which has been worked at different times from the period of the Roman Empire. Since 1787, considerable copper has been produced by the cementation process, from the water issuing from the old workings. In 1828, the amount of iron used was 150,000 pounds, and 112,500 pounds of copper were produced. In 1833, the yield was about 140 tons.

A number of English Companies have taken hold of the Linares Mines, which yield lead and copper, and the amount of the latter metal produced will probably continue to increase.

The most reliable estimates give the quantity of copper raised in the whole kingdom at 450 tons for the year 1849, and it probably somewhat exceeds that at present.

Italy.—The only copper mines of any note in Italy are in Tuscany, and their produce is very limited. The interesting cupriferous ores of Monte Catini occur in contact deposits, and, contrary to the usual rule, have been found to grow richer the deeper they have been worked.

Turkey.—This country exports considerable copper of a good quality, but I have been able to obtain no particulars of its mode of occurrence or geographical situation.

Algiers.—A mine of copper has been wrought for some time near the foot of the Mouzaia Pass. The veins are composed of spathic iron and gray copper ore, and are contained in rocks which are very high up in the geological series, belonging partly to the gault and partly to the middle tertiary.

East India.—Copper is found in the Ramghur Hills, about 150 miles from Calcutta. The locality is said to be valuable, and might produce largely, were foreign labor and capital applied to its development. The natives smelt the ore with charcoal, although mineral coal is said to exist in the neighborhood.

Japan.—Large quantities of copper have been sent from Japan to China and Holland, that being one of the principal articles of which the exportation is permitted. The quality of the metal is very superior. There is, of course, but little information in regard to its mode of occurrence. The quantity exported is about 1000 tons per annum, as near as can be ascertained.

Australia.—The cupriferous deposits of this country were very celebrated before they were thrown in the shade comparatively by the gold discoveries. The first mine of copper was opened in 1836, on a location thirty-four miles distant from Adelaide; and in 1848 more than twenty companies were at work, or preparing to commence operations, in that region, of which twelve had shipped ore, although only one had been very successful. This was the celebrated Burra-Burra Mine, which was opened in September 1845, and at once began to produce largely.

The ore is chiefly the red oxide, mixed with the green carbonate; and it seems to form a powerful vein in a rock analogous to the killas of Cornwall, both in a mineralogical and geological point of view. The distance of the mine from Adelaide, eighty-six miles, gives employment to a great number of persons in conveying the ore, which has to be transported over a very bad road. The greatest drawback to the prosperity of the mine, is the want of wood for timbering the works and for smelting the ore. But so rich is the ore and so abundant, that large profits have been realized. Only £6 had been paid in on a share; and in January 1850, the stock was quoted at £157. The last dividend was paid in March 1853, of £5 per share on 2464 shares; the whole amount of dividends paid since the opening of the mine, was £135 per share. At the time of the discovery of gold in Australia, the produce of this mine was enormous; but the gold excitement drew all the miners away. The produce for the six months ending September 30, 1851, immediately preceding the gold discoveries, was 10,732 tons of ore, on which the profit was £49,506. The prudence of the managers in declining to declare any more dividends until the arrival of additional labor, although with a balance of profits of £73,539, cannot be too highly commended. As labor gradually flows into Australia, the working of this extraordinary mine will be resumed, and there can be no doubt of its capacity to yield largely for a long time to come.

The following are the returns of ore raised from the mine up to the time of the gold discoveries:—

| | Tons of ore. | | Tons of ore. |
|---|---|---|---|
| 1846, - - - - | 6,359 | 1849, - - - - | 7,789 |
| 1847, - - - - | 10,794 | 1850, - - - - | 18,692 |
| 1848, - - - - | 12,791 | | |

There were 1003 persons employed in and about the mine, in 1849 and '50. The amount of ore shipped to England may be seen by referring to the sales at Swansea, at the close of this chapter; the amounts there given for Australia being almost entirely from this mine. The yield of the ore shipped to England was usually from 24 to 28 per cent. In 1852, a magnificent 250 horse-power engine was erected at the mine; cylinder 80 inches diameter, and 11 feet stroke; the cost of running it amounted to £200 per week, so great is the scarcity of fuel. At the last meeting of the proprietors, held at Adelaide, October 19, 1853, it appeared that in spite of the continued attractions of the gold-fields the mine had begun to resume work; 1780 tons of 18 per cent. ore having been taken out in the preceding six months, and the number of under-ground men having been increased to eighty.

*Kapunda Mine.* This is next in importance to the Burra-Burra. It shipped to England, in 1846, 1386 tons, and in 1847, 1382 tons of ore, varying from 20 to 50 per cent. in richness.

After 1849 the shipments of ore to Swansea began to fall off rapidly, smelting works having been erected at several points in Australia, although the cost of fuel is enormous.

The amount of metallic copper furnished to commerce by Australia was, as near as can be estimated:—

| Year. | Tons. | Year. | Tons. |
|---|---|---|---|
| 1844, | 20 | 1849, | 2500 |
| 1845, | 450 | 1850, | 3300 |
| 1846, | 850 | 1851, | 3500 |
| 1847, | 1600 | 1852, | 3250 |
| 1848, | 1500 | 1853, | 2650 |

The effect of the gold discoveries was very sensibly felt in the amount of ore raised from the mines in the years immediately following 1850; but as a considerable quantity of ore had already accumulated, or was on its way to market, the table above does not indicate as great a falling off as actually took place.

New Zealand.—A copper mine is worked to some extent by a Scotch Company at Kaw-aw. Their sales of ore have not been over a few hundred tons annually.

South America.—Although cupriferous veins have been found along the whole chain of the Andes, from one extremity to the other of South America, yet the mines now extensively worked are chiefly confined to Chili. There are, however, in Peru, some veins which are worked in spite of the great disadvantages under which the mineral industry of that country labors.

According to M. Crosnier,* the law of the geological distribution of the copper ores in Peru is similar to that in Chili. A great number of auriferous and cupriferous veins are found in the granitic and igneous rock; but in this formation, whatever the nature of the ores may be, they never contain silver. In the stratified rocks, on the other hand, there are, in addition to numerous veins of silver ores and

* Ann. des Mines (5), ii. 10.

argentiferous galena, many others bearing ores of copper, which always contain more or less silver in combination.

Near Antarangra a vein is worked to some extent, which furnishes rich cupriferous ores, giving at one operation a regulus yielding fifty per cent. of metal, which is sent to England to be refined. Near Colcabamba is another furnace, smelting ores which occur in a segregated mass in the granite. There are numerous veins of argentiferous copper ores in the districts of Niñobamba and Castra Vireyna, which were formerly wrought very extensively by the Spanish: recently, the attention of the Peruvians has been turned to them, and several companies have been formed for the purpose of taking them up again. The whole amount of copper produced in Peru is very small, and cannot be given with exactness. We imported into this country, in 1852, 157 tons from this source.

Chili.—In spite of the very great difficulties which the mining interest of this country has had to contend with, such as the entire want of facilities for communication, and the high price of materials of every kind, it has been developed rapidly within the last few years. Copper is the most important product, although the amount of silver raised is very considerable.

Among the important mines are those of Carrisal, north of the valley of Huasco. They are in a feldspathic rock. Those of San Juan and La Higuera, between Huasco and Coquimbo, also furnish a large quantity of rich ore, of from twenty to twenty-five per cent.: it consists principally of copper pyrites and oxide of copper. Numerous mines are worked in the vicinity of Copiapo. Here was obtained, from the upper part of the veins, an immense quantity of gold, up to the close of the last century. As the gold began to fall off, copper was found; and in 1845 there were more than fifty mines at work in the department of Copiapo, producing an ore of from twenty-three to twenty-five per cent. of metal, part of which was smelted there, but the larger portion was shipped to England. At the Cerro del Cobre Mines, the ore, red oxide and pyrites, is mixed with magnetic oxide of iron. The principal vein of the Cerro Blanco district was worked for silver to a depth of 600 feet below the summit of the mountain. Here the ore changed gradually, and consisted principally of gray copper and galena: at a still greater depth the gray ore gave way to pyritous copper, and the proportion of lead increased.

No country, unless it be England, exceeds Chili in the extent of its production of copper; and these two countries together supply more than half of the amount consumed in the world. The yield of Chili for 1849 and a few of the previous years, as near as can be determined, was as follows:—

| Years. | Tons. | Years. | Tons. |
|---|---|---|---|
| 1840, . . . . | 9,000 | 1847, . . . . | 11,850 |
| 1845, . . . . | 13,270 | 1848, . . . . | 12,275 |
| 1846, . . . . | 13,800 | 1849, . . . . | 12,450 |

Since 1849, the production has probably increased somewhat, although it has not been possible to obtain any reliable information as to its amount, the statistics of the South American mines being always exceedingly imperfect. The following table will show at once the great and increasing importance of our imports of this metal from Chili.

VALUE OF IMPORTS OF COPPER INTO THE UNITED STATES FROM CHILI (YEARS ENDING JUNE 30TH).

| | 1849. | 1850. | 1851. | 1852. |
|---|---|---|---|---|
| Ores, . . . . . | $102,817 | $155,002 | $52,164 | $230,497 |
| Pigs, bars, and old, | 779,376 | 1,008,044 | 1,367,191 | 1,294,481 |
| Sheathing, . . . | | | 13,656 | 221 |
| Sundries, . . . | | 8,579 | 178,108 | |
| | $882,193 | 1.171,625 | 1,611,119 | 1,525,199 |

CUBA.—The copper mines of this island have been of great importance; but of late have somewhat fallen off in their yield. The ores do not seem to be in regular veins, but in beds and masses, subordinate to igneous rocks, especially greenstone and serpentine. The gangues are quartz, dolomite, and carbonate of lime. The yellow sulphurets are associated with hydrated oxide of iron; near the surface considerable native copper, red oxide and carbonate, and other oxidized ores, have been found. The ore appears, thus far, to have been entirely exported from the island to be smelted; and, with the exception of a small amount sent to this country, it is sold at Swansea. The table of the sales of ores at Swansea, therefore, in which those furnished by Cuba are included under one head, will serve to give a pretty good idea of the progress of the copper mines of the island. The yield of the ores sold at Swansea has usually been from 15 to 18 per cent.

There are two principal companies working on the island, both of which are English. They are the Cobre and the Royal Santiago.

*Consolidated Copper Mines of Cobre Association.*—This is the principal Cuban company, and has been in operation since 1834. Their affairs are not made public, but they are well known to have been highly successful. They have 12,000 shares, on each of which £40 has been paid in, and which now stand at a small premium. They have divided £61 12*s.* per share, up to January, 1854. Their sales of ore have been nearly as follows:—

| Years. | Tons of ore. | Years. | Tons of ore. |
|---|---|---|---|
| 1836, | 2,077 | 1843, | 16,433 |
| 1837, | 5,346 | 1844, | 17,744 |
| 1838, | 6,758 | 1845, | 22,741 |
| 1839, | 10,732 | 1846, | 14,755 |
| 1840, | 16,699 | 1847, | 14,711 |
| 1841, | 22,094 | 1848, | 21,761 |
| 1842, | 22,544 | 1849, | 19,772 |

The average yield of the ore shipped is about 16 per cent.

The principal depot for copper ores is at Port du Salle, four miles from Santiago. Thence there is a railroad to Cobre, nine miles distant.

*Royal Santiago Mining Company.*—This company was formed in 1837, and has 17,000 shares, on each of which £13 has been paid in. From 1840 to 1845, 55,440 tons of ores were raised, of the value of £577,533, at an expense of £247,043, the net profit realized being £48 per ton. Up to 1848, £33 4*s.* per share had been paid in dividends; but since that time the mines have been worked at a loss, and in 1853 an assessment had to be called.

The mine at the present time is about 110 fathoms deep.

The total amount of copper produced in Cuba appears to have been nearly as follows for the years indicated:—

| Years. | Tons. | Years. | Tons. |
|---|---|---|---|
| 1835, | 681 | 1849, | 3,594 |
| 1840, | 4,139 | 1850, | 3,239 |
| 1845, | 6,532 | 1851, | 3,300 |
| 1846, | 4,092 | 1852, | 2,500 |
| 1847, | 3,288 | 1853, | 2,200 |
| 1848, | 3,867 | | |

The amount of copper imported into this country from Cuba has never been large. It was, in the last four years, in value as follows:—

| | 1849. | 1850. | 1851. | 1852. |
|---|---|---|---|---|
| Sheathing, . . . . . . | $188 | | | |
| Pigs, bars, &c., . . . . . | 27,266 | $38,848 | $8,740 | $24,820 |
| Ores, . . . . . . . . . | 3,980 | 1,294 | 2,331 | 1,793 |
| | $31,434 | 40,142 | 11,071 | 26,613 |

JAMAICA.—The attention of English capitalists has been recently drawn to this island, and several companies have been formed for working copper veins which are known to exist there.

The *Jamaica Copper Mining Company* has several localities where valuable lodes are reported as having been found. Among them are the Mount Vernon, Washington, and Bloxburgh mines.

MEXICO.—Ores of copper are found scattered through Mexico in considerable abundance, but they are not worked to any extent as yet. The following localities are mentioned by Humboldt: the mines of Ingaran, a little south of the volcano of Jorullo, and San Juan Guetamo, in the province of Valladolid. The ores are vitreous copper and the red oxide, as also native copper.

---

## SECTION III.

### GEOGRAPHICAL DISTRIBUTION OF COPPER IN THE UNITED STATES.

NEXT to gold and iron, copper is the most important metal to the United States; since, although the value of the lead raised in this country may exceed that of the copper at present, yet the large amount of capital invested in mining for the latter metal, the certainty of its increasing production, the extent and number of the mines from which it is obtained, and the fact that general attention has been called to its occurrence, within a few years, by the discovery of a new and interesting cupriferous region in this country, all combine to give to this metal a predominating interest. Of copper only can it thus far be said, in speaking of the United States, that there are large, productive, and permanent mines.

The copper mines, and localities where ores have been discovered, in the United States, may be classed according to the following scheme, which is formed partly on a geographical and partly on a geological basis.

### I. LAKE SUPERIOR COPPER REGION.

Native copper, in true veins, in the trappean rocks and associated conglomerates and sandstones of lower Silurian age; extensively worked.

### II. COPPER DEPOSITS OF THE MISSISSIPPI VALLEY.

Ores of copper, chiefly pyritous, in the unaltered limestones and sandstones of lower Silurian age; mostly occurring in gash-veins, or contact-deposits: principal localities, in Wisconsin and Missouri; not worked at present to any extent.

### III. CUPRIFEROUS DEPOSITS OF THE ATLANTIC STATES.

The deposits of this division are not to be separated geographically; but, geologically, they may be described in three groups, as follows:—

*a.* The copper-bearing veins of the Appalachian chain, in rocks of the metamorphic palæozoic age; ores chiefly pyritous; deposits mostly in the form of segregated veins; localities numerous, extending along the flanks of the great Appalachian chain, from Vermont to Tennessee. Worked in numerous places.

*b.* Deposits in the sandstones and associated trappean rocks, of the formation commonly called the New Red Sandstone: ores, carbonate, oxide, and native copper, principally; contact-deposits, usually of limited depth. Localities numerous, in Connecticut and New Jersey; formerly extensively worked, but now abandoned.

*c.* Veins traversing the new red sandstone and the older

metamorphic rocks, and bearing principally ores of copper in the sandstone. Locality confined to Montgomery and Chester Counties, Pennsylvania, and there extensively worked.

We will now proceed to the consideration of each of the above-mentioned divisions in order, beginning with Lake Superior.

### THE COPPER REGION OF LAKE SUPERIOR.

The occurrence of native copper on Lake Superior has been known since the time of the earliest explorations of the Jesuit Fathers, who in the latter half of the seventeenth century travelled extensively in that region, and whose accounts are filled with the most exaggerated and extravagant stories of the abundance of the metal.* Other travellers through the great northwestern lakes added their testimony, and embellished the previous accounts with stories of gold and precious stones. The first actual mining operations within historical times were commenced near the Forks of the Ontonagon, in 1771, by Alexander Henry. Having worked without success for a while at this point, searching in the clay bluffs which line that river for masses of native copper, operations were transferred in the next year to the north shore of the Lake; but, as might have been expected under the circumstances, they proved entirely abortive.

In 1819, Gen. Cass, accompanied by Mr. H. R. Schoolcraft, made a journey along the southern shore of Lake Superior to the Mississippi, and visited the famous mass of native copper lying in the west branch of the Ontonagon. A few years later, in 1823, Major Long, acting under the orders of the War Department, passed along the north side of the Lake on his return from a scientific expedition to the Mississippi and St. Peter's Rivers. Professor Keating, the geologist of the party, mentions that they had seen numerous boulders of native copper strewed over the valley of the Mississippi. All the early explorers seemed to agree that copper might be found in abundance; but that, so great was the distance of this region

* See Foster and Whitney's Report on Geol. of Lake Superior, Part I. p. 6.

from a market, and so wild and unsettled the country, there would be little prospect of any mines being worked with profit.

Their observations, however, could not fail to draw attention to the region, and not long after this portion of the country came into the possession of the State of Michigan, an exploration was commenced by Dr. Douglass Houghton, State Geologist, and his official report did more than anything else towards awakening an interest in that region, and directing towards it the attention of explorers. In 1841 he published an account of his observations, in the form of an annual report to the Legislature of Michigan, in which the first definite information with regard to the occurrence of native copper, in place, on Lake Superior, was given to the public. Afterwards he undertook to carry on a linear survey of the northern peninsula of Michigan, comprising the whole extent of the copper and iron-bearing districts, in connection with a geological exploration; while engaged in the execution of this survey, and before it was fairly commenced, he was drowned during a snow storm and gale, near Eagle River, on the 13th of October, 1845. But already mining operations had commenced in that region, and explorers and speculators were flocking to it from all quarters.

The cession by the Chippeways to the United States of the district extending from Chocolate River, west to the Montreal, and southerly as far as the boundary of Wisconsin, which was ratified March 12, 1843, was the signal for the opening of the Lake Superior mineral region to the pioneers of the West. During the following summer (1843), several miners crossed over from Wisconsin by land, and selected numerous tracts of land, including many of those now occupied by the best mines in the country. These tracts, at first three miles square, and afterwards reduced to one mile, were leased by the War Department to the persons applying for them, in virtue of an Act of Congress made in reference to the lead lands of Illinois. By the terms of the leases the applicant was required to work the mines he might discover on the tract selected by him with all due diligence, and render to the United States six per cent. of all the ores raised.

In the summer of 1844, as soon as it became generally known that the Lake Superior country was opened to settlement, numerous persons visited that region, and the first mining operations were commenced, on leases secured the year before. Many loose masses of native copper, some of which contained silver, and were of large size, were picked up, and discoveries of veins and deposits of copper in the rock were made. When these facts were reported in the eastern cities, of course with many exaggerations, a great excitement or "copper fever" was the result, and in 1845 the shores of Keweenaw Point were whitened with the tents of speculators and so-called geologists. Many hundred "permits," or rights to select and locate on tracts of land for mining purposes, were issued by the Department, and three hundred and seventy-seven leases were granted. Most of the tracts covered by these were taken at random, and without any explorations whatever; indeed, a large portion of them were on rocks which do not contain any metalliferous veins at all, or in which the veins, when they do occur, are not found to be productive.

In 1846 the excitement reached its climax; the speculations in stocks were continued as long as it was possible to find a purchaser, and a serious injury was inflicted on the mining interests of the country by the unprincipled attempts to palm off worthless property as containing valuable veins. But every such mania must have an end, and in 1847 the bubble had burst, and the country was almost deserted. Only half a dozen companies, out of all that had been formed, were actually engaged in mining.

The issue of permits and leases having been suspended in 1846 as illegal, Congress passed, in 1847, an Act, authorizing the sale of the mineral lands, and a geological survey of the district. In the mean time, while this survey was going on, the companies which had continued their operations made considerable progress, new ones were formed, and lands were purchased by them after *bona fide* explorations and discoveries of veins, the position and character of the really metalliferous rocks began to be known, and confidence was gradually restored. At the time of the completion of the geological survey, in 1850, and the publication, in the following year, of maps of the whole region, on which the range and extent of the geological formations were laid down, copper mining in the Lake Superior district had become established on a firm basis, and was rapidly developing.

It is not possible here to describe the "ancient mining" on Lake Superior, which is of so much interest at the present day to the antiquarian, and even to the explorer, as a clue to the richness of the veins. It may be stated, however, that throughout the whole extent of the copper region, from the extremity of Keweenaw Point to a considerable distance beyond the Ontonagon, and on Isle Royale even, numerous excavations, made for the purpose of procuring copper, have been found, and which must evidently have been made at a very remote period. Some of this ancient mine-work is on a quite extensive scale; in one instance, at least, attaining the depth of fifty feet in the solid rock. The principal tools found in the excavations are the so-called stone-hammers, or small boulders of rock around which a groove has been cut for the purpose of fastening on a handle. These are accumulated in some of the ancient pits in great quantities; with them, fragments of charcoal have been often noticed, indicating that the method pursued by the miners was similar to that still employed in some of the European mines, in the use of fire to attack the rock. A few tools of copper and wood have been also discovered; but no remains of habitations, or burial-places, which might furnish a clue as to the race by which this work was done, have yet been found. It is known, from the size and age of the trees growing over many of these excavations, that they are several hundred years old, at least. The present Indians have no traditions as to copper mining in the region, and have not now any conception of how, or by whom, the work could have been done.

Having thus given a short sketch of the progress of the Lake Superior region, opening the way to the development of its mineral resources, it will be proper to add a few words with regard to its geological structure, before proceeding to consider the mode of occurrence of the copper.

The basin of Lake Superior, the largest collection of fresh water known, occupies, for the most part, a great synclinal trough, caused by a depression in the sandstone which appears to form its bed. From each side of the Lake, the dip of the strata is towards its centre. The northern and southern shores, which for a considerable distance are 160 miles apart, are very different in character and appearance. On the north, around the deep bays, which extend inland from 30 to 40 miles in some instances, almost perpendicular cliffs rise from the water's edge, sometimes to the height of more than 1000 feet, presenting scenes of picturesque grandeur unrivalled in the Northwest. The southern shore, on the other hand, is comparatively low, only occasionally rising to a height not exceeding 200 feet above the Lake.

The reason of this difference in the aspect of the two shores is easily perceived, upon noticing their geological structure. On the east and north, the sandstone which originally existed there has been worn away, until the more enduring granite and trappean rocks only were left, presenting an effectual barrier against the farther encroachment of the Lake; only here and there limited patches of the sedimentary rocks remain, where they were sheltered from the action of the waters, standing as outliers in the small islands along the coast, and behind Isle Royale.

This sandstone appears along nearly the entire southern shore of the Lake, from Saut Ste. Marie to Fond du Lac, its continuity being interrupted in only a few points, where the trappean or granitic ranges have been for a short distance denuded of the sedimentary beds which were originally deposited upon them. The general trend of the southern shore is east and west, but at a nearly equal distance from each end of the Lake, the regularity of its outline is broken by a projecting point of land, which extends for sixty or seventy miles in a northeast direction, gradually curving round to the east. This is Point Keweenaw, the locality where, by the present generation, the copper-bearing veins were first opened and worked.

The sandstone of Lake Superior, in regard to the geological position of which there was formerly some disagreement, has now been satisfactorily determined to be of Lower Silurian age, and probably the equivalent of the Potsdam Sandstone, the lowest fossiliferous rock recognized in this country.* Above it, as we proceed southward from any point between Saut Ste. Marie and the Pictured Rocks, we find the upper members of the Silurian system cropping out in succession, with a slight southerly dip. Along this portion of the Lake the sandstone lies nearly horizontally, and is made up of rounded grains of quartzose sand, but slightly colored by iron, and having little coherence, and the whole thickness does not seem to exceed 300 or 400 feet. Where it comes in contact with the older azoic rocks, as may be observed in the vicinity of Carp and Chocolate Rivers, it is seen resting unconformably upon them, having been deposited nearly horizontally on their upturned edges. On Keweenaw Point, however, its character is entirely changed; it has increased greatly in thickness, is tilted up at a considerable angle, and is associated with very heavy beds of conglomerate and trappean rock. On tracing into the interior the ranges which approach the Lake at the extremity of Keweenaw Point, they are found to extend in a general southwesterly direction along the whole line of the Lake, at a distance of a few miles from it, and, gradually becoming less conspicuous in Wisconsin, they finally disappear before reaching the Mississippi.

They form a series, usually of two, but sometimes of three or more parallel ridges, having an average height above the Lake of about 500 feet, and presenting steep mural faces toward the south, while dipping at a moderate angle in the direction of the Lake to the north. Along this line of elevations, which is familiarly known in the region as the "Trap Range," the copper mines of the southern shore of Lake Superior are situated, the metalliferous belt occupying in Michigan a length of over 120 miles and a breadth varying usually from two to six miles.

The rocks of which the trap range is made up are somewhat varied in their mineralogical character, but they belong mostly to the igneous class, and it is apparent, from their

* F. & W. Report on Geol. of Lake Superior, Part I. p. 99.

mode of formation and position, that they were poured out from the interior of the earth, at the time the deposition of the sandstone was going on, from a series of fissures which extended along the line now occupied by the metalliferous formation. In the more elevated and central portion of the range, the igneous rocks predominate, containing intercalated beds of conglomerate, of very inconsiderable thickness, between heavy masses of trappean rock. As we recede from the line of igneous action in either direction, we find that the belts of trap become thinner, the conglomerate predominates, but gradually disappears, and is succeeded by the sandstone with its normal character. Thus the appearance of the conglomerate is seen to be allied with, and subordinate to, that of the igneous masses, and it appears to have been a result of the combined action of the two classes of agencies by which the trap and sandstone were formed. The whole system of the bedded trap and the interstratified masses of conglomerate is developed on a grand scale, some of the single beds acquiring a thickness of several thousand feet.

The lithological character of the trappean rocks is quite varied, and their mechanical structure is liable to important changes. The usual mineral components of the trap are labradorite and augite, with a smaller proportion of various other minerals, among which magnetic oxide of iron, chlorite, and epidote are the most abundant, with smaller quantities of the zeolitic minerals and calc. spar, as accidental ingredients. The feldspathic and augitic portions are usually finely granular, and form a compact homogeneous paste, in which the others are embedded; and the recognized differences in the characters of the different trappean beds do not seem to depend so much on chemical composition as on mechanical structure; as they seem in most cases to contain the same or very similar mineralogical components, in very different conditions of mechanical aggregation; at least, chemical analysis has not yet given a clue to the nature of those changes in the character of the rocks which are easily perceived by the eye.

There are certain varieties of the trappean rock which are universally recognized in the Lake Superior region, and which have a marked influence on the character of the veins as they pass through them. These changes of character are most distinctly perceived on Isle Royale and Keweenaw Point, where the bedded structure of the trap is more apparent than in the other districts. There are two kinds of trap especially well-marked on Keweenaw Point, the amygdaloid, and the crystalline trap or greenstone. Of these, the former is the productive, the latter the unproductive rock. The veins seem to be best developed and richest in metallic contents in a rock which is neither too crystalline and compact, nor too soft and porous; in the beds which have a very vesicular structure they are liable to lose their regularity and form floors of vein-stone, with but little metallic matter, or too irregularly distributed to be worked with profit: in the very compact and finely crystalline rock, on the other hand, they are pinched up and barren.

The most remarkable feature of the Lake Superior district is the character of its metalliferous product, which is not an ore of copper, but exclusively the native metal, which previous to the opening of this new mining region had never formed the object of persistent mining operations. Native copper does indeed occur in many veins, but usually in small masses, which were found near the surface, and evidently resulted from the decomposition of the sulphurets. In this extraordinary district, the veins, in those rocks in which they are most productive, carry exclusively native copper, with a small amount of native silver intermixed, and there has been no change observed in the character of their contents at any depth which has yet been reached. The occurrence of native metal characterizes the veins in that part of the trappean rocks in which a distinct bedded structure may be observed; the same may be said of the conglomerate, except that in this latter rock the quantity of metalliferous matter is small, and very unequally distributed through the vein-stone. In those parts of the Lake Superior district where the trap is not distinctly bedded, it ceases to bear native metals, but contains sulphurets of copper, zinc, lead, &c., of which traces only have been found in the true native copper-bearing veins. Thus in the Bohemian or Southern Range of Keweenaw Point, which appears to have been protruded at a late epoch, and under different conditions, and to have tilted up the system of the bedded trap and interstratified conglomerate which lies to the north, the veins bear only sulphuret of copper; and on the north shore, where the trappean rocks are most developed they appear to be of the same unbedded character, and they are traversed by powerful veins bearing the sulphurets of copper, zinc, and lead.

In describing the mines of the Lake Superior district, they will be divided into four groups, each of which is characterized by its geographical position, and by a mode of occurrence of the copper somewhat differing from that of the others, and peculiar to itself. These groups are:—

1. Keweenaw Point.
2. Isle Royale.
3. Ontonagon.
4. Portage Lake.

Before proceeding to a particular notice of the mines of each district, some general remarks will be made in relation to the peculiar features of its geological structure, and the occurrence of its metalliferous deposits.

### KEWEENAW POINT MINING REGION.

This district embraces a large number of mines, some of which are extensively worked, and extends over a space of about thirty-six miles in length and two to three in breadth. Its geological features are strongly marked. From its eastern extremity a belt of metalliferous trap extends through it in a nearly east and west direction, but gradually curving in its western prolongation towards the south; this belt is made up of a variety of trappean rocks, which differ from each other in their structure and in the character of their metalliferous contents. There are, through nearly the whole extent of the Point, two well-marked ranges of elevations, known as the Greenstone Range and the Southern or Bohemian Range. The former comprises a line of bluffs rising sharply from the valleys of the two streams, Eagle River and the Montreal, which drain the district, rising near its centre, and flowing longitudinally through it in opposite directions. The greenstone ridge is made up of a compact, crystalline, homogeneous trappean rock, analogous to dolerite in composition. It forms a powerful bed, which has a thickness of several hundred feet, and dips to the north at an angle of from 20 to 30 degrees. Its northern limits are not sharply defined, but on the south its extent is well known, as a marked change takes place in the character of the veins at this point. Between this bed and the next inferior there is a stratum of conglomerate, accompanied by a thin deposit which seems to be a consolidated volcanic ash, or mud deposited on the floor of the ocean; beneath these lies the great southern metalliferous belt of Keweenaw Point, in which numerous mines have been opened, but in none of which have the workings been found profitable when extended through the "slide" or "cross-course," as the thin belt of conglomerate and trappean ash is called, into the greenstone above. The relative position of these beds will be seen by the annexed section (Fig. 19), made at a point

Fig. 19.

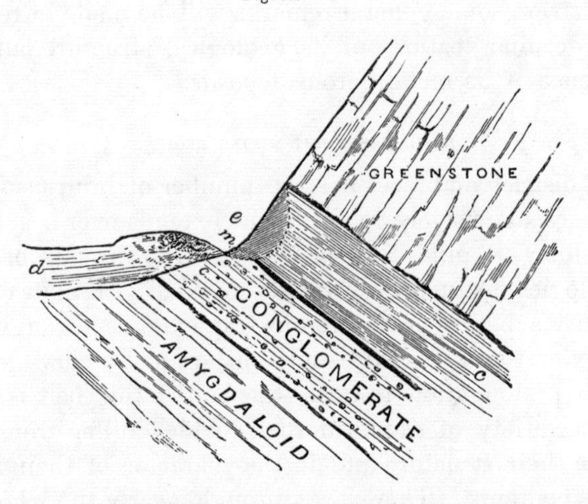


*c.* Bed of chloritic and ashy matter. *d.* Drift, and rubbish thrown out of *e.* *e.* Excavation made by ancient miners. *m.* Cupriferous stratum.

now known as the Waterbury Mine, where a shaft has been sunk at a point where quite an excavation had been made by the ancient miners in search of copper.

The belt of conglomerate, which, at the eastern end of the Point, is 30 or 40 feet in thickness, gradually thins out and finally disappears near the Cliff Mine, although the other and main features of the distinction between the crystalline and amygdaloidal rock remain as well-defined as before, and continue so for several miles farther. At the Eureka Mine the following section is presented between the greenstone and amygdaloid.

Greenstone; thickness, 500 to 800 feet.

| | |
|---|---|
| 1. Seams of quartz, irregular, . . . . . . . | 3 feet. |
| 2. Conglomerate of a red color, 14 inches, . . . . | |
| 3. Conglomerate of a greenish-white color, 3 inches, . . | |
| 4. Indurated clayey matter containing specks of copper, 6 inches, . . . . . . . . . | |
| 5. Indurated clayey matter, 6 inches, . . . . . | |
| 6. Trappean ash, 7 inches, . . . . . . . . | |

Amygdaloid; thickness undetermined.

The bed of rock between the conglomerate and the greenstone often contains thin sheets and particles of copper, and the conglomerate itself is not without frequent indications of the same metal.

To the south of this thin belt of conglomerate, the amygdaloid extends for from two to three miles, but as it lies in the low ground, occupied by the valleys of the Eagle River and Montreal, it has only been explored, in general, by underground working in the mines, nor have these been extended sufficiently far to show the thickness of the beds of which it is made up.

On the north of the conglomerate belt, the greenstone occupies a width on the surface of from a quarter to half a mile, and gradually becomes less crystalline and compact. At length, by an imperceptible change, the rock is found to have become amygdaloidal, resembling what it was on the other side of the conglomerate. From the point where this change takes place, to the appearance of the first belt of sandstone, is a space of from a mile to a mile and a half, which is occupied by a variety of trappean beds, some of which are more and others less metalliferous; but, together, they constitute the "northern metalliferous amygdaloid belt," in which several important mines are opened, and the particulars of whose structure will be noticed under the description of the Copper Falls and Phœnix Mines. Still farther to the north is a series of alternating belts of amygdaloid and sandstone of moderate thickness, varying from 50 to 500 feet, and these are succeeded by a heavy belt of conglomerate, which occupies an extent on the surface of nearly a mile. Beyond this is still another bed of a very amygdaloid rock, of about 1500 feet in thickness, succeeded again by conglomerate, which forms the northern portion of the point, from its extremity as far west as Agate Harbor.

The mines of Keweenaw Point, almost without exception, are worked on metalliferous deposits which have all the characters of true veins. They cross the belts of rock in a direction nearly at right angles to the strike of the formation, and have, in many instances, been traced through beds of both the aqueous and igneous formations, from the southern amygdaloidal belt across the greenstone, the northern metalliferous beds, the alternating beds of sandstone and conglomerate, to the Lake shore. It has not yet been proved by direct observation, that the same veins extend across to the southern range, and there bear sulphurets: as the valley which separates the greenstone and the porphyritic or southern range, and which is underlaid by the amygdaloid, is too much covered with drift, and too low, to admit of the veins being traced across it on the surface. It seems probable, however, that such would be the case. But we have pretty satisfactory evidence that the veins do not extend, for any great length, into the sandstone. Only one vein is known to have been worked on both sides of the greenstone in the northern and southern belts; this is the Copper Falls vein, which is the same as that opened in the Northwestern Mine. In their passage through the different belts of rock, the veins exhibit marked changes in their characters. In the conglomerate their gangues are mostly calcareous; the copper is usually concentrated into large masses, but they are of rare occurrence. In one case black oxide of copper has been found in this rock; but this seems to have been a solitary instance.

In the true copper-bearing rock, which is, as before remarked, of a fine texture, not too crystalline, and with occasional amygdules scattered through it, the veins appear with a gangue made up of quartzose matter, mixed in a greater or less degree with calcareous spar and the zeolitic minerals, of which Prehnite and Laumonite are the most common. The most favorable veinstone is one which contains considerable crystallized and drusy quartz, with Prehnite and granular carbonate of lime intermixed; where this variety of gangue occurs, copper may be looked for with confidence. Datholite, in one exceptional instance, the Hill Vein, forms a considerable portion of the veinstone: in no other place has it been found in this part of the district. Many of the veins have a more or less brecciated character; that is, they appear to be made up in a considerable degree of fragments of the adjoining rocks, cemented together by the same veinstone as occurs in other veins. Numerous smaller veins or strings are made up almost wholly of Laumonite, and carry very little copper. There are instances in which veins with almost wholly calcareous gangues have been worked in the trap, but these are now regarded as quite worthless, especially when the carbonate of lime is in the form of coarsely-crystallized spar. Such veins, moreover, have not the course of the true metalliferous veins.

The width of the productive veins is usually from one to three feet; they sometimes widen out to ten feet or even more, but rarely continue to hold those dimensions for any considerable distance. The wider the vein, as a general rule, the richer it is in metallic contents.

There is but one system of veins in this district, so far as has been ascertained, and they are remarkably regular in their course, which in the productive ones, as before remarked, is nearly at right angles to the line of bearing of the formation; when they pass from one belt of rock into another of a different character, they are sometimes shifted to one side or the other, for a few feet, as if there had been a sliding of the beds on each other after the formation of the vein-fissure; but there are no regular cross-fractures, or counter lodes, intersecting the main ones and heaving them, as in many other metalliferous districts, where there are sets of fissures of different ages, and extensive faults which interrupt the continuity of a whole series of veins of an earlier epoch. In fact, the general parallelism of the productive lodes of Keweenaw Point is one of their most marked features, and they do not appear to have a tendency to unite together, and form what the Cornish miners call "champion lodes."

The dip of most of the veins in this district is nearly perpendicular, and generally pretty regular, the underlay, or deviation from a vertical line, being rarely more than 8 or 10 degrees. The selvages are usually well marked, the vein being separated from the wall-rock by a thin layer of red clay or flucan, and the walls themselves striated and polished, sometimes extending for a hundred feet or more in an almost perfectly straight line.

Throughout the whole of the region the occurrence of the copper in the veins is marked by the same characteristics. It is found mixed with the veinstone in pieces of every size, from almost microscopic particles up to masses of from one to two hundred tons in weight. Some of the veins carry only the former variety, while in the most productive ones heavy masses are frequently met with, and form a large part of their produce. There are three heads under which the copper is classed, according to the size of the pieces in which it occurs. These are, mass copper, barrel-work, and stamp-work.

1st. *Mass copper.*—Where, as is frequently the case, a mass of copper is met with in the vein which extends along for several feet, sometimes 20 or 30, the rock is stoped away from one side of it, and it is then detached from the wall on the other side by heavy charges of powder inserted behind it. It is then divided up into pieces of such a size that they can be drawn by tackle or otherwise moved to the shaft, so as to be raised to the surface. This cutting up of the masses is often a tedious and costly process, when they are of great size, and sometimes requires several months before a single one is entirely removed from the mine. It is performed by the aid of chisels, with a cutting edge of about $\frac{1}{4}$ of an inch in width, and varying in length according to the thickness of the mass to be divided. Two persons are required, one of whom holds the chisel and guides it, moving it so as to prevent its becoming wedged in the groove, while the other strikes it with a heavy sledge-hammer, so that chips are gradually taken out of a length equal to the distance which is to be cut across, and of a thickness of about an eighth of an inch. A repetition of this process at length completely severs the mass. The expense of this operation is generally about $6 per square foot of surface exposed on one side of the cut. The pieces raised to the surface often require farther subdivision, after detaching all of the rock and veinstone adhering to them, so as to be reduced to a weight not too great to be transported to the shore of the Lake. As thus prepared for shipment, the mass copper contains 70 or 80 per cent. of the pure metal, and sometimes is almost wholly free from foreign matter, yielding from 90 to 95 per cent. when melted down in the furnace.

2d. *Barrel-work.*—This includes the smaller pieces, weighing usually a few pounds, which are too large to go under the stamps and too small to be sent away without barrelling. In the productive mines a considerable quantity of this form of copper is obtained in breaking and preparing the stamp-work. The pieces are cleaned as well as possible from the rock by hammering, and as thus prepared contain 60 or 70 per cent. of pure copper.

3d. *Stamp-work.*—This forms a large part of all the veins; and each mine requires, therefore, when fully under way, a set of stamps of a number of heads proportionate to the amount of productive veinstone taken from the workings, and the extent to which the mine is expected to be opened. The stamp-work is prepared to go under the heads by being calcined and broken into small fragments. For this purpose it is piled on a layer of billets of wood, which are fired, and in burning, raise the heap of rock upon them to a high temperature. Care has to be taken to distribute the heat as uniformly as possible, and not to allow any part of the copper to become fused and oxidized, as a loss would thus ensue.

Having thus noticed some of the more general facts connected with the Keweenaw Point veins, such mines in that district as are worthy of notice from the amount of work done upon them will be described as concisely as possible, in a geographical order, beginning at the east and proceeding towards the west; taking first those of the northern metalliferous belt, and then those south of the greenstone and lying in the southern belt of amygdaloid.*

*New York and Michigan.* T. 58, R. 28, Sect. 12, southwest quarter. A mine which had been worked at this place in 1846–7, was resumed by this company in 1852, at which time the shaft was 84 feet deep. It was carried down to about 150 feet and then stopped, in September 1853. A drift was run off to the north for 100 feet or more, and the vein found to be in some places 20 inches wide; veinstone, quartz and considerable Prehnite, carrying a small percentage of native copper. Four small masses, weighing about 1800 lbs., were shipped in 1852. The position of this vein within the unproductive belt of greenstone was found to be fatal to its being worked with profit, and up to this time it does not appear to have been traced into the more productive belt of rock to the north.

*Clark Mining Company.* T. 58, R. 28, Sect. 9, northeast quarter. This company commenced operations in the autumn of 1853, on a vein running north 10° west, a foot to eighteen inches wide, and well filled with copper. It has been opened at several points, from the edge of the greenstone northwards, and found to be remarkably well-defined, and promising favorably for successful working.

*Washington Mining Company.* Has several veins in the vicinity of Mosquito Lake, on parts of Sections 4, 5, 8, and 9 of T. 59, R. 29; operations not yet commenced.

*Agate Harbor Mining Company.* This company, which is not yet fully organized, has been exploring during the past winter, on parts of Sections 5, 6, 7, and 8, of T. 58, R. 29, and of Sections 1 and 12 of T. 58, R. 30. Several powerful veins are represented as having been discovered, which have a course of about south 19° east.

On the adjacent section, 11, a vein has been opened to some extent, called the "Keliher Vein;" and a company was partially organized to work it under the name of the *Lake Superior Mining Company*, but no work has as yet been done. The vein bears about north 25° west, and is from 2 to 8 feet in width, averaging, perhaps, 3 feet. It is made up of a breccia of rock and veinstone,

* In describing these mines the location of the works will be given by township, range, and quarter section. The ranges are numbered westward from the principal meridian of Michigan, which passes through Saut Ste. Marie, and the range-lines run north and south at a distance of six miles from each other: the town-lines are also six miles apart, but run east and west, being numbered from the north towards the south, commencing at a base-line drawn from Lake St. Clair due west to Lake Michigan. The townships, bounded by the town-lines on the north and south, and the range-lines on the east and west, are subdivided into thirty-six sections, of one square mile each, whose numeration begins from the northeast corner of the township; the first tier being counted from the east towards the west, the second from the west to the east, and so on alternately.

with considerable copper associated with it, and has been opened over an extent of 2,500 feet. It presents favorable indications for successful working. This is the most important vein on the property, but several others have been found in the vicinity.

*Native Copper Mining Company.* T. 58, R. 30, Section 10, northeast quarter. Mining operations have been carried on here since 1852; a shaft had been sunk 120 feet, and a cross-cut driven, at 10 fathoms depth, to the vein, on which a winze was sinking at the time I last visited the mine (Sept. 1853). The vein is wide, but made up in a considerable degree of a breccia of rock mixed with argillaceous matter, and not productive of copper. This vein bears a few degrees west of north; another, with a direction of north 10° to 15° east, has been explored to some extent, and found marked by a line of ancient excavations.

*Eagle Harbor Mining Company.* This company has made some explorations on Section 9, T. 58, R. 30, but without arriving at any very satisfactory results.

*Copper Falls Mining Company.* T. 58, R. 31, Section 14. The old "Copper Falls Company," to whose rights and property the present company succeeded, was formed in October 1845, and commenced mining on the vein now known as the "Old Copper Falls Vein," in 1846. At this time nothing was known of the geology of the country, and the workings were commenced and carried on in a belt of trap of only 170 feet in thickness measured at right angles to the dip, and 430 feet across on the line of the adit, enclosed between two beds of sandstone. The workings were continued down to the underlying bed of sandstone, and a shaft was sunk through it, a distance of 53 feet. So long as the workings were in the trap, the vein was quite rich, and one mass of over seven tons in weight was obtained, as well as numerous other small ones. On entering the sandstone, the vein contracted rapidly in width, and became split up, so that it was not attempted to follow it; but the shaft was sunk perpendicularly through the sandstone, and then cross-cuts were driven in each direction for a distance of 40 feet, without recognizing the lode. This, however, cannot be considered as evidence that the fissure does not extend through the sandstone, the cross-cuts having been driven too near the junction of the two rocks, where it would have been most disordered. The same vein has since been traced on the surface for a considerable distance, to the south of all the belts of sandstone.* From this mine copper to the amount of $15,000 in value was taken, and $100,000 expended in mining and surface improvements, including $11,060 97 paid for the lease.† The Directors remark, in regard to this work, as follows: "With the present mining experience on Lake Superior, a mine would not be commenced in a belt of trap of so little thickness as that in which all the works which have been described were situated. It should be remembered, however, that these mining operations were among the earliest undertaken on

* See Foster and Whitney's Report, Part I. p. 136, where a section of the old mine will be found.

† Third Annual Report of Directors, Boston, 1852.

Lake Superior: that they were begun at a time when there were but few inhabitants in the country, . . . and, what is of more importance than all the rest, when hardly anything was known of the geology of the country."

In order to remedy this error, Mr. S. W. Hill, who had been recently connected with the Geological Survey of the Lake Superior Land District, was selected to make a thorough geological examination of the location, and in the course of his explorations, several veins were discovered to the south of all the belts of sandstone, in the great northern metalliferous belt of amygdaloid. Two of these veins only have, up to the present time, been worked to any extent. These are the Copper Falls and Hill Veins, so called. On the former vein, mining was commenced in December 1850, and has been continued uninterruptedly since that time. The Hill Vein was opened about a year later.

The several veins on this location are nearly parallel with each other, their course being north 22° to 25° west. They have, in almost every case, been traced across the whole width of the belt of trap north of the greenstone, a distance of more than a mile.

The Copper Falls Vein is opened by an adit of 1250 feet in length to shaft No. 1, where it strikes the vein, at a depth of 200 feet below the surface, and is continued on it to the next shaft south, a distance of 750 feet. There are seven shafts, extended along a line of about 2350 feet in length, the deepest of which (shaft No. 1) is down to about 12 fathoms below the adit.

The Hill Mine, upon the vein of that name, is opened upon the most extensive plan of any mine in the Lake Superior region. The deep adit starts from a point only about 50 feet above the level of Lake Superior, and is driven on the vein, and will drain it at the present principal working shaft (No. 4) at a depth of 520 feet, and at No. 7 at 700 feet below the surface. There are seven shafts opened, in all, and the deepest (No. 4) is now down about 200 feet. The deep adit must be driven 2320 feet to intercept Shaft No. 1, to the south of which there is a length of over 6400 feet on the vein, within the metalliferous formation.

The veinstone of the Copper Falls Vein contains considerable quartz and calcareous spar, a good deal mixed with argillaceous matter near the surface, but becoming finer and less mixed with rock-breccia in descending. The most abundant accidental minerals observed in this vein are analcime, often beautifully crystallized, and Leonhardite, a variety of Laumonite. The Hill vein, where it is richest and widest, contains a good deal of datholite, in addition to the other usual minerals. This vein sometimes widens out to several feet, and is then almost always much richer than when narrow. Both these veins are not uniformly productive throughout their whole extent, but show alternate belts of rich and poor ground, which dip, with the formation, about 26° to the north. In working the mines great attention should be paid to this fact, so that as little money as possible should be sunk in opening the unproductive ground.

In these mines a remarkable feature has been discovered during the past winter, which is unlike anything yet noticed on Point Keweenaw.* This is

---

* S. W. Hill's Report, dated March 1, 1854.

the occurrence of a metalliferous bed included in the formation and parallel with it. This stratum, which has been intersected in the works of both mines, is about 100 feet thick; and between it and the overlying bed, a bluish granular trap, there has been a movement or sliding, so as to produce a fissure along the line of contact, which is filled with veinstone. This east and west vein is distinctly marked on the surface at the Copper Falls Mine, where it contains small bunches of copper. Underlying it is the bed of volcanic ash, as it is considered by Mr. Hill, which is of a brownish color, quite soft, and everywhere filled with copper. From it sheets and lumps of pure copper have been taken, from a few ounces up to twenty or thirty pounds in weight. Near the Copper Falls Mine there are extensive ancient workings on this metalliferous belt. In the Hill Mine, where it is thinner than in the Copper Falls Mine, it contains an intercalated bed of sandstone 18 inches thick, which carries enough copper to be excellent stamp-work. Mr. Hill remarks of this cupriferous belt as follows: "By some it is believed that it contains 1 per cent. of copper, but by others it is thought to be much richer. It is perfectly clear, from what can now be seen of it, that many thousand tons of mixed rock and copper will be taken from it in opening the mines. It will require no calcining to stamp and wash easily, and can be cheaply excavated. So little has been done in testing the value of the bed in question, that great caution should be observed in giving an opinion in regard to it; but metalliferous beds have been, and are now, mined in the Ontonagon district with some success, and on Portage Lake with prospects decidedly flattering."

From Mr. Hill's Report, dated March 1st, 1854, we learn that up to that date, in both mines, 543 fathoms of vein had been stoped, which yielded 469,863 lbs. of 70 per cent. stuff, equal to 865 lbs. of pure copper per fathom. Some parts of the ground stoped in the Hill Mine have yielded more than one ton per fathom. The amount of stoping ground, open in both mines, at the same time, was over 4,000 fathoms of productive ground, besides 7,000 not considered worth removing.

During the past year, a steam-engine and stamps have been erected, and are now nearly ready to go into operation, all the copper thus far shipped from this mine having been in masses and barrel-work. The engine is of six feet stroke, and eighteen inches diameter of cylinder. It is calculated to drive forty-eight stamp-heads, of which twenty-four are intended to be put in use immediately.

From the various Reports of the officers of this company, it appears that for the year ending March 1st, 1854, the average cost of sinking the shafts was $14 04 per foot; that of driving the levels, $5 44; and of stoping, per fathom, $14 26.

The sum of $180,000 has been paid in on assessments on the stock of this company, and $37,000 received from sales of copper, which, together, are expected to prove sufficient to put the mines in a paying condition. The shipments of copper have been as follows:—

| | 1852. | 1853. |
|---|---|---|
| Weight of barrel-work, . . . . . . . | 13,000 lbs. | 42,113 lbs. |
| " masses, . . . . . . . . | 4,662 " | 96,407 " |
| Amount pure copper produced, . . . . . | 12,651 " | 91,737 " |
| Percentage yield, . . . . . . . . . | 72 " | 64 " |

The shipments for 1854 are expected to reach 400 tons.

*Humboldt Mining Company.* T. 58, R. 31, Section 21. This company was organized, and commenced explorations, in the summer of 1853, on a tract of land situated between the Copper Falls and Phœnix locations. Several veins have been discovered, but not opened as yet to a sufficient extent to make it possible to decide upon their value.

*Meadow Mining Company.* T. 58, R. 31, Sec. 20, northeast quarter. This Company was organized in 1853, and is engaged in working several veins, which have a course of north 28° west. According to Mr. Kelsey, the superintendent, there are four veins: the principal mine, at present, is opening in the most eastern, the "Kelsey" vein. An adit has been driven on a branch of the lode for 293 feet, and there intersects shaft No. 1, at a depth of 38 feet. Shaft No. 2 is 287 feet farther south. The vein is said to average two feet in width, and to be well filled with copper.

*Phœnix Mining Company.* T. 58, R. 31, Sec. 19, southeast quarter. This company, as originally constituted February 22, 1844, was possessed of seven three mile square leases on Keweenaw Point. It was the first organized company of the Lake Superior region, and was called the "Lake Superior Copper Company." Its stock was divided into 1200 shares, of which the proprietors of the leases received 400 unassessable for their interest. The first superintendent was C. H. Gratiot, who had been previously engaged in digging lead in Wisconsin. The seven locations, embracing over 40 square miles, were nearly all situated in the very richest portion of the mineral region.

During the summer of 1844, Dr. C. T. Jackson examined several veins which had been discovered on the property by C. C. Douglass and others, and under his direction work was commenced October 22, 1844, on Eagle River, near the place now known as the "Old Phœnix Mine," and carried on through the year 1845, and a stamping mill and crushing-wheels, of a kind suitable for grinding drugs, were erected, but soon proved to be entirely unserviceable. Up to March 31, 1849, when the Phœnix Company was organized and took possession of the Lake Superior Company's property, the latter company had expended $105,833 40, of which about half was probably for actual mining work, but they had done little or nothing towards developing the value of the property. The principal shaft was sunk on a "pocket" of copper and silver, without any signs of a regular vein, which soon gave out entirely. In 1846, after this fact had been ascertained, a drift was run off from the shaft, at the depth of 90 feet, towards the river, to find a vein, and directly under the river the workmen came upon a crevice, evidently worn out by the stream and afterwards filled up with gravel and water-worn accumulations. In this old channel, workings were carried on for some time in search of a vein which once had occupied the space excavated by the river, and a large pot-hole was discovered filled with rounded pieces of native copper and silver in considerable quantity.

From this pot-hole principally, and from the washing of the gravel taken out of the other excavations, 18,000 lbs. of nearly pure copper were obtained, besides considerable silver, most of which was stolen by the miners; but one piece weighing 96·8 ounces troy came into the company's possession. Finally the vein itself was struck, and sunk upon for about 90 feet; and in that state the works were, when they were suspended, and not resumed until the Phœnix Company commenced operations, in the fall of 1850. Mining on the vein discovered under the river, and now called the Phœnix vein, was carried on from the spring of 1851 to 1st March, 1853, and resulted in demonstrating the fact that it was too poor to be worked with profit, at least in that part of its course. The principal shaft was sunk to the depth of about 264 feet, and three levels driven between five and six hundred feet. The vein was found to be very regular, but too narrow for working, being usually but a few inches wide; towards the river it was wider and richer, but the influx of water from that part of the mine made it necessary that those workings should be dammed up. The veinstone consisted of calc. spar with jaspery and argillaceous matter, and contained copper disseminated through it in fine particles, and masses up to 1200 lbs. weight, with considerable silver; but, as before remarked, it was too narrow, and since the abandonment of the mine the efforts of the company have been very properly directed to exploring the vein in longitudinal extent, it being now a well-ascertained fact that the same vein may be too poor to be worth working in one belt of rock, while in another more congenial it may be rich in copper.

Besides the Phœnix, three other veins have been worked to some extent by this company. On the Armstrong vein a shaft was sunk 110 feet and a level driven 400; but it is poor in copper; the same may be said of the Ward vein, on which a shaft was sunk 75 feet. The New Phœnix vein is situated on the northeast quarter of Section 19. From it a mass of copper weighing 2,390 lbs. was taken almost at the surface.

The shipments of this company amounted in 1852 to a little over 13 tons, and in 1853 to 5,729 lbs. The entire amount expended here cannot fall much short of $200,000. Hardly a location on Lake Superior shows more abundant indications of copper, and there seems still good reason to suppose that some one of the numerous veins upon it will be found capable of being profitably worked in some one of the belts of rock which it traverses.

Having thus briefly noticed the principal mines working on the northern metalliferous belt of Keweenaw Point, we will next take up those which are situated on the southern side of the greenstone bluffs. They have in all cases quite a similar position, which is one presenting great facilities for working. In no case have they been worked to any extent in the greenstone; but they have been found fully as productive as anywhere else, if not more so, close up to the conglomerate belt which divides the amygdaloid from the greenstone. In the neighborhood of the Cliff Mine, the position of the rocks is such that no considerable part of the productive portion of the vein can be drained by an adit level. Towards the eastern extremity of the Point, the amygdaloid rises, so that at some of the mines in that vicinity from one hundred to two hundred feet of back can be had above the adit level. Frequently the amygdaloidal rocks are so covered by drift, that the veins can only be discovered by tracing them down from the greenstone, and opening shode-pits on their supposed course. Hence it was, that so many ineffectual attempts were made to mine in the greenstone, conglomerate, and other rocks which are more easily explored on account of their elevated and exposed situation.

*Keweenaw Point Copper and Silver Mining Company.* T. 58, R. 28, Section 13, southwest quarter. A company formed in England to work the veins on the property previously known as the "New Lac la Belle." Working was commenced last autumn, but has not yet been prosecuted to sufficient extent to develop the veins, which are four or five in number. The location was visited and reported on favorably by Mr. De Bussy, of London, in 1853.

*Star Mining Company.* T. 58, R. 28, Section 9, east half. At the old workings, two shafts were sunk to the depth of sixty or seventy feet, and then abandoned. There were two veins, which were supposed to unite at a distance of about three hundred feet farther south. Near the supposed point of junction a shaft was sunk one hundred and six feet in depth, but afterwards discontinued. During the summer of 1853 a new vein was discovered, about one-third of a mile east of the old mine, running about north 10° or 15° west. It had not been opened sufficiently, at the time I saw it, to enable me to form an opinion of its value; but it is said to be promising well.

*Manitou Mining Company.* T. 58, R. 28, Section 8, southwest quarter. Work was commenced here in the autumn of 1852. There are two veins. The eastern one had been opened at several points on the face of the hill, which rises perhaps two hundred feet from the valley to the greenstone on the north. In the drift near the base of the hill, which had been carried in two hundred feet last September, the vein was poor at the entrance, but widened out and improved in character towards the end, showing several inches of good veinstone, well filled with copper. No. 2 vein, forty rods west of the one just described, is wide, but somewhat intermixed with rock; clear veinstone two or three inches wide on foot-wall. The vein is well-defined, but poor in copper.

*Cape Mining Company.* T. 58, R. 28, Section 7, southwest quarter. A shaft has been sunk on the west vein 150 feet, and levels extended both ways at 90 and 150 feet. From the lower level there is a cross-cut 84 feet to the east vein; there are numerous bunches and strings of zeolitic minerals and quartz, containing lumps of copper of considerable size, but no regular well-defined vein, so far as I could observe. South of the adit, the rock is a brecciated mass of trap, very amygdaloidal and much mixed with sandy matter resembling indurated clay, as if the beds had been broken up and the interstices filled with mud, which afterwards became consolidated.

*Iron City Mining Company.* T. 58, R. 29, Section 14, northeast quarter. Much money has been wasted here on an unproductive, barren calc. spar vein. Two shafts have been sunk, at a distance of 257 feet apart, and connected by levels. Shaft No. 1 is 288 feet deep. The vein is wide and regular, but entirely barren of copper; the gangue being calc. spar cleaving into large rhombohedral pieces, and hence called by the miners "block-spar." This kind of veinstone, in the trappean rocks of Lake Superior, when unmixed with quartz and the zeolitic minerals, seems to be always unaccompanied by copper. The work has been abandoned, and ought never to be resumed. On the south half of section 11, 968 and 1368 feet north of the south line of the section, two shafts have been sunk to the depth of 60 or 70 feet on a narrow vein carrying some copper. The rock is the crystalline trap, or greenstone, and is coarse-grained and crystalline. Farther north, beyond the limits of this unproductive rock, the vein may be found of some value.

*Bluff Mining Company.* T. 58, R. 29, Section 15, northwest quarter. This company is engaged in working a very regular and well-defined vein, which varies from a few inches to eighteen in width. In August 1853, the principal shaft (No. 2) had been sunk about 159 feet. From it two levels have been driven to the north about 260 feet, at the depth of 50 and 114 feet respectively. In the excavations already made, the vein is usually too poor to pay for working, although it has produced some good stamp-work and a few small pieces of copper. There is a space of some 600 or 700 feet, between the end of the present drifts and the conglomerate, to the north, which may prove a more congenial ground, and should be carefully explored by sinking or driving upon the vein in that part of its course.

*Northwest Mining Company of Michigan.* T. 58, R. 30, Section 15. This company was incorporated and organized in 1849, and took the place of an association called the "Northwest Copper Mining Company," which had been working in a small way since 1847. Mining on an enlarged scale was commenced in 1849, and has since been continued, until the works have become so extensive as to be exceeded by few in the region.

Three veins have been opened near the centre of Section 15, two of them quite extensively. They are called the Stoutenburgh, Hogan, and Kelly veins; but the latter has not been worked to a sufficient extent to be worthy of notice. The principal mine is on the Stoutenburgh vein. This has a course of north 16½° east. The Hogan runs north 19° west; and the two would intersect at a point 320 feet south of the mouth of the adit-level on the Stoutenburgh, according to their courses as thus stated. The third, or Kelly vein, has a course nearly parallel with that of the Hogan, but is not very well defined or of much importance.

*The Stoutenburgh Vein* has been opened by four shafts, as will be seen by reference to the section (Plate I.), which represents the state of the workings at the end of February, 1854. The engine or D shaft is now sinking below the seventy-fathom level, its entire depth being five hundred feet from the surface. C shaft is down fifty fathoms below the adit. The longest level is driven about 1000 feet. The whole amount excavated by sinking shafts is 1130 feet; by driving levels, 5450 feet; and the number of fathoms stoped is about 2780. The vein averages about a foot in width, varying from six or seven inches to eighteen. The lode consists of argillaceous and sandy matter, with occasional strings of calc. spar, and has not a promising appearance. It contains but a small amount of stamp-work, and that of a poor quality; but masses are frequently met with which in some degree compensate for the poverty of the vein in fine copper. These weigh from a few hundred pounds up to several tons. One of eleven tons was taken from the twenty-fathom level. A great deal of the copper stuff brought out of the mine is nothing but amygdaloid trap, with shot copper scattered through it; and it may frequently be observed in the mine that the rock in the vicinity of the vein is richer in copper than the lode itself. Indeed, the large masses appear to make outside of the lode, and in close proximity to it; and when they occur, they impoverish it for some distance in every direction, and near them there are many lateral fissures running off into the country.

There are several "amygdaloid floors," designated on the section by the letters $a^1$, $a^2$, &c., which are beds of a few feet in thickness of very amygdaloidal trap, dipping with the formation. These have a marked influence on the character of the vein in their vicinity; thus it will be seen that in the upper part the workings have hardly in any case been extended to the south of the amygdaloid belt marked $a^2$, and in the upper portion of the mine, the productive ground lay between $a^2$ and $a^4$, until a new belt $a^3$ came in, which seems to have thrown it still farther to the north, and leads to the expectation that in that direction, near to the conglomerate, the vein will be found rich in copper. The 30 fathom level has been opened through the conglomerate, which was found to be 23½ feet in thickness, at right angles to its dip (25°). According to S. W. Hill, Esq., a very little copper was met with in the soft shaly rock resting on the conglomerate, as at the Waterbury mine. The vein has not been traced into the greenstone, but has probably been shifted to one side or the other, at the junction of the two belts of rock.

To the south of $a^2$, in the ends of all the levels, the vein has been found split up and mixed with clayey matter, but it has not been sufficiently explored in this direction, and there is no reason why productive ground may not be found on the south of $a^1$. This, however, will be determined by sinking the shafts C and D, which pass rapidly through the different floors, as the beds of rock dip at an angle of 23° only.

The works on the Hogan vein are much less extensive than those on the Stoutenburgh. They are represented in the annexed section (Fig. 20). The principal shaft (No. 2) was about 250 feet in depth in March 1854, and 278 fathoms of ground have been stoped. The veinstone of this lode is much more "kindly" looking than that of the Stoutenburgh, being less mixed with argillaceous matter, and more crystalline. It has produced small masses, and

Plate I

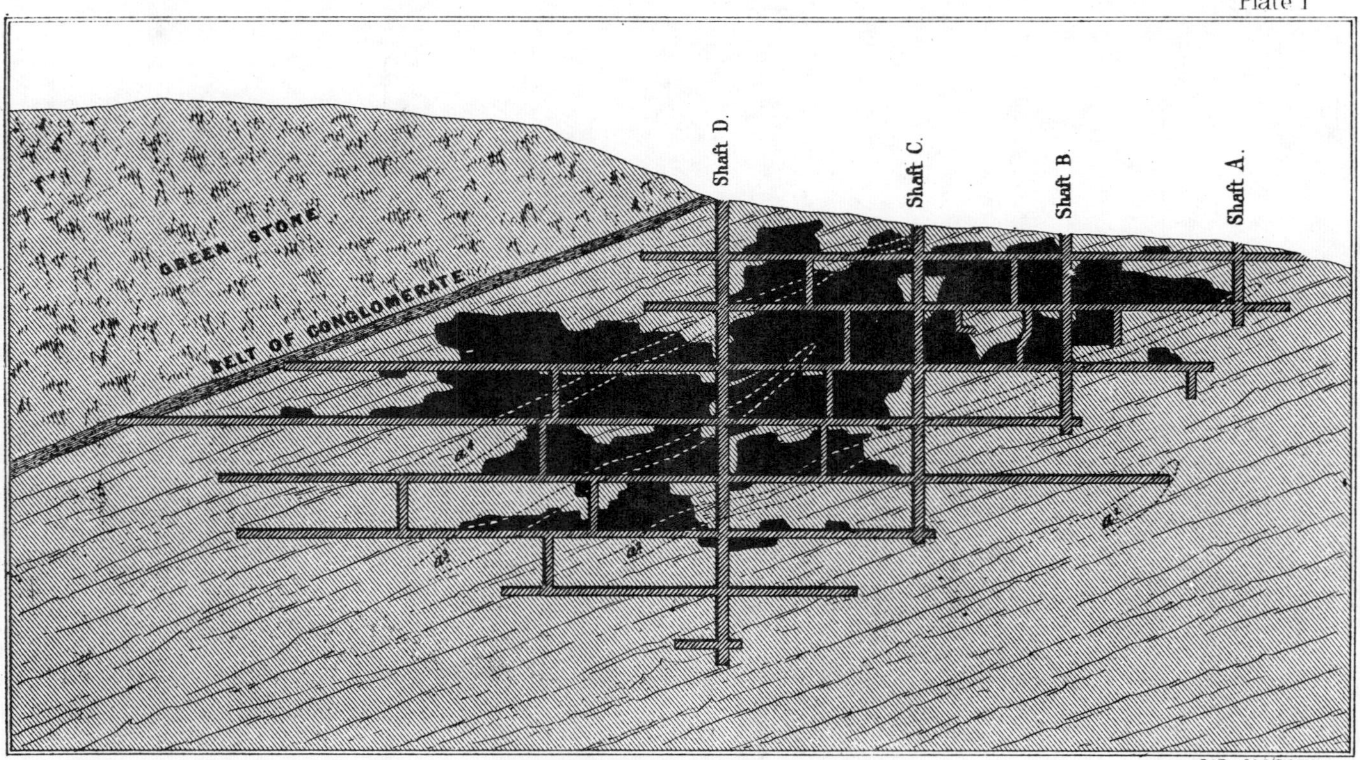


**SECTION OF THE NORTH WEST MINE.**

Feb 28. 1854.

barrel-work, as well as good stamp-work. In the ten-fathom level, the vein was from 10 to 12 inches wide, but it is now represented as being 18 inches wide at the bottom of the shaft, and well filled with copper.

Fig. 20.

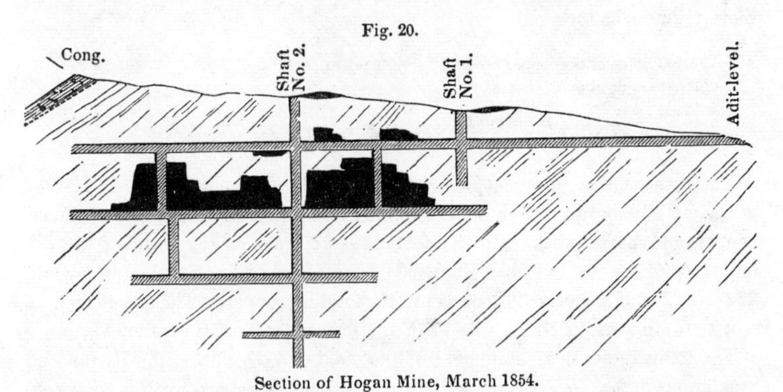


Section of Hogan Mine, March 1854.

From the forty-fathom level in the Stoutenburgh Mine a cross-cut was driven west for some distance, in the expectation of intersecting a vein which was formerly opened on the surface, and called the Clark vein, but it has not been found below.

There are two steam-engines in operation at the mine. One, of 17 inch cylinder and 5 foot stroke, winds and pumps at D shaft. The other, of 12 inch cylinder and 4 foot stroke, is used for driving the stamps. The following returns of the amount of rock stamped, and its yield of copper, for three months, in 1853, furnished by William Petherick, Esq., former superintendent of these mines, will be read with interest, as this is the only instance on the Lake where accurate and reliable accounts have been kept of the operations of the stamps.

| 1853. | Rock stamped. | Yield of copper. | Percentage. | Copper in ton of rock. |
|---|---|---|---|---|
| April, . . . | 440 tons. | 12,194 lbs. | 1·38 | 27·7 |
| May, . . . | 505 | 13,996 | 1·38 | 27·7 |
| June, . . . | 506 | 13,503 | 1·25 | 26·6 |
| | 1,451 | 39,693 | 1·34 | 27·4 lbs. |

The following statement shows the actual cost of raising the stamp-work from the mine, after the ground has been laid open for stoping, and of all the operations necessary for preparing the copper for market, for the above three months:—

| | Per Ton. |
|---|---|
| Stoping, . . . . . . . . . . . . . . . . . . | $2 00 |
| Filling, landing, and wheeling, . . . . . . . . . . . . . . | 26 |
| Tramming to kiln, . . . . . . . . . . . . . . . . | 06¼ |
| Burning, including wood and labor, . . . . . . . . . . . . | 35 |
| Dressing, . . . . . . . . . . . . . . . . . . | 68¼ |
| Steam-engine and labor at stamps, . . . . . . . . . . . . | 76 |
| Miscellaneous charges, carpenter, smith, &c., . . . . . . . . . . | 30 |
| | $4 42 |

The value of the copper produced is reckoned at 30 cents per pound, and the stamp-work is estimated to yield 90 per cent. of pure metal, making 27 cents per pound value of the stuff shipped, less $\frac{3}{4}$ cent for freight and smelting; there remains as the

| | |
|---|---|
| Actual value of the copper produced from 1 ton of rock, . . . . . . | $6 97 |
| Expenses as above, . . . . . . . . . . . . . . . . | 4 42 |
| Profit per ton, . . . . . . . . . . . | $2 55 |

The returns of a shipment in 1853 of 106,903 lbs., smelted by J. G. Hussey & Co., at Pittsburgh, which was estimated to consist of 57 per cent. of mass copper and barrel-work, and 43 of stamp-stuff, gave as the yield of the different varieties, as follows: Mass, 73; barrel-work, 68; No. 1 stamp-work, 92; No. 2 stamp-work, 59 per cent. of metallic copper. The proportion of No. 1 stamp-work to No. 2, was 77 of the former to 23 of the latter.

The following tabular statement of the yield of both mines, the Hogan and Stoutenburgh, has been compiled from the printed reports of the company, and from information furnished by the secretary, John Fausset, Esq.:—

| Year. | Amount shipped. | Copper produced. | Percentage. | Value. | Expended. |
|---|---|---|---|---|---|
| 1849, . . . . | 44,196 lbs. | 34,322 lbs. | 77·7 | $5,672 71 | $12,321 51 |
| 1850, . . . . | 270,873 | 195,020 | 72· | 35,786 66 | 39,906 88 |
| 1851, . . . . | 434,993 | 293,199 | 67·5 | 53,360 46 | 79,696 78 |
| 1852, . . . . | 380,429 | 269,174 | 70·7 | 53,071 58 | 74,983 06 |
| 1853, . . . . | 304,223 | 229,077 | 75·3 | | |

The whole produce of metallic copper, per fathom of vein removed, in the workings of both mines, is, approximately, 225 lbs.

The company own a very extensive tract of land, extending for three miles along the southern metalliferous belt, and occupying 4320 acres, a large part of which yet remains comparatively unexplored. A new vein was discovered about a quarter of a mile east of the present mine during the last summer, and has been opened during the winter, with what success I am not informed.

*Connecticut Mining Company.* This company has made some explorations on Section 16, next west of the Northwest location, and several veins have been discovered; none of them had been opened last summer to any extent.

*Waterbury Mining Company.* T. 58, R. 30, Sec. 17, southeast quarter. By referring to figure 19, on page 256, the geological position of this mine will be seen. It is opened in *c*, a chloritic mass, interposed between the conglomerate and the greenstone, which contains thin sheets and spangles of native copper scattered through it, and especially concentrated along its planes of contact with the rocks lying below and above. Two shafts have been sunk, each about 100 feet deep, going down upon the conglomerate, which dips at an angle of about 25°. A drift was also commenced between the shafts. This bed has not been found to contain copper enough to be worth working, and it is not probable that it will grow richer in depth; as deposits of this kind, which were formerly opened to some extent in the Porcupine Mountains, only grew poorer as they were worked downwards.

At several other mines along the southern edge of the greenstone, this layer of cupriferous rock, commonly known as the "Waterbury Vein," has been intersected by pushing the levels to the north; but I know of no instance in which any considerable amount of copper has been met with. There is no reason *a priori* why masses should not be found at the contact of the conglomerate and the greenstone; but this does not seem to be the mode of occurrence of this metal on Keweenaw Point.

*Summit Mining Company.* T. 58, R. 30, Section 19, west half. This company was organized in the winter of 1852–3, and commenced operations early in the latter year. Two veins had been opened to some extent in September 1853. The western vein, No. 1, underlays to the east, and is from 1 to 2½ feet wide. A shaft had been sunk upon it to the depth of 85 feet. The veinstone is chloritic, with Prehnite and quartz, containing some copper. Vein No. 2 is about 200 feet farther east, and was opened to the depth of 50 feet. It appears to underlay towards No. 1, so that the two will probably unite at no great depth. The gangue of these two veins is similar, being much mixed with what appears to be chlorite; the east vein, however, is more brecciated and argillaceous in its composition, resembling the Copper Falls Vein. It is well filled with copper, in fine particles and small lumps.

*Dana Mining Company.* T. 58, R. 31, Sect. 24, northeast quarter. This mine was first opened in April 1851, although but little work was done previous to 1852. In that year one of the principal veins, No. 1, was opened by three shafts, and an adit-level driven for about 270 feet on the vein. Shaft No. 1 was sunk 30 feet below the 10 fathom level; No. 2 was carried a few feet below the adit; No. 3, near the conglomerate, was just commenced. At the surface the vein was from 18 to 20 inches wide and well charged with fine copper and some barrel-work; but as it seemed, according to the Superintendent, Mr. S. W. Hill, to grow poor in depth, it was abandoned in 1853. During the past winter, another vein has been opened and worked on to some extent, with favorable results, as is reported. The "Old Copper Falls Vein" must cross this location at some point, and explorations were directed to finding it last summer, but at the time of my last visit had not been successful. The veins appear to become split up and barren of copper as they pass from the amygdaloid into the greenstone, and their continuity is often interrupted by a slide or heave to one side or the other. Towards the eastern extremity of the Point their courses seem to suffer a permanent deflection at the junction of these two rocks. Hence they have but rarely been traced across the greenstone, and discoveries are usually made quite independently in the metalliferous belts on each side of it.

*Northwestern Mining Company of Detroit.* T. 58, R. 31, Sect. 24, west half. Organized in 1845, under the name of the "Northwestern Mining Company;" reorganized in 1848, under the above name. The land owned by the company formed a part of one of the leases originally granted to the "Lake Superior Company." Mining operations were commenced in 1845, but only a small force was employed during the first two years. The workings have been hitherto confined to one vein, which is the same one wrought in the Copper

Falls Mine, affording the only instance thus far known in the Keweenaw Point district of a vein extensively worked on both sides of the belt of crystal line trap which extends through the region. The annexed section (Fig. 21), will convey an idea of the extent of the workings, in February 1854. The gangue of this vein is a light-colored quartzose material, mixed with chlorite; and it is characterized by an abundance of crystallized analcime, in both mines which are worked upon it, but the specimens found in the Northwestern Mine are particularly fine. It is associated with delicate crystallizations of reddish feldspar. The width of the lode is quite variable, and when it is widest it is much the richest in copper. At the surface its appearance was not very favorable; and it has required no little energy on the part of the Directors, and the Superintendent, John Slawson, Esq., to carry forward the work. The vein, however, has continued to improve in depth, and in some parts of the mine it now appears wide and rich.

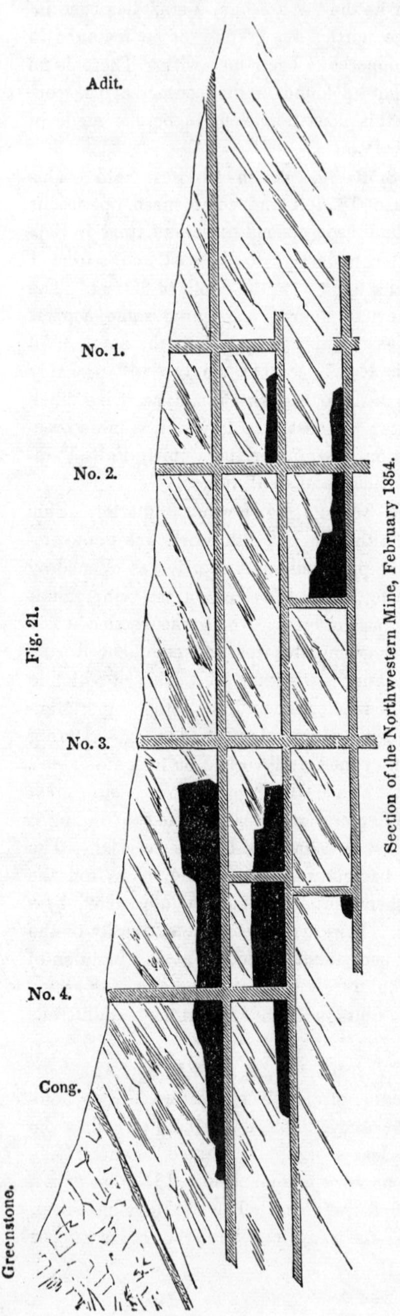


Fig. 21. Section of the Northwestern Mine, February 1854.

From a letter of the Superintendent, dated February 28, 1854, I learn that the average width of the vein in all the parts of the mine now working, is three feet; in the twenty-fathom level, north of the winze, between shafts 3 and 4, it is four feet wide, and rich in stamp and barrel-work. In the ten-fathom level, driving to the north of shaft No. 4, the vein is of the same dimensions; and a mass of copper, about twelve feet in length and three feet high, has been discovered here. In the adit-level, north of shaft No. 4, the vein is also very wide, being in some places over ten feet, and well filled with copper.

Two steam-engines have been erected at this mine. One of them is used for stamping, and is capable of running twenty-four heads. Eight are now at work, and the number will be increased to sixteen immediately. The other engine, of twelve inches diameter of cylinder and four feet stroke, is intended for winding and pumping at shaft No. 4.

The shipments from the mine were as follows:—

| | 1852. | 1853. |
|---|---|---|
| Ore shipped, . . . . . . . . | 13,836 lbs. | 61,165 lbs. |
| Produce, fine copper, . . . . . . | 8,622 | 44,166 |
| Percentage, . . . . . . . . | 62·3 | 72·2 |

*Winthrop Mining Company.* T. 58, R. 31, Sect. 23, southwest quarter. This location is supposed to have upon it the Hill Vein, but it has not yet been developed to any extent. Nothing was doing there at the time I visited it, in 1853; but work has since been resumed on what is supposed to be the Hill Vein, with fair prospects of success, as is stated.

*Eagle River Mining Company.* T. 58, R. 31. This company is engaged in exploring on sections 28, 32, and 33; no mine-work of any extent having been yet commenced.

*Pittsburgh and Boston Mining Company. The Cliff Mine.* T. 58, R. 32, Section 36, southeast quarter. The discovery and opening of this mine formed an era in the history of Lake Superior, and are also of high interest to the country, as it was the first mine in the United States, those of coal and iron excepted, systematically and extensively wrought, and at the same time with profit. Besides this, it has a peculiar importance as being opened on a vein bearing copper exclusively in the native state, a feature entirely unknown in the history of mining previous to the discoveries on Lake Superior.

The history of the discovery of the Cliff Mine and the previous operations of the company is briefly as follows:*

During the summer of 1843, a Mr. Raymond made certain locations in the Lake Superior region, for which he obtained leases, three of which he disposed of to parties in Pittsburgh and Boston, who commenced mining in the summer of 1844. The first location made was at Copper Harbor, where the outcrop of a cupriferous vein, on what is now called Hays's Point, was a conspicuous object, known to the *voyageurs* as "the green rock," and had given a name to that beautiful harbor long before it became the centre of the copper excitement. A little work was done here in the autumn of 1844; but on clearing away the ground on the opposite side of the harbor, where Fort Wilkins now stands, numerous boulders of black oxide of copper were found, evidently belonging to a vein near at hand, which was discovered in December and proved to be a continuation of the one before worked on Hays's Point.

Mining was commenced here immediately; two shafts were sunk, about 100 feet apart, and considerable black oxide of copper taken out, mixed with the silicate. This was very remarkable, as it is thus far the only known in-

* See Foster and Whitney's Report, Part I. p. 128; also letter of Thos. Jones, in a Report of the Committee of the Stockholders. Boston, 1847.

stance of a vein containing this as the principal ore, or in any other form than as an impure mass, mixed with the sulphuret of copper, and oxides of iron and manganese, and resulting from the decomposition of the common ore, copper pyrites. This proved, however, unfortunately, to be only a rich bunch in the vein, of limited extent, and which gave out at a depth of a few feet, although the fissure continued. The workings were entirely confined to the conglomerate, which at that time was supposed to be as favorable to the development of the vein as any other rock. The gangue associated with the black oxide was principally calc. spar, and some argillaceous and quartzose matter intermixed. Fine crystals of analcime were found connected with it. Crystallized red oxide and native copper were also obtained in fine specimens. About 30 or 40 tons of black oxide were obtained in all, and sold for $4500. The main shaft was continued down 120 feet, and levels driven each way for some distance, without striking another bunch of ore, so that in 1845 the attention of the company began to be turned to exploring their extensive property, and in August of that year the Cliff Vein was discovered, by a party of explorers under the direction of a Mr. Cheny.

This vein was first observed on the summit and face of a bluff of crystalline trap, rising with a mural front to the height of nearly 200 feet above the valley of Eagle River at its base. The break or depression made by it in the back of the ridge was quite distinct, and has since been traced to the Lake and found marked by ancient excavations. At the summit of the bluff, as I saw it, a few days after its discovery, it appeared to be a few inches wide, and contained native copper and specks of silver beautifully incrusted with capillary red oxide, with a gangue of Prehnite. Half-way down the cliff, it had expanded out to a width of over two feet, and consisted of numerous branches of Laumonite, with a small percentage of metallic copper finely disseminated through it. Of course, at this time, nothing whatever was known of the varying character of the lode in different belts of rock, nor had the trap been supposed by the miners to be the principal metalliferous rock. It is now known that the vein could not be worked with profit in the rock in which it was discovered, namely, the crystalline trap or greenstone, as no vein has yet proved sufficiently metalliferous in that belt of rock to be profitably mined.

Without knowing anything of the entire change in the character of the rock which takes place at the base of the cliff, where there was a heavy accumulation of fragments of rock dislodged from above, and suspecting as little as any one else the unprecedented discoveries about to be made in the metalliferous bed beneath, I advised the clearing away and opening of the vein at as low a point as possible, because it appeared to widen out and improve in depth. A shaft was sunk a few feet, a little below the edge of the bluff, and a level driven into the greenstone a short distance, but nothing was done of importance, until the talus at the base of the cliff was cleared away, and the vein traced into the amygdaloid. A level was then driven in upon it, and, at a distance of 70 feet, the first mass of copper was struck, a discovery of the greatest interest, since it revealed the presence of a metalliferous belt whose existence had not before been suspected, and showed the extension of the

Plate II.

SECTION OF THE CLIFF MINE.

Feb 1854

lodes of Lake Superior into belts of rock of different lithological character, and the variations in richness attendant on such transitions.

On inspecting the accompanying section of the mine, Plate II., it will be seen that the beds of rock dip at an angle of about 28° to the north, consequently the extent of ground in that direction, the mine being opened to the south of the greenstone, is constantly increasing as each successive level is extended northwards from a greater depth. To the south the mine is limited by the line of the Company's property, the exact position of which, singular as it may appear, had not been determined last summer. The longest level has been extended 600 feet to the south of shaft No. 1, and in the same level (No. 6, 244 feet below the adit), the supposed distance to the crystalline trap is 900 feet, and each level is extended about 100 feet farther than the one 66 feet above. In fact, the addition to the interval from the shaft to the greenstone should be a little more than a hundred feet, from level to level, since the belts of rock form parts of great curves, not sensibly differing from straight lines in a short distance, but whose divergence must be quite perceptible in a perpendicular descent of 500 feet.

The mine was, until recently, worked chiefly through two shafts which were less than a hundred feet apart, thus very much retarding its progress. During the last winter, a third shaft (No. 3) has been extended from level No. 1 up to the surface at the upper edge of the bluff, a distance of 138 feet, and a winding engine is to be placed there, which will, for the present, suffice to elevate the stuff from that part of the mine. The engine-shaft is now sunk to the ninth level, 444 feet below the adit, and the distance from the collar of shaft No. 3 to level No. 9, will be 630 feet.

Owing to the want of the necessary shafts and winding machinery, the mine is opened but little in advance of the stoping, and much loss must have been incurred for want of room to carry on the sinking and driving without interrupting the stopers. The remarkable and uniform richness of the vein may be inferred from the fact, that no part of it is so poor as not to be worth taking down, and so far as the work has been carried, hardly a fathom of ground has been left standing in it. On calculating the number of fathoms of the vein removed in the drifts, shafts, and stopes, I find it to be, approximately, 8270; and there has been produced an average amount of 761 pounds of copper per fathom, a result which is truly astonishing, when it is considered that the *whole* of the vein has been taken down.

The vein is remarkably regular in its course, which is about north 27° west, and its underlay is about 10° to the east. In the lower levels its dip is more varying. Some parts of it expand to three or four feet in width, other portions are pinched up to a few inches, but its average width is probably from 15 to 18 inches. The veinstone is principally quartz, calc. spar, and the zeolitic minerals, and is characterized by an abundance of finely-crystallized minerals, of which drusy quartz is the most common, and found at this mine in finer specimens than anywhere else on the south shore of the Lake: it is associated with apophyllite, Prehnite, and calc. spar, in various crystalline forms.

The metallic contents of the veinstone consist exclusively of copper and silver, both only in the native state. The copper occurs here, as in some other mines of this region, in masses of great size, from a few hundred pounds up to nearly 100 tons, and the vein is not only rich in these, but also furnishes a large quantity of stamp-work, containing an unusually high percentage of copper. The relative amount of the various kinds of mineral raised from the mine, for five years in succession, may be seen from the following table:—

| Years. | Masses est. at 60 per cent. | Barrel-work at 50 per cent. | Stamp-work at 5 per cent. |
|---|---|---|---|
| 1848, . . . . . . . | 1,209,852 lbs. | 486,487 lbs. | 3,879,392 lbs. |
| 1849, . . . . . . . | 1,077,884 | 566,314 | 5,584,500 |
| 1850, . . . . . . . | 710,046 | 482,322 | 6,145,000 |
| 1851, . . . . . . . | 836,409 | 515,462 | 6,114,000 |
| 1852, . . . . . . . | 877,789 | 705,421 | 4,761,000 |

No winding-engine was erected until Sept. 1850, a fact which sufficiently explains the falling off which took place in that year, the mine having become choked up with rubbish. Two such are now at work, or will be at an early period. There is also a fine engine, of 24 inch cylinder and 6 feet stroke, which drives 30 heads of stamps and does the pumping.

The operations of the mine may be seen by reference to the annexed table, in which is given as complete a statement of them as can be made out from the published reports and such additional information as the Secretary has very politely furnished me:—

| | Expended. | Stuff shipped. lbs. | Yield. Per cent. | Yield. Amount lbs. | Amount received. | Dividends. Per share. | Dividends. Amount. |
|---|---|---|---|---|---|---|---|
| 1844, . . . | $3,066 76 | | | | | | |
| 1845, . . . | 25,574 40 | 33,171 | 60 | 19,903 | $2,968 70 | | |
| 1846, . . . | 66,128 05 | 108,774 | 38·1 | 37,625 | 8,870 95 | | |
| 1847, . . . | 89,409 16 | 729,848 | 56·3 | 410,783 | 70,977 32 | | |
| 1848, . . . | 105,278 52 | 1,655,304 | 60·2 | 996,467 | 166,407 02 | | |
| 1849, . . . | 106,968 77 | 2,285,050 | 56·1 | 1,282,127 | 155,227 04 | $10 | $60,000 |
| 1850, . . . | 118,784 14 | 1,521,391 | 46·9 | 714,643 | 177,044 36 | 14 | 84,000 |
| 1851, . . . | 127,623 51 | 1,528,465 | 55·3 | 846,486 | 174,931 96 | 10 | 60,000 |
| 1852, . . . | 112,929 81 | 1,660,330 | 49·9 | 829,356 | 161,917 08 | 10 | 60,000 |
| 1853, . . . | . . . . . | 2,062,958 | 45 | 929,615 | | 15 | 90,000 |
| 1854, . . . | . . . . . | . . . . . | . . . | . . . . . | . . . . . | 10 | 60,000* |

The amount of silver obtained from the Cliff Mine has sometimes been quite considerable. Early in the history of its operations, a great excitement was raised on the subject of the occurrence of this metal in the vein, a rich pocket of it having been met with; but experience has uniformly shown that in the Lake Superior veins these bunches of silver are of limited extent, and cannot be depended on as a permanent source of revenue. The mode of occurrence of the silver is precisely similar to that of the copper; like that metal it is found only in the native state. It occurs intimately united with the copper, being, as it were, soldered to it, or forming blotches and specks within it; but the two metals are never found alloyed together, each being almost chemically pure, even when associated thus closely.

The most usual mineral accompanying silver in the veins is a greenish, magnesian substance, which appears to be talc; in the Hill Vein, where this metal is found in considerable quantity, talc is one of the prominent vein-mine-

* First semi-annual dividend.

rals. At this latter mine were found the only crystals of silver which have come under my notice from this region; they were cubes, of about one-eighth of an inch across their faces. The argentiferous portion of the lodes seems to be, in general, near the plane of contact of two beds of different lithological character. The largest deposit of this metal yet noticed was in the old Copper Falls Mine, in the upper level in the trap, at a distance of a few feet only from the sandstone; but, according to Mr. Hill, in the lower levels no silver was noticed. He remarks also that the same is true of the Hill Mine, the quantity observed at some depth being much less than above. Hence he infers, that the veins of this region cannot be successfully mined for silver. A small percentage of this metal is found in the smelted copper, a few ounces to the ton, but not enough to justify its separation.

The silver rarely forms lumps of more than a few ounces in weight, although some pieces weighing several pounds, and nearly pure, have been obtained. Unfortunately, such pieces are often looked upon by the miners as their especial property, and the amount received by the companies from this source is considerably less than it ought to be.

The annexed is a statement of the amount of silver obtained at the Cliff Mine; it is mostly picked out by hand from the coarse metal which is taken out from under the stamp-heads.

| Years: | 1846. | 1847. | 1848. | 1849. | 1850. | 1851. |
|---|---|---|---|---|---|---|
| Silver obtained, in lbs. troy, . . . . | 24·16 | 32·5 | 81·25 | 24·75 | 239· | 34·83 |

*North American Mining Company.* Old Mine, T. 57, R. 32, Section 2, east half; New or South Cliff Mine, on Section 1, northeast quarter.

This is also one of the oldest companies on Lake Superior, and has mined extensively at two localities. The "Old North American Mine," so called, was opened in 1846, and worked until the spring of 1853, at which time the workings had reached the depth of 415 feet. The course of the principal vein is north 58° west, and it is therefore not parallel with the productive lodes of the vicinity, which are found to be nearly at right angles with the line of bearing of the formation. There were, however, three lodes here, which were supposed to unite, No. 2 falling into the main lode near the 155 foot level, and No. 1 at a depth of 275 feet. Never having had an opportunity to make a thorough examination of this mine, I am unable to state all the facts connected with it. But it appears that the lode was from the beginning very irregular and variable in width, and that although, in some places, it carried copper, occasionally in masses of considerable size, yet it never presented a sufficiently favorable appearance to encourage any strong hopes that it could be worked with profit. The produce, however, was sufficient to justify the managers in going on, in the hope of meeting with a main vein, or champion lode, having the regular course of the veins of the country, into which the other veins should fall as branches.

The produce of the mine was as follows:—

| 1849. | 1850. | 1851. | 1852. | |
|---|---|---|---|---|
| 77,000 | 256,000 | 257,000 | 77,000 lbs. | Yielding on an average, 66·8 per cent., or 446,000 lbs. fine copper. |

The entire sum expended on this mine, including, however, the cost of very extensive surface improvements, and of the land purchased by the company, 2400 acres, was about $200,000.

It is a curious fact, and one which shows that confidence in a mining region is a plant of slow growth, that although this company owned the line of the famous Cliff Vein up to a point within a stone's throw of where it was producing most abundantly, yet it was not until the summer of 1852 that the attempt was made to open it upon the North American Company's territory. One reason of this delay was, undoubtedly, the belief in the very great thickness of the drift in the valley of the Eagle River, through which it would be necessary to sink before the vein could be reached; nothing was known with certainty respecting it, but it would not have been a difficult operation, at least, to bore down and settle the matter. On actually sinking a shaft, it was found that it passed through from 8 to 12 feet of sand, and then through quicksand, until the stratum of *hard-pan* was struck lying directly on the rock, which, itself, was 52 feet from the surface. All difficulty in sinking here, from water, would have been obviated by constructing the shaft of boiler-iron, and in sections. On striking the rock, a shaft was sunk 22 feet, and on cross-cutting at that depth, a vein was found at a distance of 14 feet from the shaft, $3\frac{1}{2}$ feet wide, and showing all the character of the Cliff Vein itself, which it was.

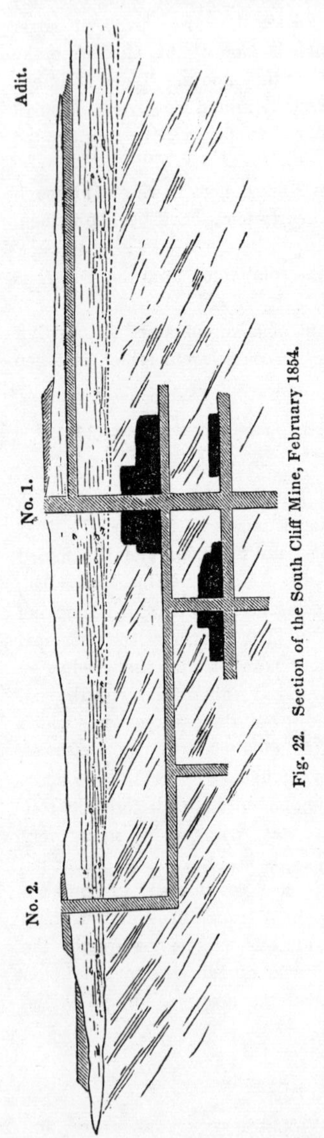

Fig. 22. Section of the South Cliff Mine, February 1854.

The annexed section (Fig. 22) represents the state of the work on this vein at the present time (Feb. 1854). The whole amount of ground opened in driving is 715 feet; cross-cutting, 76 feet; sinking in rock, 438 feet; stoping, 106 fathoms. From these workings the extraordinary amount of 506,000 lbs. of stuff, yielding $67\frac{1}{4}$ per cent. of copper, was taken.

The largest mass ever yet observed on Lake Superior was thrown down in this mine on July 4, 1853; it was about 40 feet long, 20 high, and supposed to average 2 in thickness. Its weight was estimated at from 150 to 200 tons. At the time of my last visit to the mines the cutting of this magnificent mass had just commenced, and was expected to occupy several months; what the precise weight proved to be I have not been able to ascertain.

The veinstone of this mine, near the surface, furnished beautiful specimens of Prehnite with crystallized copper. An interesting occurrence of the former mineral in radiated nodules in perfectly pure metallic copper, as represented in the annexed cut (Fig. 23), is well worthy of notice, as throwing light on the origin of the metal in the veins.

Fig. 23.

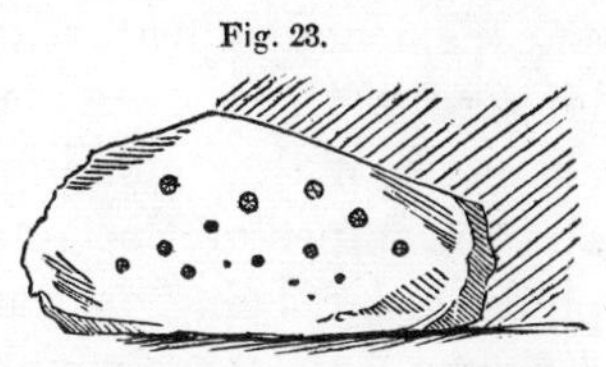

Mass of copper containing nodules of Prehnite.

As the workings of this mine are pushed to the south of the greenstone farther than in any other mine in the same geological position, it will be a matter of great interest to notice the changes which take place in the character of the vein as it is opened in that direction. At shaft No. 1, it was unprecedently rich; on driving towards No. 2 the ground became poorer, and continued so for some distance, but at the time of my last visit, it was improving rapidly. In the 2d level, the vein is said to be much better than above.

Two steam-engines are in operation, one of which is employed in pumping and winding, and the other in stamping. The company has also an extensive farm, on the products of which a profit was made, in 1853, of about $3500.

*Albion Mining Company.* T. 57, R. 32, Section 11. This company is at present located on Portage Lake; but operations were carried on by them at this Point for several years. The vein is in the same geological position as that of the Cliff Mine, but is barren of copper. A shaft was sunk in the greenstone, and continued through the conglomerate, which it intersected at 115 feet in depth, into the amygdaloid, where the vein was found to have been thrown 70 feet to the east. The shaft was sunk to nearly 200 feet below the adit, without any encouragement; the vein, which was about 2½ feet wide, being of the most unpromising character. The work was abandoned in 1852.

*Fulton Mining Company.* T. 57, R. 32, Section 33, southeast quarter. This was formerly known as the Forsyth Mine, and was abandoned in 1847, without having been sufficiently proved. The work was resumed early in the season of 1853, by the present company, and has been pushed with uncommon energy. The vein has a course of north 9° to 11° west, and dips to the east at an angle of about 75°. Its width, where best developed, is from one to two feet. The adit-level is driven in about 500 feet, at which distance it intersects shaft No. 3, at a depth of 88 feet below the surface. Above the adit a few fathoms of ground have been stoped out, and produced largely. To the south of shaft No. 3, the vein is intersected by a slide, which dips to the north, and has thrown the vein out of its course, so that it has not yet been found in that direction. Shafts No. 2 and 3 have been sunk to the ten-fathom level, and a rich block of ground will be opened between them. Between shafts 3 and 4, above the adit, the lode was found to be very rich, producing nearly a ton of copper to the fathom. Several masses, weighing from 500 lbs. to a ton, have been taken out.

The veinstone, unlike most of those of Keweenaw Point, contains considerable epidote, which is mixed with calcareous spar. An unusually large amount of silver was obtained from this mine in the superficial workings.

Beyond the Fulton Mine, the distinction between the crystalline trap, or greenstone, and the amygdaloid, which is so conspicuous a feature of the geology of the Point, can no longer be traced. A marked change takes place in the character of the metalliferous deposits within a few miles, and the mines of Portage Lake, which are next in geographical order, being only about twelve miles distant, are quite different from those which have hitherto been described.

Before passing to the next division, however, it may be mentioned that there are also, on Keweenaw Point, some important veins which have not yet been the object of exploration by companies; they are usually known each by the name of its discoverer, or of the preemptor of the quarter-section on which it is situated. One or two of the principal ones are mentioned below.

*Manhattan Location.* T. 58, R. 28, Sections 14 and 15. Two lodes have been opened on the surface in Section 14, with a fair show of copper. No regular mining has been commenced. The western lode is three feet wide, and consists of the usual veinstone of the productive veins of the region, with small particles of copper finely disseminated through it.

*Montreal Location.* T. 58, R. 28, parts of Sections 8, 9, and 17. This has not yet been organized into a company. There are several veins upon the location, but none of them have been opened to any extent. The "Montreal Vein" appears to be the continuation of the "Black Oxide Vein," formerly worked at Copper Harbor. It is from one to two feet in width, and has an excellent-looking veinstone, carrying considerable fine copper disseminated through it. On the northeast quarter of Section 8, is another vein, which has been opened at various points along a line 1800 feet in length, showing a width of 12 inches, and a small percentage of copper. This property is worthy of a thorough trial. The "Clark Vein" is near the centre of Section 8, and has been opened in the greenstone, where it is two feet wide, and well filled with copper: pieces of several pounds weight having been found.

## ISLE ROYALE.

Having thus given an account of the most important localities or mines of copper on Keweenaw Point, we will next take up Isle Royale, which is, in many respects, the counterpart of the region just described. The ridges of trap traverse the island longitudinally, and this rock, with occasional intercalated belts of conglomerate, forms the whole island, with the exception of a part of its southwestern end. The trap all belongs to the bedded class, and contains the same metalliferous products as on Keweenaw Point. The strata have, however, a dip which is just the reverse of that of the rocks on the other side of the Lake, and their mural faces are turned to the north. In one respect, there is a great difference between the bedded trap of this island, as well as of Michipicoten, and that of the southern shore; it is in much thinner masses, there being none of those thick beds in which the great mines of the other side are worked. As a consequence, the metalliferous lodes are liable to frequent changes, passing through a variety of different beds, and exhibiting a different character in each. This has formed the great drawback to the development of the cupriferous veins, which are very numerous upon the island.

The most extravagant notions formerly prevailed with regard to the richness of Isle Royale in copper, and soon after the opening of the Lake Superior region, nearly the whole island was taken possession of by different companies, and operations were commenced at numerous points. All the mines opened were on or near the shore, as the physical difficulties of exploring the interior are such as to render it almost impossible that important discoveries should be made at any distance inland.

There are numerous veins which, like those of Keweenaw Point, traverse the formation nearly at right angles to it, and others which are parallel with it, resembling in their main features the veins of the Ontonagon district, but occasionally dipping in the opposite direction to that of the beds of rock. Epidote belts occur here which are filled with fine particles of native copper; but they have not been found sufficiently persistent in their metalliferous character to be worthy of being worked.

On visiting the island in the summer of 1853, I found nearly all the mines abandoned. At two of them only, operations were still continued on a small scale.

*Siskawit Mining Company.* T. 66, R. 34, Section 13, southwest quarter. This mine has been quite extensively worked, and yielded considerable copper in the productive rock in which it was first opened. The annexed section (Fig. 24), although not strictly accurate, will convey a sufficiently clear idea of the extent of the mine, and the position of the rocks in which it is situated. It represents nearly the state of the workings in the summer of 1853. In the bed of rock indicated by *a* on the section, the vein yielded considerable copper, in stamp-work and masses, as indicated by the stoping; but, unfortunately, the workings could only be extended with profit by carrying them under the lake, as the dip of the rocks is in that direction; and on sinking but a short distance, a stratum of hard, compact, basaltic trap (*b*), 15 feet thick, was encountered, on entering which, the lode contracted to a mere fissure, and was entirely barren of copper. The principal shaft has been sunk to the 60-fathom level below the adit, during the past winter, in the hope of intersecting a more productive belt of rock. The vein in the bed *b* was so pinched up as to be hardly visible, but the walls remained good, so that it could be followed. Below, in *c*, the rock was softer and less crystalline, with occasional amygdules of calc. spar. At the point indicated on the section as the bottom of the shaft, the vein was about one foot wide, of which three inches were good veinstone, consisting of quartz and chlorite, and carrying a little copper in fine scales. I learn from the Superintendent, M. Curnow, Esq., that in sinking the last 120 feet the vein has improved considerably, but that farther change for the better would be requisite to put the mine in a paying condition.

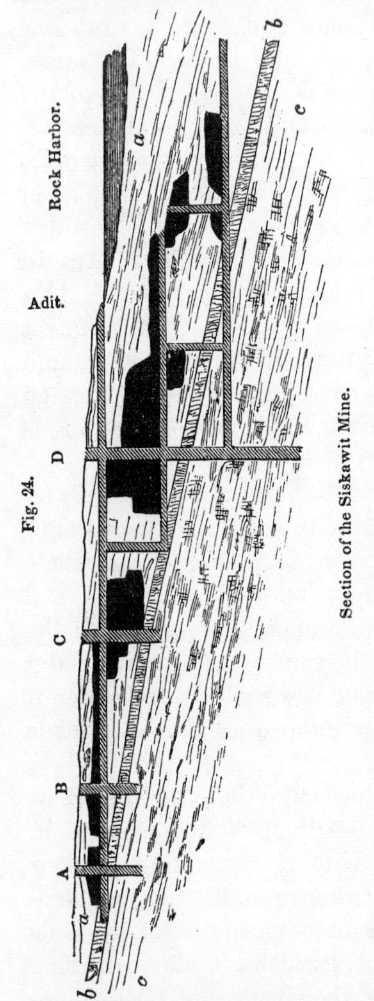

Fig. 24.

Section of the Siskawit Mine.

The North Vein, so called, a short distance from the one just described, averages about 14 inches wide, and is well filled with copper.

There is a steam-engine at the Siskawit Mine, which pumps, and drives 12 heads of stamps. It was employed during the last summer in stamping the rock formerly rejected as halvans, working up 350 tons per month. From this quantity of rock about 1 ton of 90 per cent., and ½ ton of 50 per cent. copper was obtained, at an estimated expense of $400.

I have been unable to ascertain the exact shipments of stuff from this mine, but, according to official information, it appears that $16,000 had been received from sales of copper up to January 1852, and, together with $68,000 raised by assessments, expended on the mine. In 1852, about 80,000 lbs. of stuff, averaging 78 per cent. of copper, were shipped; and in 1853, 38,437 lbs., which was probably of about the same richness.

*Pittsburgh and Isle Royale Mining Company.* T. 65, R. 36, Section 12, northwest quarter. The principal workings of this company have been on a vein, near the lake, which is contained in a hard, crystalline trap, and is quite narrow, but rich in copper, considering the nature of the rock in which it occurs. It bears north 20° east, and has a width, when best developed, of about 18 inches. Shaft No. 2 is 135 feet deep, not having been continued any farther on account of the miserably defective machinery for removing the water. Shaft No. 1 is of the same depth, and the two are connected by a level 113 feet in length. Above this level, considerable ground has been stoped out, and also in the vicinity of shaft No. 2; and from 14 to 15 tons of stuff, mostly in sheets, and nearly pure copper, sent to market.

A shaft has also been sunk, on the hill about one mile south of the present mine, to the depth of 223 feet, on a wide vein, but destitute of copper.

## ONTONAGON MINING DISTRICT.

Following a purely geographical division, the district of Portage Lake should have been taken up next to that of Keweenaw Point, but as the metalliferous deposits of that region are quite distinct in character, and form a group by themselves, they will be discussed after those of the Ontonagon; in this way a regular gradation in the character of the mode of occurrence is made the basis of the classification, each district succeeding the one with which it has the greatest analogy in this respect.

The region in which are situated the mines which are about to be described, takes its name from the principal river which drains it, the Ontonagon. This stream has three branches, one of which flows from the east, another from the west, and the third from the south, and, uniting nearly at the same point, they cross the trap range at right angles to its course, and furnish a tolerable means of communication between the mines and the Lake, which might be much improved at a small expense. The mines are situated on the trap range, and are worked at various points for a distance of twelve miles on each side of the river, making the whole length of the district about twenty-four miles. Between the most northeasterly mine of the district, the Douglass Houghton, and the mines on Portage Lake, there is a distance of twenty-five miles, in which no mining operations are now carried on, with, perhaps, one exception. The trap range, in that part of its course, is much broken up into small knobs, and is almost entirely concealed by drift. The explorations which have been made in that part of the range, seem to indicate that there are no veins there which can be profitably worked. To the west, the limits of the really valuable part of the range are not yet defined; I have seen no veins which seemed to me sufficiently well-developed to be profitably worked, west of the line between townships 41 and 42; and, indeed, in the Geological Report on that region, the lands were not marked as mineral beyond a point three miles west of the line between townships 40 and 41. Farther explorations may reveal the presence of metalliferous deposits of value in the region extending west to the Montreal River, but such remain still to be discovered.

The mode of occurrence of the cupriferous deposits of this region differs materially from that exhibited by the veins of Keweenaw Point. They are characterized by a constant parallelism with the line of strike of the formation, although sometimes occurring in deposits in which no particular direction can be perceived.

The character of the trappean rocks is somewhat different in this district from what it is on Keweenaw Point. The varieties of rock are more numerous, and epidote becomes a frequent associate both of the rock and the veins, almost always occurring where copper is found. West of the Ontonagon, a large part of the range on the north is made up of a reddish quartzose porphyry, which appears to be entirely barren of copper in any form. Intercalated in the trap are frequent beds of conglomerate, which are usually quite thin, and to the north it is flanked, as in Point Keweenaw, by heavy beds of this rock. There is no marked belt of unproductive crystalline rock extending through the Ontonagon region, and the position of the veins and deposits is not so well ascertained with reference to any fixed line of upheaval.

Copper is apparently quite as abundantly diffused through the Ontonagon district as anywhere else; but thus far, the workings have not been attended with so great a degree of success as on Point Keweenaw, mainly because there seems to be less concentration of the metal within limited spaces, and into workable veins.

The copper of this region occurs in four forms of deposit.

1st. Indiscriminately scattered through the beds of trap.

2d. In contact deposits between the trap and sandstone or conglomerate.

3d. In seams or courses parallel with the bedding of the rocks, and having the nature of segregated veins.

4th. In true veins coinciding in direction with the beds of rock, but dipping at a different, and usually a greater angle, in the same direction with the formation.

Deposits of the first-mentioned class occur frequently, and have been worked in several places, but with poor success. Were the copper, which is thus scattered through the rock without law or rule, collected into a reasonable number of veins, it might be profitably worked, but, as it is, it seems almost a hopeless case at present. Masses of many hundred pounds weight have been repeatedly found in the trap, without any connection with a vein-fissure, and sometimes unaccompanied by veinstone, although this is not usually the case. The particles of copper, when smaller, often fill amygdules in the trap, and are usually somewhat more concentrated along the line of junction of two beds of different character, indicating a tendency to an arrangement of the character of the second class above indicated.

Contact deposits have, in some instances, become of considerable importance in this district, although their permanent value, as it seems to me, can only be tested by deep workings. When they occur between the sandstone and the trap, as illustrated in the annexed section (Fig. 25), which represents a locality formerly worked by the Isle Royale Company, in the Porcupine Mountains, they appear to be of little value. In such cases they are soon found to lose their metallic contents, on being opened at any depth. This was the case at the point illustrated by the figure, in which A represents a seam of blue plastic clay and chlorite, with fragments of the walls, the whole forming a sort of breccia, and containing, near the surface, considerable native copper. At the Union Mine, in the vicinity, a chloritic bed having a similar position was worked to a considerable depth, but its metallic contents, which were not large in quantity at the surface, diminished rapidly in depth.

Fig. 25.

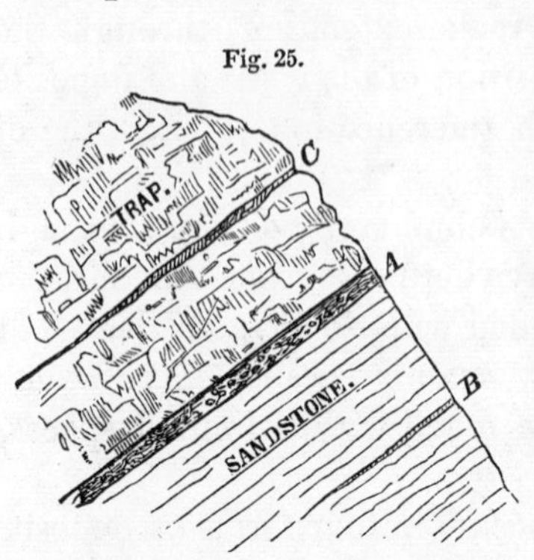


A, Contact deposit between trap and sandstone. B, Parallel deposit in the sandstone. C, Seam of calc. spar in the trap.

The deposits of copper which occur between the trap and the conglomerate in this district, are much more difficult to classify. They appear to belong to the contact deposits, but have some of the characteristic features of true veins. Immense masses of copper are accumulated in this position near the surface, and even at a considerable depth, and it remains to be seen what degree of persistence they will exhibit at a still lower point.

The third class of deposits, that in seams parallel with the bedding of the rocks, or in segregated veins, is one peculiar to this district. In several of the mines the vein is wont to be found irregular in its course, suddenly heaved to one side or the other, or disappearing altogether, and lacking throughout some of the principal characteristics of true veins; in such cases, I regard the metallic matter as being accumulated, with more or less veinstone, in parallel courses, which coincide with the bedding of the rocks, but are irregular in respect to their extent and the distribution of the metalliferous and mineral matter in them. This mode of occurrence may be illustrated by the annexed section (Fig. 26), which was taken near the Douglass Houghton Mine. The parallel layers of veinstone are represented by *a*, *a*, *a*, and the rock adjacent contains a large amount of epidote, indicated by the oblique lines. These metalliferous belts are not always strictly parallel, but occasionally run into each other, both horizontally and vertically, and thus give rise to the so-called feeder-veins; frequently they diminish to a mere seam, destitute of veinstone or metal, and then the vein is supposed to have been thrown to one side or the other, and on cross-cutting another seam is struck, and perhaps found to be well filled with copper, and it is then supposed that the same vein has been recovered. I am not prepared to say that deposits of this kind may not in some instances be profitably worked, but great caution must be observed in expending money upon them.

Fig. 26.

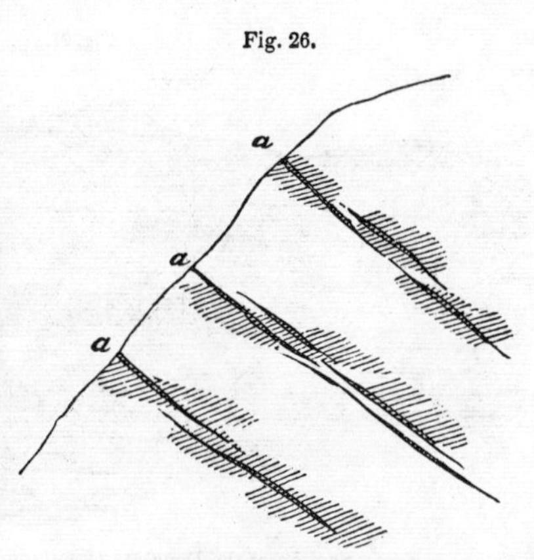


Parallel layers of cupriferous veinstone, near the Douglass Houghton Mine.

Metalliferous deposits occur in this district which have most of the characteristics of true veins; they do, indeed, coincide with the line of bearing of the rocks, but in their downward direction they are found to be wholly independent. Such veins are often rich in copper, and may be worked with almost certainty of success; unfortunately, their number in this district appears to be quite limited, although their longitudinal extent is considerable.

The above enumerated forms of deposit will be sufficiently illustrated in speaking of the particular mines, to which we now turn our attention.

*Algonquin Mining Company.* T. 52, R. 37, Section 36. Considerable work has been done here at various times, and in various places on the Section.

In 1848, there was at one point an open cut, about 35 feet in length and 10 feet deep, from which a considerable quantity of copper had been taken.

Fig. 27.

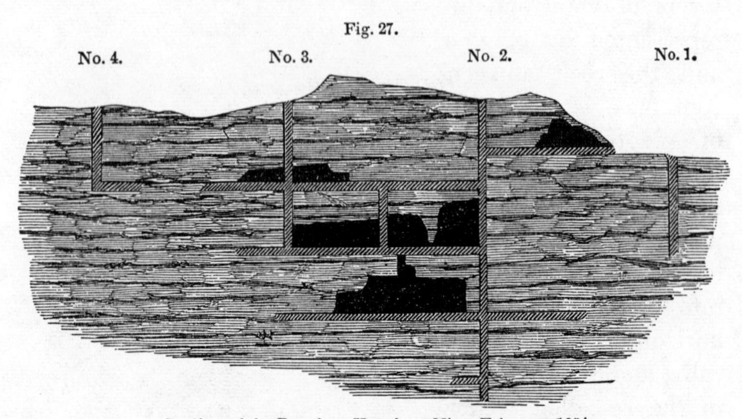


Section of the Douglass Houghton Mine, February 1854.

*Douglass Houghton Mining Company.* T. 51, R. 37, Section 15, northwest quarter. Operations were commenced here on a small scale in 1846, but it was not until 1850 that the work was vigorously pushed. The extent of the mine will be seen by referring to the annexed section (Fig. 27), which represents the state of the works, February 1st, 1854. The vein, at the surface, appeared between two and three feet wide, with a quartzose veinstone, and was quite well filled with copper. At that point, it had two perfectly defined walls, separated from the rock by selvages of argillaceous matter, and a gangue distinct from the rock. In the workings of the mine, however, it has not so much regularity as would be desirable: in some places it is two feet in width, and well charged with copper; and in others it becomes entirely lost, and can with difficulty be traced. There is a break or fault intersecting the vein vertically between shafts 2 and 3, and displacing it to the amount of 14 feet. Two slides have also been found to traverse the rocks, and have shifted and deranged the vein along their course. During the past winter, the appearance of the mine is said to have improved materially, and I learn from Mr. Coulter, the Superintendent, that there were in February last 2500 fathoms of ground ready to be stoped, in most of which the lode is wide, and well filled with barrel and stamp-work.

There is a stamping-mill near the mine, with eight heads, moved by water power.

About 5 tons of barrel-work and mass copper were shipped in 1853, and 120 barrels of stamp and barrel-work, and four masses, amounting to 25 tons in all, were at Ontonagon, ready for shipment, in February 1854.

*Toltec Consolidated Mining Company.* T. 51, R. 38, Section 25, south half. This is a consolidated company, formed on the Farm and Toltec locations. Mining was commenced in 1850, and at the present time a large amount of ground is opened for stoping, and the excavations are quite extensive, as will be observed by examining the annexed section (Fig. 28), which represents the amount of work done up to March 1854. The deepest shaft is down 210 feet. The No. 1 shaft, on part of the vein formerly belonging to the Farm Mine, was found to have been sunk on what is thought to be a "feeder" to the main vein, and the level between that shaft and No. 2 is driven on the same. A cross-cut, 66 feet in length, from this level, intersects the main vein, as it is considered. The character of the lode varies very much in different parts of the mine. At shaft No. 4, it dips 64°, and is about 3 inches wide; at No. 3, it has an inclination of 56°, and is 20 inches wide, and well filled with copper, a mass of a ton in weight having been found here. At other points it is from 2 to 3 feet wide, and very variable in richness. Indeed, so irregular is the distribution of the metal in the vein, that it would be quite hazardous to make any assertions as to the future prospects of the mine. The work thus far has been executed with judgment, and, if sufficient care is taken to avoid excavating the poor ground, and not to allow the rich part of the vein to escape notice, it seems not unlikely that operations may be conducted with profit.

Fig. 28.

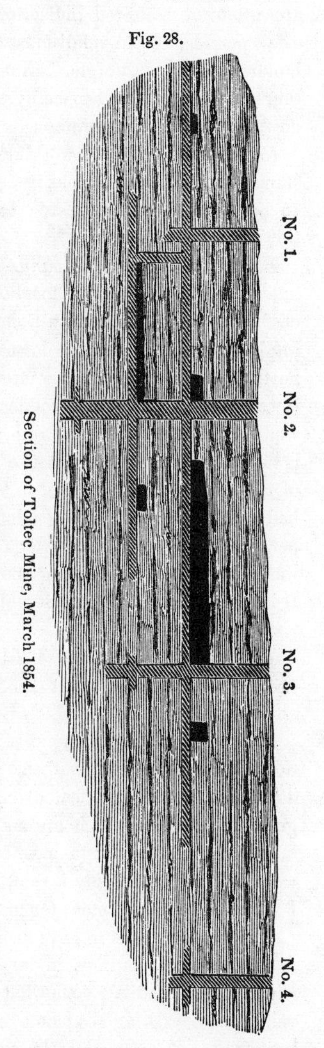


Section of Toltec Mine, March 1854.

The gangue of the vein is almost exclusively quartz, often well crystallized, and occasionally associated with fine specimens of Prehnite.

Stamps have not yet been erected, but it is understood that they soon will be. A small shipment of masses and barrel-work was made in 1853, amounting to about 11,000 pounds.

*Algomah Mining Company.* T. 51, R. 37, Section 30, west half. This location adjoins the one just noticed on the east, and it is said that the same vein has been opened upon it during the past winter, and found to be well-developed. There was nothing doing on the property at the time I was there, in 1853.

*Aztec Mining Company.* T. 51, R. 37, Section 31, west half. There is no regular vein or appearance of one at the point where workings have been carried on by this company. A large excavation has been made in a bluff of a very irregular shape, about 20 feet high, and descending at a steep angle, but it was filled with water at the time of my visit. The bluff was found to have been very extensively worked over by the ancient miners, and the copper appears to be scattered indiscriminately through the rock, mostly in lumps of a few pounds, and in small masses, sometimes, however, amounting to several hundred pounds in weight. Finely crystallized minerals are found in connection with the copper, especially calc. spar, in various forms, among which the dog-tooth is the most common.

As the result of these workings at random in the rock, about 4 tons were shipped in 1852 and 10 tons in 1853, of stuff which would probably yield about 75 per cent. of pure copper. At the time I visited the place, in 1853, operations had been suspended.

*Bohemian Mining Company.* T. 51, R. 37, Section 31, east half. This is known as the Piscataqua location, but it is worked by a company under the charter of the old Bohemian Company. In September 1853, a shaft was sinking here, and was already down 63 feet, inclining at an angle of 33° to the north. A seam of epidote, mixed with quartz and calc. spar, was observed in it, carrying in places a good deal of copper. A drift has been extended on what is supposed to be the vein, both east and west from the adit-level, but it is difficult to trace the existence of any regular vein. A cross-course or slide of clayey matter intersects it to the east of the adit, and beyond that the vein had not been found. The epidote seam, when it can be traced, is rich in copper. An amount of stuff which would produce about 1½ tons of pure copper was shipped in 1853, leaving about 20 tons of good stamp-work on the surface. It is understood that considerable copper has been taken out here during the past winter.

*Ohio Mining Company.* T. 51, R. 38, Section 36, east half. A shaft had been sunk here, in August 1853, 69 feet, on a worthless seam of quartz and epidote.

*Adventure Mining Company.* T. 51, R. 38, Section 36, west half, and Section 35, northwest and southeast quarters. A large amount of work has been done here, and a good deal of copper taken out, from extensive but very irregular excavations, which are not upon any vein, but ramify through the bluff, in which copper seems to have been distributed with so much abundance that, were it concentrated into a vein, it would be well worthy of working. As it is, I see little encouragement for prosecuting the works any farther at this point.

The copper seems to have been scattered through the bluff promiscuously, and is found sometimes in large masses, not unfrequently accompanied by silver. This rock is a somewhat crystalline and compact trap, and bunches of veinstone, consisting of epidote, beautifully crystallized calc. spar, feldspar, &c., are found in the amygdaloidal cavities of the rock.

In 1853, the old workings were abandoned, and a drift was carried in on a belt of rotten veinstone, between two beds of rock at the base of the bluff, dipping at an angle of 48°, and carrying some copper: it did not appear to be very promising.

From the above workings, irregular as they are, a considerable quantity of copper has been obtained, as will be seen by the following statement:—

| | 1851. | 1852. | 1853. |
|---|---|---|---|
| Shipped, . . . . . . . . | 21,073 | 54,074 | 23,527 lbs. |
| Yield of pure copper, . . | 12,478 | | 11,988 |
| Percentage, . . . . . | 59 | | 50 |

The silver found here occurs in vugs, generally in masses of calc. spar; one specimen is said to have weighed 5 lbs., and to have contained about 4 lbs. of the pure metal.

*Ridge Mining Company.* T. 51, R. 38, Section 35, southwest quarter. A considerable amount of money has been expended here, so far as I can see, without any sufficient grounds. Two shafts have been sunk, at 300 feet distance apart. The deepest, Clark's shaft, was, in Dec. 1853, 12 feet below the 45-fathom level. Four levels have been extended between them, but no stoping had been done in 1853. During the last summer, an engine and stamps were erecting, but I was unable to find in the stuff taken from the mines anything which would pay for stamping.

According to the Report of the Directors, there had been expended, up to December 1st, 1853, $53,583 77, and $5817 received for copper sold. Occasional masses of copper, of some size, are met with; one of over a ton in weight was taken from the 45-fathom level. Such specimens are, unfortunately, too rare in the mine to warrant the expectation of profitable working.

*Evergreen Bluff Mining Company.* T. 50, R. 38, Section 6, northeast quarter. Mining operations have been commenced here, during the past winter, on a vein which is represented by the agent as 2½ feet wide. From it, near the surface, a number of small masses have been taken.

At several points in this vicinity, mines are said to have been opened during the past winter, with what appearances of success I am unable to state with any confidence.

*Minnesota Mining Company.* T. 50, R. 39, Section 15. The property originally purchased by this company, being the north half of an original 3-mile square lease (location No. 98), consisted of about 3000 acres, but portions of it have since been set off and organized into companies, called the Rockland, the Flint Steel, and the Peninsula Mining Companies.

The Minnesota Company now owns 2035 acres of land, a considerable quantity of additional wood and farming land having been recently purchased. The vein, which has, until now, been by far the most productive one in the Ontonagon region, and only second to the Cliff Vein on Lake Superior, was discovered in the winter of 1847–8, at a time when there was very little confidence felt in this portion of the Lake country, and nothing in reality had been done towards developing its hidden wealth. It is to Mr. S. O. Knapp that the credit of this discovery belongs; and it was indeed an important one, as it immediately turned the attention of the public in the direction of the Ontonagon district, at that time almost abandoned.

The line of the vein was very strongly marked by ancient excavations, which were quite perceptible, even under a covering of three feet of snow. Early in the summer of 1848 one of the principal ones was opened, and found to be 26 feet deep on the vein. It was filled with an accumulated mass of clay, sand, and mouldering vegetable matter; but on penetrating to the depth of 18 feet, a mass of nearly pure copper, weighing over six tons, was met with, which had been raised about five feet from its native bed by the ancient miners, secured there on timbers placed under it, and abandoned, apparently, on account of the difficulty of raising it to the surface. This mass had been hammered over its whole surface until it was entirely smooth, and by the side of it were found large quantities of stone-hammers and a few other tools. Of course, here was pretty strong evidence of the existence of copper in considerable quantity, and operations were immediately commenced by Mr. Knapp, and carried on with much perseverance and energy.

By referring to the accompanying section of the mine (Plate III.), which represents the state of the works on the main lode in January 1854, a good idea may be obtained of their position and extent. Four principal shafts are opened on the vein, following its inclination, which is, at shaft No. 3, 64° to the north, and at various other points from 52° to 55°. The dip of the beds of rock in the vicinity is about 44°; thus it will be seen that the vein must cross them, at a small angle, in its downward course, a fact of the highest interest, since it makes it evident that this is a true vein, although it coincides in its line of bearing with the range of the formation. Its course is north 65° east, and the altitude of shaft No. 4, above the river, at the saw-mill of the Forest Company, is given by Mr. Merriwether, by whom it was levelled up, as 645 feet.

The annexed section (Fig. 29) will serve to represent the dip of the lode, and its position with regard to a conglomerate and sandstone belt a short distance south of the mine. According to the dip of the belt, as observed in the adit, it should intersect the lode at about 40 fathoms in depth, and this point has already been reached in one of the shafts, as I am informed, and an immense mass of copper has been found there.

Fig. 29.

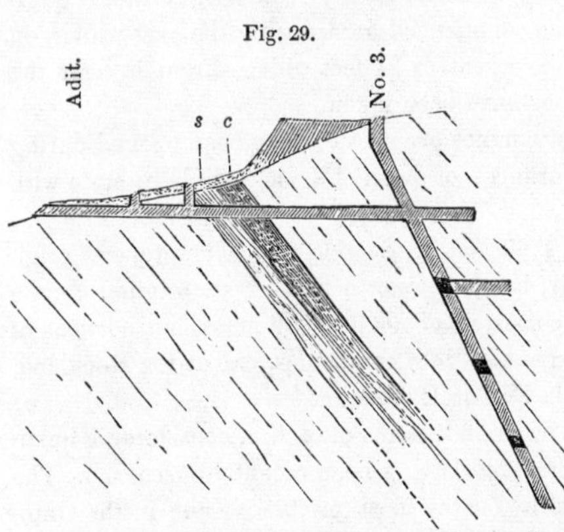


Section of adit and shaft No. 3, Minnesota Mine. *s*, Sandstone; *c*, Conglomerate.

The gangue of this vein is chiefly quartz, calc. spar, and epidote. Fine crystallizations of the two former minerals are obtained. Large crystals of quartz, terminated with the usual pyramid at both ends, have been frequently met with; some of them weigh several pounds. They are usually incrusted with delicate crystallizations of feldspar. Calc. spar occurs in large dog-tooth and rhombohedral forms, in great abundance, sometimes associated with crystallized copper, so as to form specimens of great beauty.

The walls of the vein are in general well-defined, although in some places not very regular; they are usually smooth, and sometimes finely smoothed and

Plate III

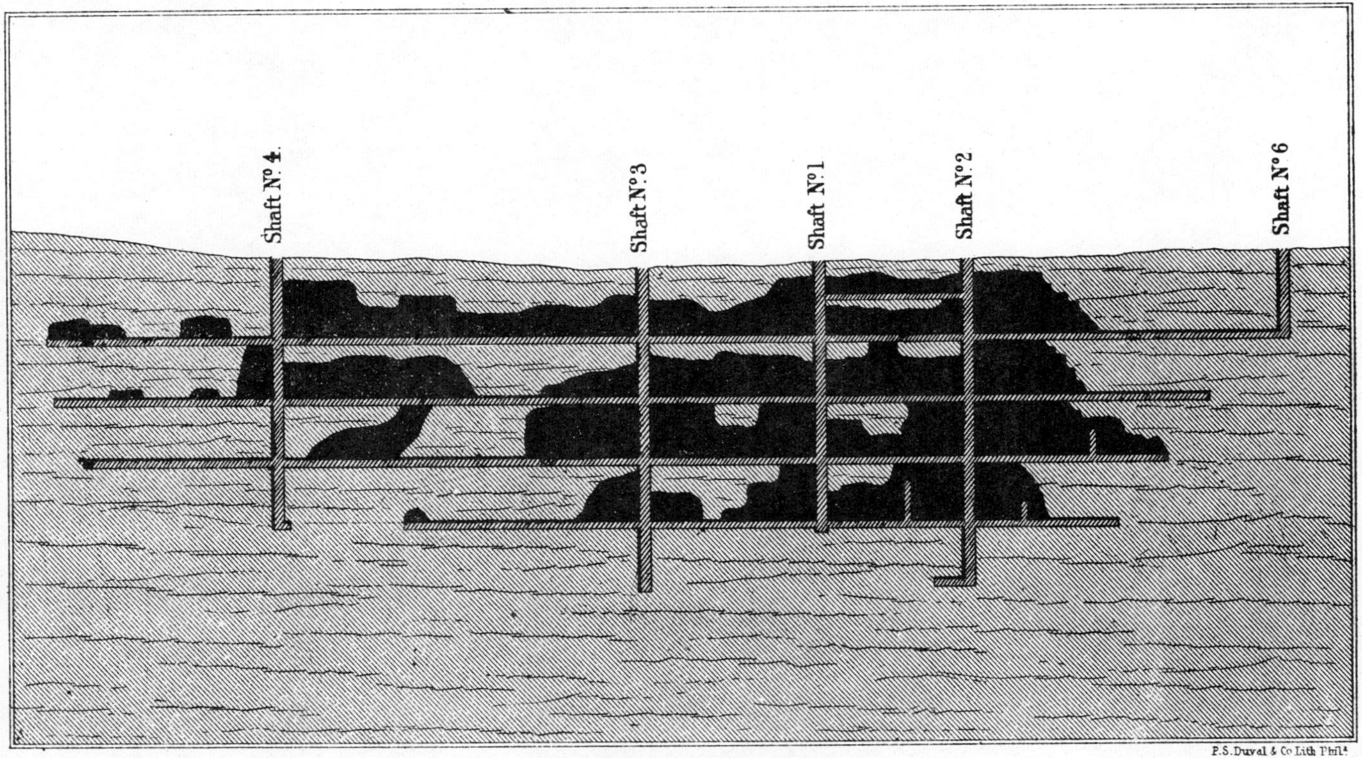

Shaft No. 4.

Shaft No. 3

Shaft No. 1

Shaft No. 2

Shaft No. 6

P. S. Duval & Co. Lith. Phil.a

**MINNESOTA MINE.**

Jan. 1854

striated, but generally destitute of selvages of flucan. Near the walls, thin lenticular sheets of mixed calc. spar and Laumonite are frequently found overlapping each other. The lines of bedding of the rock are frequently observed inclining at a small angle with the lode, and are troublesome to the miners, as they make it necessary to break more ground than would otherwise be required.

At the time I last visited this mine, September 1853, its appearance was as follows: 100 feet east of shaft No. 4, below the adit, was a mass of several tons, 20 feet long, 10 high, and 1 thick; in the bottom of the adit, 150 feet east of No. 4, the lode was very large and rich; in the 10-fathom level, 230 feet west of No. 4, the lode was rich, but poor for 150 feet east of that point; east of No. 4, vein large and rich, and large masses at 100 feet distance in that direction. In the 20-fathom level, 90 feet west of No. 4, vein small, but good stamp-work; 100 feet east of No. 4, lode 5 to 6 feet wide, rich in stamp and barrel-work, with some masses; 156 feet from the shaft, in the same direction, vein very wide, and carrying masses; in the same level, 100 feet east of No. 2, the lode is heaved to the north 9 feet; farther east, very large and carrying masses; the vein contains at this point considerable jaspery matter; at 220 feet from the shaft, an almost continuous mass of copper was observed for the distance of 40 feet, and 20 feet high. In the 30-fathom level, 100 feet east of No. 2, lode large and rich; 50 feet west of No. 2, lode rich, from 6 to 10 feet wide, and carries masses; 30 feet farther in same direction, a mass was cutting up; west of No. 1, the lode was from 6 to 8 feet wide, and contained masses.

Fig. 30.

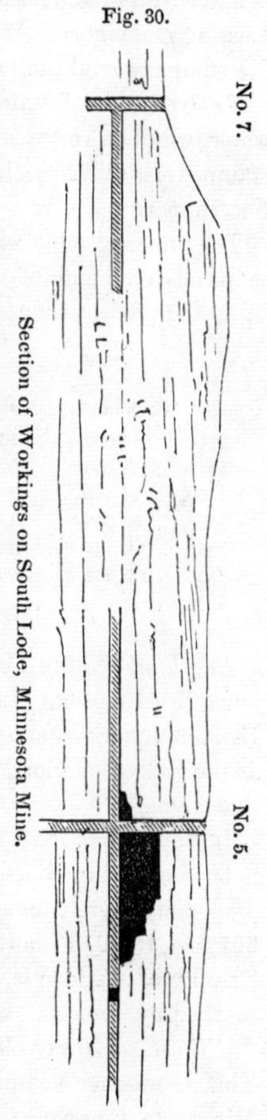


Section of Workings on South Lode, Minnesota Mine.

The whole number of fathoms of vein removed in the working of this mine amounted, in January 1854, to 4152; and the yield of copper was, approximately, 582 lbs. per fathom.

*South Lode.* During the summer of 1852, workings were commenced on Section 16, adjoining on the west, by the National Company, in a vein lying between a belt of conglomerate and the trap, from which a large quantity of copper was taken. As this conglomerate band crosses the Minnesota location a short distance south of their mine, as seen in Fig. 29, it was immediately opened by this company, by driving a cross-cut, from the adit-level south, at a point 200 feet west of shaft No. 4; and the conglomerate bed was intersected at a distance of 96 feet from the level.

On driving by the side of the conglomerate, large masses of copper were found, and preparations were immediately commenced for working this vein, by sinking two shafts, Nos. 5 and 7, as shown on the annexed section (Fig. 30), which represents the excavations on this "South Lode," as it is called.

Near shaft No. 5, there was, in September 1853, a lode, in some places 5 feet wide, and filled for a distance of 40 feet with an almost continuous mass of copper. By recent letters from the mine, dated March 15th, 1854, I learn that in drifting between shafts No. 5 and 7, the lode is found to be rich, and carrying masses of copper, with considerable silver; one piece of the latter metal weighed 3 lbs. Stoping has also been commenced near shaft No. 7.

The operations of this mine have been retarded by the want of the necessary machinery and fixtures. It was only in September last that a winding-engine was set to work, and instead of laying down tram-roads in the shafts, the stuff is hauled up in kibbles, at a great loss of power, and wear and tear of the hoisting machinery. With so rich a vein open over so great an extent of ground, the company ought to be producing more largely than at present.

Twelve heads of stamps are at work, but are frequently delayed by want of water, which is very scarce in and about the mine in dry weather. The stuff stamped is said to yield about four per cent. of work, worth 60 per cent. of fine copper.

The annexed table will give as correct a view as can be made out, from the materials on hand, of the operations of the company from its commencement up to the present time:—

| | | | Yield. | | | Dividend. | |
|---|---|---|---|---|---|---|---|
| | Expended. | Stuff shipped. | Per cent. | Amount. | Amount received. | Per share. | Amount. |
| 1848, | $14,000 | 13,288 lbs. | 75 | 9,950 lbs. | Not stated. | | |
| 1849, | 28,000 | 102,679 | 75 | 77,000 | " | | |
| 1850, | 58,000 | 198,614 | 75 | 150,000 | " | | |
| 1851, | 88,000 | 513,599 | 72 | 369,800 | " | | |
| 1852, | 123,144 31 | 626,000 | 74 | 467,250 | $51,209 62 | | |
| 1853, | 168,244 34 | 1,490,000 | 72 | 1,075,633 | 339,719 63 | | |
| 1854, | | | | | | $30 | $90,000 |

*Rockland Mining Company.* T. 50, R. 39, Section 11. This property adjoins the Minnesota Mine on the east, and was set off from their location in 1853, and work commenced in the same year. It was, until recently, known as the Lake Superior Mine, the company having purchased an old charter of that name.

In September 1853, a large and rich lode was opened here, which appeared to be identical in course and mineralogical character with the Minnesota Vein. Two shafts were commenced; also an adit, which was intended to be driven 307 feet, from the north, and to intersect the vein 152 feet from the surface. The dip of the lode is, near the surface, 48½°, and its whole line of outcrop is marked by numerous ancient excavations.

*Flint-Steel River Mining Company.* T. 50, R. 39, Sections 11 and 12. This is another company formed on the location originally purchased by the Minnesota Company. The vein was poor when I examined it last summer, but since that time large masses of copper are said to have been found. The principal shaft has been sunk about 70 feet, and connects with the adit at that depth.

*Peninsula Mining Company.* T. 50, R. 39, Sections 15 and 16 (a small fraction at the southern corner of these two Sections). This company was organized in 1850, and a vein was worked to some extent in 1851. It appears now to be abandoned, as there was little encouragement to go on with it. 6140 lbs. of mass and barrel-work, averaging 75 per cent, were taken from it near the surface, but it was very poor at some depth.

*National Mining Company.* T. 50, R. 39, Section 16. This company was organized in 1852, and commenced operations in the summer of that year. Ancient mine-work was discovered to have been carried on extensively in a vein lying between the belt of conglomerate, seen on the section of the Minnesota adit (Fig. 29), and the overlying trap. A shaft had been sunk here in former times, to the depth of about 50 feet, on the lode. It had become filled up with clay, sand, and vegetable matter; but, on cleaning it out, the remains of stulls, or timbers forming a scaffolding across the shaft, were found, and a nearly continuous sheet of copper down its side. Workings were vigorously prosecuted during the next winter, and a shipment of 34,908 lbs. of masses, and 46,406 of barrel-work, averaging 72 per cent. of pure copper, was made during the next year.

From the Report of the Secretary, dated January 28th, 1854, it appears that shaft No. 1, at that time, was 169 feet, and No. 2, 157 feet in depth. They are connected by levels 39 feet in length. The first level is at 72 feet from the surface, and above it most of the ground has been stoped out, and yielded richly in copper. The second level is about 95 feet below the first. Two adits are driving on the lode from the west; the lower one will give a back of about 200 feet. The whole amount of ground stoped, January 1st, 1854, was 206 fathoms; ready for stoping, 2307 fathoms.

The position of this vein is remarkable, and, up to the time of its discovery, entirely unprecedented on Lake Superior. It bears all the marks of a true vein, so far as regularity is concerned, but lies between two dissimilar formations. The amount of veinstone connected with it is small; for a considerable part of the distance the lode was almost a solid sheet of copper, of 3 or 4 inches in thickness. The distance through which this productiveness will continue can only be ascertained by the development of the mine, but as large masses have been found in the same position at the Minnesota Mine, there is every reason to expect a considerable extent of the same richness, in depth as well as length.

To the north passes the line of the Minnesota Vein, and a cross-cut has been extended in that direction 183 feet; at 130 feet a vein of spar, quartz, and epidote, 11 feet in width, was intersected, and was supposed to be the one sought for, but it was found poor in copper at the point of intersection. It appears that there was some uncertainty as to the identity of the lode, as the cross-cut has been continued still farther to the north.

*Forest Mining Company.* T. 50, R. 39, Section 30, southwest quarter. Very extensive excavations had been made on this tract by the ancient miners, and quite large masses of copper were found near the surface in cleaning out the old workings, of which there were four parallel rows along the bluff, extending in a direction of about north 64° east. One mass weighed 1800 pounds. Work was commenced here in the winter of 1849–50, and several tons of copper were taken from near the surface, where the vein appeared quite regular and well-defined. The principal shaft is now down about 225 feet, and two others have been sunk to nearly the same depth. The principal levels are from 600 to 700 feet in length. In the eastern part of the mine, the lode is quite irregular, and the greater part of it is not worthy of being stoped; but, as the workings have been extended westward, its appearance has much improved. Evidently, if the mine is to be worked, the openings should be pushed to the west. Here there is a single wall, against which the copper is accumulated, sometimes in considerable quantity, and through a space several feet in width, but in the opposite end of the mine it is not possible to say that any regular wall exists. The copper occurs chiefly associated with epidote and calc. spar, and sheets of from one to three inches in thickness, are occasionally met with. On the whole, the appearance of the mine has much improved in the last year.

Eight heads of stamps were, until recently, in operation, but the stamping-mill and engine were destroyed by fire in March last. A larger engine and an increased number of stamps will be immediately erected, and the Superintendent expects to make a handsome shipment of copper during the present year. In 1853, 73,702 lbs. of masses, barrel, and stamp-work were sent to market. The average cost of sinking shafts, in 1853, was $11 12½ per foot; of drifting, $7 84; and of stoping, $24 12 per fathom.

*Glen Mining Company.* T. 50, R. 39, Section 31, west half. This tract formed a part of the original purchase of the Forest Company, and was set off in 1852. A nearly vertical vein has been opened on it, which is about 12 inches wide, and carries some copper in a gangue of calc. spar and quartz. But little has yet been done here or on the other half-sections on the west, which are known as the *Shirley*, *Devon*, and *Tremont* locations.

*Norwich Mining Company.* T. 49, R. 41, Section 2, southeast quarter. A precipitous trap bluff crosses this tract, in a nearly east and west direction, rising to the height of 300 feet above its base, and nearly 400 above the west branch of the Ontonagon, which is about ¼ mile south. The existence of the vein here has been known since 1846, but it was not worked to any extent until 1850. The following account of its appearance at the surface was written by me in 1847.* "The course of the vein is nearly east and west, and it dips from 47° to 49° to the north. It has been traced for a considerable distance along the brow of the cliff, and found to vary much in width and character. In one place it measured 12 inches in width, and consisted of partially decomposed chlorite and epidote; wall-rock well-defined, generally separated by selvages of chlorite or argillaceous matter. In a few feet distance, the vein narrows down to three inches, and consists of quartz, with radiated epidote, and native copper and red oxide, with a curious vermilion hue, and much stained with carbonate of copper. Farther on, the vein is almost entirely quartz; here they have taken out a mass of copper weighing 110 lbs. The occurrence of beautifully-radiated epidote seems to be characteristic of this vein."

The present state of the workings may be seen by referring to the annexed

* Documents accompanying the President's Message, 1849–50, vol. iii. p. 718.

section (Fig. 31), which represents the amount of excavation on the lode, in February 1854.

Fig. 31.

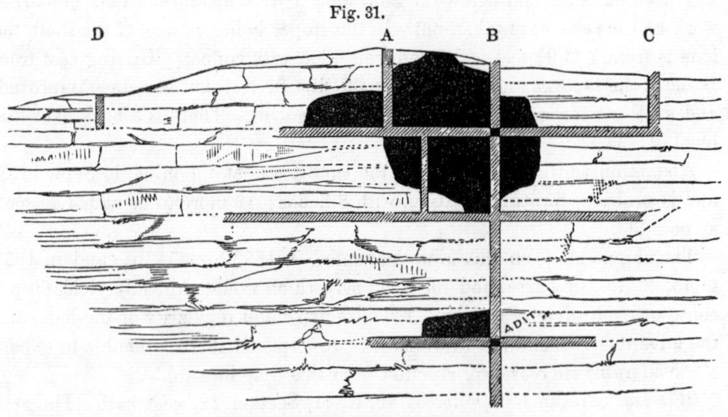


Section of Norwich Mine, Feb. 1854.

The deep adit which intersects shaft B in the lower level, as shown in the figure, is driven from the base of the bluff, a distance of 439 feet. The length of a straight line drawn from the upper adit to the lower, at their points of intersection with the vein, is 180 feet, and its dip is 61°. The underlay of the lode gradually decreases in this distance from 34° to 23°, so that it forms an arc of a circle, of which the above imaginary line is the chord. The adit-level commences in the conglomerate which flanks the trap range on the south, and has afforded a most interesting and instructive section of the junction of the two formations. For 24 feet from its mouth, the rock excavated is conglomerate, without any recognizable bedding; then follows sandstone for 55 feet, which shows in the most marked manner the mechanical action by which the trap was uplifted. It is broken and crushed into fragments, which afford the most evident proofs of having been rubbed against each other with immense force, at the time the upheaval took place. The next 52½ feet are occupied by trap, the southern portion of which is distinctly observed to dip to the south, although evidently much crushed and dislocated during its uplift, while the northern half has a steep inclination to the north. Nothing can be clearer than the evidence which is here afforded of a fracturing of the strata and an immense mechanical force exerted in the upheaval of their northern portion. At about 130 feet from the entrance of the adit, is another belt of sandstone, which is 13 feet thick, and has apparently the regular dip of the beds of rock in which the mine is wrought, about 44°. At the mean inclination of the lode, which in B shaft is 61°, it would intersect the bed of sandstone, which dips at 47°, at 900 feet below the deep adit, measured on its own course; at the inclination which it has at the adit, 65°, it would meet it at a depth of 700 feet.

The vein has been found, in many places, rich in copper, in masses and sheets, and also furnishing good stamp-work, in a gangue of epidote and quartz. By a communication recently received from the Superintendent, A. C. Davis, Esq., dated February 21, 1854, it appears that the main shaft, B, has been sunk 58 feet below the adit, and, in that distance, 4 tons of barrel-work and masses were taken out. In the stopes going on west of the shaft, the lode is from 1 to 2½ feet wide, and well filled with copper. Driving east from B shaft, the lode was not so good for the first 80 feet, but has since improved, and mass copper is in sight at the end of the drift. There is a back over this level of 134 feet.

A stamping-mill was erecting at the time I was at the mine, in Sept. 1853, and is probably now in operation, with 8 heads; 12 more to be added as soon as possible.

The shipments from this mine amounted in 1852 to 9054 lbs., and in 1853 to 45,325 lbs. of barrel and mass copper, which would probably yield 60 per cent., or over, of pure copper. From the size and regularity of the lode, and the advantageous manner in which it is now opened, it is reasonable to expect a considerably increased production from this time forward.

*Windsor Mining Company.* T. 49, R. 41, Section 12, west half. The vein here is in the same position as the one just described, and appears to be a continuation of it. It is opened by three shafts, the deepest of which is down about 210 feet. A deep adit is also driving from the base of the bluff, and will intersect the lode at nearly 300 feet from the surface. The appearance of the vein was quite favorable when I examined it, some portions being from 1 to 2 feet wide, with good stamp-work and some masses. Since that time, as I learn by a letter from the Superintendent, Mr. Plummer, it has improved considerably, and a mass of 2 or 3 tons has been found.

In driving in the adit-level, a promising vein was intersected 470 feet south of the one previously worked on, and a shaft is now sinking to prove it. The appearances here are decidedly favorable for successful mining.

*Ohio Trap-Rock Mining Company.* This company commenced operations in 1846, and up to September 1853, about $90,000 had been expended with but little returns of copper. Several different veins or courses of epidote, carrying some metal, have been opened. From the Superintendent's Report, dated July 6, 1853, it appears that there were at that time 3 shafts on the "Indian Diggings Lode," the deepest of which was down 234 feet, and the two others each 100 feet. The character of the lode is said to be very variable, and it appears that but little excepting stamp-work had been taken from it. "Park's Lode" has one shaft on it 100 feet deep. The "New Lode," which is represented as 2 feet wide and carrying good stamp-work, is opened by two shafts recently commenced, at a distance of 320 feet apart. The company intended to put up 24 heads of stamps, to work out the copper from the large amount of stamp-stuff now at the surface.

*Sharon Mining Company.* T. 49, R. 41, Section 9, southeast quarter. I have not examined this location; but, according to Mr. Davis, the Superintendent, a shaft has been sunk 90 feet on a good lode from 6 to 18 inches wide, and well filled with stamp-work, with a few pieces of barrel-copper. Another shaft is down 26 feet, at a distance of 150 feet from the first mentioned, and a level driving to connect them at 90 feet depth. Work has been recently commenced on another vein, 560 feet farther south, which is said to be from 2 to 4 feet wide, and to contain a fair amount of copper.

There are numerous other localities west of the Norwich Mine where mining operations have been commenced on a small scale, but as I have never visited them, and as they have, as yet, made no shipments, I will not attempt to give an account of them. Several of the veins are described as presenting favorable indications for successful working, but the remoteness of their locality, and the bad state of the roads, will operate strongly against these mines, until the companies shall have acquired sufficient confidence in their value to unite in building a good plank-road to the mouth of the Ontonagon, a distance of about 25 miles.

### PORTAGE LAKE DISTRICT.

The opening of this district of the Lake Superior region is by no means a recent event, as mines have been worked there since 1846; but it was not until 1852 that the attention of the public was called to it in any marked degree. As yet, the mines are not sufficiently developed to enable a very decided opinion to be formed as to their probable permanent success, especially as the mode of occurrence of the copper is quite peculiar to the district. There appear to be few, if any, regular veins, but the metal is found disseminated, mostly in small masses and barrel and stamp-work, through certain metalliferous beds, which run with the formation, and differ very slightly in composition from the other trappean beds with which they are associated. Such a mode of occurrence is not wholly without analogies in the Lake Superior district, since, as has been seen in the description of the Ontonagon region, there are localities there where the copper is irregularly disseminated through certain beds of rock, without any appearance of a vein-fissure, or, in some cases, even of veinstone. In the vicinity of Portage Lake, however, these metalliferous beds appear to have a much greater degree of regularity, and are neither broken up nor deranged in their course, and their metalliferous contents are more uniformly distributed through them than on the Ontonagon. In some instances, the same bed has been distinctly traced for a mile or more, by a line of ancient excavations, and, wherever opened, found to contain copper disseminated through it.

A large number of mines have commenced work within the past year, and, according to the statements of those interested, several of them will make considerable shipments of copper during the present season.

There are at least twelve companies now at work in this district, employing a force of about 400 men, and expending from $30,000 to $40,000 monthly. With this energetic development of its metalliferous deposits, the question of their importance will soon be settled.

*Isle Royale Mining Company.* T. 54, R. 34, Section 1, northwest quarter. Operations were commenced here in May 1852, and I visited the place about three months later. At that time, a row of ancient pits had been traced for a mile on the line of the metalliferous bed, which bears about north 62° east. A shaft had been sunk for a few feet, from which a considerable quantity of copper had been taken. It appeared to lie in an amygdaloidal belt of rock, of a light-brown color, with patches of veinstone, principally epidote, quartz, calc. spar, and Prehnite, scattered through it. According to Mr. Hill, who examined it during the past winter, this bed is distinctly defined, being flanked on both sides by a gray trap. The dip is about 60°, and it is as even in its bearing and thickness as a belt of sandstone under the same circumstances would be.

The two principal shafts were, at that time, down to the second level, 150 feet from the surface; they were connected by a level 200 feet in length, and stoping was going on above it. Three other shafts have also been commenced, whose aggregate depth was, in February last, 490 feet, and 802 feet had been drifted.

From the shafts sunk and levels driven, a large quantity of copper has been taken, mostly in the form of barrel-work, or small pieces which could be cleared of the rock. Shipments were made last season of 31,783 lbs., which yielded 18,738 lbs. of pure copper, or 58·9 per cent. This amount has been shipped without any stamping having been done; but preparations are now making to erect an engine capable of driving 48 heads.

If the copper is found to hold in sinking on this bed, it will be capable of furnishing a very large amount of that metal, and we may soon expect to see it rivalling with the most productive mines of Lake Superior.

*Portage Mining Company.* T. 55, R. 34, Section 36, southwest quarter. According to C. H. Palmer, Esq., this company is working principally on the "Portage Vein," which runs parallel with the "Isle Royale Vein," at a distance of 200 feet west of it. They are in every respect similar to each other. The metalliferous bed is from 10 to 15 feet in width.

Six shafts have been begun, 4 on the Portage Vein, and 2 on the Isle Royale; their aggregate depth amounted, in February last, to 478 feet, and 510 feet of drifting had been done at the same time. Mining was commenced in November 1852, and during the season of 1853, a shipment of about 10,000 lbs. of barrel-work was made, and several hundred tons of stamp-work are at the surface.

*Montezuma Mining Company.* T. 55, R. 34, Section 35, southeast quarter. Work was commenced here in September 1853. Two shafts are sinking, and an adit driving from the Lake on the course of the vein.

*Albion Mining Company.* T. 55, R. 34, Section 36, southeast quarter. Operations were commenced here by the company in August 1853, they having discontinued their mine near the Eagle River, and removed to this place. The Isle Royale and Portage Veins extend through a corner of the quarter section. Three shafts have been commenced on the latter, whose aggregate depth was, in February last, 237 feet. I have no information as to the amount of copper contained in the vein at this point.

*Quincy Mining Company.* T. 55, R. 34, Section 26. Work was first commenced here as far back as 1846 or 1847; but little was done until 1851. In 1852, numerous excavations had been made, but had not succeeded in developing anything of value. A vein, bearing about north 25° east, had been driven on for a distance of 120 feet, at a depth of about 80 feet, the vein being pretty distinctly marked, and well filled with copper, but narrow. Another shaft was sunk 90 feet on the plane of contact of two amygdaloidal beds of rock, near which considerable copper was found disseminated.

According to Mr. Palmer, valuable discoveries have been made on this location during the past winter.

In the foregoing sketch of the different mining districts of the southern shore of Lake Superior, most of the working mines have been noticed; and, in order to present at one view the progress of the development of this region, the following table is appended, in which the product, in pure copper, of all the mines for each year, is given in tons (2240 lbs.), from 1845 to the end of 1853.

From all the principal mines pretty accurate returns have been received, both of the amount of the shipments and the yield of metallic copper; and although, in the case of some of the less important ones, it has been necessary to estimate the percentage of the mixed copper and rock sent to the smelting-works, yet it is believed that the table can be relied on as a near approximation to the truth.

## TABULAR STATEMENT OF THE PRODUCT OF THE LAKE SUPERIOR MINES.

| | 1845. | 1846. | 1847. | 1848. | 1849. | 1850. | 1851. | 1852. | 1853. | Total. |
|---|---|---|---|---|---|---|---|---|---|---|
| Cliff, | 8·88 | 16·79 | 183·38 | 444·85 | 572·38 | 319·04 | 377·89 | 370·25 | 415· | 2708·46 |
| Minnesota, | | | | 4· | 34· | 66· | 165· | 208·50 | 480·19 | 957·69 |
| North American, | | | | | 22·90 | 76·20 | 76·50 | 22·90 | 112·32 | 310·82 |
| Northwest, | | | | | 15·32 | 87·06 | 130·89 | 120·17 | 102·27 | 455·71 |
| Copper Falls, | | | 20· | 10· | 10· | 2· | | 5·64 | 40·95 | 88·59 |
| Northwestern, | | | | | | | | 3·85 | 19·72 | 23·57 |
| Siskawit, | | | | | 14· | 13·80 | 16·68 | 27·80 | 12·87 | 85·15 |
| Phœnix, | | 4· | 5· | | | | | 7· | 1·53 | 17·53 |
| Forest, | | | | | | 3· | 1·50 | 4· | 19·25 | 27·75 |
| Adventure, | | | | | | | 5·57 | 13· | 5·35 | 23·92 |
| Norwich, | | | | | | | | 2·83 | 15·38 | 18·21 |
| Aztec, | | | | | | | | 3· | 7· | 10· |
| Peninsula, | | | | | | | 2· | | | 2· |
| National, | | | | | | | | | 26·13 | 26·13 |
| Isle Royale, | | | | | | | | | 8·36 | 8·36 |
| Pittsburgh and Isle Royale, | | | | | 1·50 | 4· | | | 5·54 | 11·04 |
| Ridge, | | | | | | | | | 10· | 10· |
| Toltec, | | | | | | | | | 3·41 | 3·41 |
| Douglass Houghton, | | | | | | | | | 2·33 | 2·33 |
| Fulton, | | | | | | | | | ·56 | ·56 |
| Ohio Trap-Rock, | | | | | | | | | 1·23 | 1·23 |
| Bohemian, | | | | | | | | | 1·44 | 1·44 |
| Portage, | | | | | | | | | 2·95 | 2·95 |
| Derby, | | | | | | | | | 1·31 | 1·31 |
| New York and Michigan, | | | | | | | | | | |
| Ohio, | | | | | | | | | ·85 | ·85 |
| Meadow, | | | | | | | | | | |
| Sundries, | 3· | 5· | 5· | 2· | 2· | 1· | 3· | 3· | 1· | 25· |
| Total, | 11·88 | 25·79 | 213·38 | 460·85 | 672·10 | 572·10 | 779·03 | 791·94 | 1296·94 | 4824·01 |

There are now about 75 mines at work, employing in and about them 2800 men, and the product of pure copper for 1854 may be estimated at 2000 tons.

The whole amount of money expended in the Lake Superior region, from its opening up to December 31st, 1853, I have estimated, from all the data which can be collected, at $4,800,000; and the value of the copper produced, at an average price of 25 cents per pound, equals $2,700,000; of this, $504,000 has been divided among stockholders, and the remainder has been applied to the development of the mines.

Of the capital thus invested in the country, a considerable portion has been expended in opening mines which may reasonably be expected to become profitable to the adventurers, and some of which will make large shipments during the present year, and probably show a balance of profits on the year's transactions; a very considerable amount was, however, irrevocably sunk during the first years of speculation and foolish excitement. But even at the present very moderate prices of Lake Superior copper stocks, their actual cash value considerably exceeds the whole amount which has been expended there. The mines of this region have a character of permanence, and there can be little doubt that their product will go on regularly increasing, as it has done in the eight years since mining operations may be said to have fairly commenced.

The trap range extends into Wisconsin, and has, at various times, been examined by the geologists of the United States Survey of that state, and by other explorers. The results, up to this time, have been entirely negative; so far as I have been able to learn, there are no veins of copper beyond the borders of Michigan which promise to become of value.

## NORTH SHORES OF LAKE SUPERIOR AND LAKE HURON.

Numerous companies have been formed at various times for working copper mines on the north shores of Lake Superior and Lake Huron, within the Canadian limits, but only one has as yet proved successful. The attempts at mining which have been made were in the trappean rocks, analo-

20

gous in geological position to those of the southern shore, and also in the rocks of the azoic period. It is only in the latter that they have, thus far, given encouraging results.

The character of the rocks on the north shore of the Lake has not yet been thoroughly investigated; but, from a hasty survey, during the last summer, of a part of the region which lies to the north of Isle Royale, it appears to me probable that the trappean rocks which form the lofty cliffs of Pie Island, Thunder Cape, and the numerous islands in that vicinity, are the counterpart of those of the South Range of Keweenaw Point. No workings are now going on here; but from 1846 to 1849 a powerful vein was worked on Spar Island and the mainland opposite, at Prince's Bay. The vein on Spar Island has a course of north 32° west, and is made up almost entirely of calc. spar, heavy spar, and quartz, and is 14 feet wide at the southern edge of the island, where it forms a conspicuous object, contrasting strongly with the dark-colored trap, and being visible from the Lake at a great distance out. It contains a little yellow sulphuret of copper and variegated ore, and two shafts have been sunk on it to the depth of 47 and 24 feet. The quantity of ore is too small for working. On the mainland the same vein appears, having a more quartzose gangue, and carrying sulphuret of zinc principally, with native silver in pockets. Splendid crystallizations of amethystine quartz and calc. spar have been obtained here. The vein has been opened on the mainland by sinking and driving, at a very heavy expense, as the body of the lode consists of crystallized quartz. A bunch of calc. spar and blende, rich in native silver, was struck near the surface, at the collar of the deepest winze, and led to unbounded expectations on the part of the shareholders, which were unfortunately destined to be entirely disappointed.

A number of localities were formerly explored and worked to some extent on Michipicoten Island and on the northeastern side of the Lake, but they are now all abandoned. The "Quebec and Lake Superior Mining Association" commenced operations in 1846 at Pointe aux Mines, Mica Bay, on a vein said to be two feet wide and rich in gray sulphuret of copper. An adit was driven 200 feet, three shafts sunk, and the 10-fathom level commenced. Smelting furnaces were erected, and, after £30,000 had been expended, it seems to have been discovered that there was no ore to smelt, and the works were abandoned.

The mines on the north shore of Lake Huron are in a formation consisting of white, and often vitreous, sandstone or quartz rock, passing into a jasper conglomerate, and interstratified with heavy masses of trap. These are supposed by Mr. Logan, the Provincial Geologist, to be of the same age as the copper-bearing rocks of Lake Superior, and he remarks that the chief difference seems to lie in the great amount of amygdaloidal trap present in the latter formation, and of white quartz rock in the former. There can be no doubt, however, of the identity in age of the sandstones and conglomerates of Keweenaw Point with those extending along the shore of the Lake on both sides of the Point, and recognized by Mr. Logan himself at Saut Ste. Marie as of the age of the Potsdam Sandstone; and the same beds are recognized in their continuation east as being deposited unconformably on the cupriferous beds of the northern shore of Lake Huron. Hence the inference is unavoidable, that these latter are older than those of Lake Superior. This is corroborated by the fact, that in the Lake Huron mines the ores are entirely sulphurets, and principally copper pyrites, associated with a gangue of quartz, differing wholly in these respects from those of the really productive part of Lake Superior.

The Bruce Mine is the only one now working on Lake Huron, and this company has been quite successful, and might have been much more so had proper discretion been used in its management. The vein was, from the beginning, worked with an entire want of mining skill; during the first year an open cut was made, which was 126 feet long and 5 feet deep; from this excavation 240 tons of ore were taken. Afterwards much money was thrown away in costly machinery not at all adapted to use in that region.

The mine is situated about 50 miles below Saut Ste. Marie, and due north of the extremity of St. Joseph's Island. The vein was discovered in 1846, and is contained in a dark-colored hornblende trap; the ore is chiefly copper pyrites, with some variegated ore. The deepest shaft was down, in 1853, 51 fathoms, and the vein was then 3 feet in width, and well filled with ore disseminated through quartz. Very expensive smelting works have been erected here, but have proved a great source of loss and embarrassment to the company, as the ore can be transported almost from the mouth of the mine to Swansea, all the way by water, without transshipment.

The entire amount of ore raised at this mine I have not been able to ascertain, but it is very considerable. During the year 1853, 1650 tons were shipped, and sold mostly in New York, netting to the company £22,586 currency. A dividend was paid of 5*s*. per share on 45,402 shares, during the past year.

This is truly a valuable mine, and should produce largely; indeed, it appears that within the past year a new order of things has commenced, and the expectations of the company that, hereafter, 300 tons of ore per month will be shipped, seem likely to be realized.

There were several other companies located on the north side of Lake Huron, but I know of none now at work except the Bruce. The success of the latter will probably induce those claiming or owning property in the vicinity, to cause more thorough explorations to be made on their territories. The Wallace Mine, 16 miles from La Cloche, a station of the Hudson's Bay Company, is said to have furnished copper pyrites resembling that of the Bruce Vein, and also nickel and cobalt ores of considerable richness. An attempt was made to resuscitate this concern, which belongs to the Upper Canada Mining Company, in 1853, but with what success I have not heard.

## COPPER DEPOSITS OF THE MISSISSIPPI VALLEY.

Under this head are included numerous cupriferous deposits which ought not to be passed over without notice, but which are of little practical importance. They occur almost entirely in the limestones of the Lower Silurian formation, or at the junction of these rocks with the azoic. Nowhere, so far as my observations have extended, do they form anything like true veins, nor have they been wrought in mines of any considerable extent. They are not unfrequently found in connection with the lead ores of the West, which they somewhat resemble in their mode of occurrence.

WISCONSIN.—The copper of this state occurring in the trap range has already been noticed; the ores included in the group now under consideration are those which have been discovered far to the south of any igneous rocks. These were considered by Dr. D. D. Owen to be of considerable importance, and he remarks in the report of his explorations made in 1839,* that were it not for the superior richness and value of the lead, the great staple of the territory, they would attract much attention. This opinion, however, could not have been based on a thorough knowledge of the nature of cupriferous deposits. From a slight examination of the region in which the principal localities of ore are situated, I cannot anticipate that they will ever become of importance.

The copper ores occur chiefly in the neighborhood of Mineral Point, on what was known as the "Ansley Tract." According to J. T. Hodge, Esq.,† who made a careful examination of this region in 1841, with a view to the development of its resources by eastern capital, the ore at that locality occupies a fissure in the limestone (Lower Silurian), 14 feet in width at the surface, and which had been traced a quarter of a mile. For about the depth of 15 feet, the fissure was found to be filled with gossan, together with lumps of sulphuret and carbonate of copper, of all sizes, up to 200 pounds in weight. Below that depth was clay, with a little ore scattered through it. About 1½ million of pounds are said to have been taken from this fissure, of which 50,000 lbs. were sent to England to be smelted, and brought in a bill of expense to the shippers. The ore was a mixture of sulphuret

* Report of a Geological Exploration of Part of Iowa, Wisconsin, and Illinois, &c. Senate Doc. 1844, p. 48.

† Sill. Am. Jour. xliii. 38.

and carbonate, resulting from the decomposition of copper pyrites.

The character of the formation in which these ores are found, and the almost certainty of the fissures terminating, or being found barren of ore, in the underlying sandstone, which is only a little more than 100 feet from the surface, are sufficient reasons, in my opinion, why these cupriferous deposits should be considered as of little value.

MISSOURI.—The copper ores of Missouri, so far as I have had an opportunity to observe, are of about as little importance as those described in Wisconsin. In travelling through a part of the metalliferous region of that state, in 1852, I was unable to find a single locality where any considerable amount of ore was being raised.

The ores of copper, as well as those of lead, cobalt, nickel, &c., found at numerous points in this state, are contained in the Lower Silurian strata, which have been deposited in previously existing depressions of the azoic rocks. The latter formation, consisting principally of granite and porphyry, often appears at the surface, as if the overlying sedimentary deposits had undergone a great amount of denudation; hence the valleys are frequently underlaid by the stratified sandstones and limestones, while the elevations and ridges consist of igneous rocks, which evidently were nearly in their present position at the time of the deposition of the overlying strata, as there is no evidence of the latter having been upheaved or metamorphosed by the adjacent igneous masses.

There are several localities known as having yielded more or less copper ore. Among them is the famous Mine La Motte, which has a historic celebrity vastly beyond what its real importance entitles it to. There are numerous so-called mines within the property, which includes 24,000 acres; but the only place where any work worth noticing was doing in 1852, was the locality called the "Philadelphia Mine." Here the sandstone and limestone are deposited upon the granite, which would be reached by sinking only a few feet, were the works to be extended in depth. The metalliferous deposit forms a bed which lies between a stratum of sandstone and another of a hard, crystalline limestone, and occupies a width of from 12 to 18 inches. It consists of a blue slaty substance, in which galena is enclosed in flat sheets, associated with ores of cobalt and nickel, in a pulverulent state. The copper pyrites occurs in patches in the fissures of the limestone, disseminated through a thickness of six or eight feet. This metalliferous stratum forms a lenticular mass, dipping at a small angle each way from its centre, and appearing to be a few hundred feet in diameter. The position of the ores in relation to each other is not constant throughout the mine. I could obtain no statistics of the yield of this mine, at which six or eight men were employed; but at all events, the copper ores will never be found to be of any value. At the time of the first discovery, in 1838, some thousands of pounds of ore were taken out, and about that time a vigorous attempt was made to raise the capital to carry on mining operations here in a scientific way, but it has had no result. At the time of Mr. Hodge's visit, in 1841, several other localities on the Mine La Motte tract were wrought to some extent, for cupriferous ores, but nothing was doing on them in 1852.

Near Fredericton, about 1½ miles south of the village, a copper mine was formerly worked in a coarse-grained crystalline limestone, containing geodes of pearl-spar and nodules of iron pyrites. There appear to be a few strings and bunches of copper ore irregularly scattered through the rock.

Mr. Hodge also visited the so-called copper region of Current River, in Shannon County,* a rough, uninhabited country, where, up to this time, no mining worthy of the name has been done. The ores appear to be contact deposits, between Lower Silurian rocks, similar to those occurring at Mine La Motte, and a reddish, quartzose porphyry, of azoic age. They are evidently similar in origin and character to those just described, and there is no reason to infer from Mr. Hodge's description of them, that they are likely to be profitably worked.

There are probably numerous localities, besides those above cited, both in this and in the other Western States,

* Sill. Am. Jour. xliii. 65.

where copper ores occur in a position similar to that above indicated; but the records of mining throughout the world will not justify the expenditure of any considerable amount of capital in developing such cupriferous deposits contained in the unaltered sedimentary rocks. They cannot be relied on as being of any permanent value.

### CUPRIFEROUS DEPOSITS OF THE ATLANTIC STATES.

The first group of this section comprehends the deposits of copper ores included in the so-called metamorphic rocks, or the crystalline schists and associated igneous masses, which extend along the eastern slope of the Appalachian chain, from Vermont to Georgia.

These deposits, wherever examined, are found to exhibit a striking similarity to each other; they are never found occurring in well-developed transverse or fissure-veins; or, at least, such have never come under my observation. They all form masses parallel with the formation, and possessing all the characteristics of segregated veins; or if, as is occasionally the case, apparently crossing the strata at an angle, such branches will be found subordinate to segregated masses, and not exhibiting, in an unmistakable manner, the phenomena of fissure-veins. The ores thus occurring are almost always pyritous, with, occasionally, a small portion of the variegated; and they do not usually appear to be oxidized to any considerable depth from the surface. Sometimes specular and magnetic oxides of iron form the outcrop of the vein, and are replaced to a greater or less extent beneath by ores of copper. On the southwestern side of the Appalachian chain, in Tennessee, this decomposition and the formation of gossan has, however, taken place on a large scale; in other respects the deposits of the ores are similar to those of the eastern slope, except that they are on a scale of greater magnitude.

But few of the very numerous localities which are known to exist have yet been worked to any extent; and it is only within the last year that the attention of the public has been called to them. The mining enterprises thus commenced are generally only in a state of preparation, and complete data for a description of what is now doing in the way of developing the mines of the Atlantic slope, scattered as they are over such an extent of country, are not easily to be procured. Most of the important localities known will be mentioned, and more or less fully described, in the following pages.

MAINE.—Copper pyrites is spoken of by Dr. C. T. Jackson, State Geologist, as occurring at Dexter,* in small quantity, in quartz veins; at Lubec, with galena, at the so-called "lead mines;" and at Parsonsfield.

None of the localities seem to present any inducements for working.

NEW HAMPSHIRE.—The localities of copper pyrites in this state are somewhat numerous, but none of them have been worked to any extent.

Orford, Lyme, and Jackson are mentioned in the Report of the State Geologist as containing localities where copper pyrites has been found, but no particulars as to the value of the deposits are given. The same ore is found in small quantity at the Shelburne Lead Mine. At Haverhill, on the estate of Mr. Francis Kimball, between the Great and Wild Ammonoosuc Rivers, occurs a vein of quartz, from one to four inches wide, containing bunches of copper pyrites.† Veins of copper pyrites occur in Franconia, said, by the State Geologist, to be "six or eight inches wide, but too narrow for profitable mining." The same ore is found in Eaton, and a vein opened for a few feet, "but the vein is too narrow, and the copper ore too rare for profitable mining." The same is said of small veins of this ore found in Canaan, and which continue into Grafton. In Bath is a vein on the estate of H. Lang, which runs north 50° west, with the formation, and is from four to eight inches wide. The locality is recommended by Dr. Jackson as worthy of being wrought. The principal locality of copper in this state is at Warren, where it occurs in a bed of tremolite, of over 50 feet in width, enclosed in mica slate. An attempt was made to work this bed about twelve years since, but it

* Third Annual Report, p. 83.

† Final Report on Geol. of New Hampshire, p. 65.

seems to have been abandoned, owing to the ore being too much scattered through the rock. Dr. Jackson remarks that he has no doubt that this mine, if wrought with economy and skill, will prove profitable.* The ore is much mixed with blende and iron pyrites, and the average yield of the bed is not known, but seems to be quite too low for working. The explorations at the time of my visit (1840) had not been sufficient to make it possible to say whether the locality was of value. There are several copper-bearing veins in this neighborhood, which should be explored. At Unity, on the farm of James Neal, is a vein of iron and copper pyrites, running with the stratification, and which has been traced for over 2000 feet, with a width of 1 to 3 feet. By my analysis the ore yielded nearly 12 per cent. of copper, and the locality may be one of some value. Nothing had been done in 1840 towards developing it.

VERMONT. — Several localities are mentioned by Prof. Adams, State Geologist, as containing copper pyrites. At Plymouth, pulverulent green carbonate is found, with particles of vitreous ore. In Corinth is a locality where copper pyrites has been found, along a line bearing north 10° west, for a distance of 200 rods. On the hill the vein is more than a foot wide, and the ore free from iron. A considerable quantity of it has been sent to the Revere Copper Company's works in Boston. Prof. Adams considers the locality as valuable, and it is understood that operations have just been commenced here by a New York company.

Rev. Z. Thompson states that in 1829, a furnace was erected at Strafford, for the purpose of smelting the copper pyrites which occurs there mixed with the sulphuret of iron. This latter mineral has been extensively worked for copperas, and its quantity is very great. If the percentage of copper is considerable, it might be worked with profit, but the undertaking proved a failure, whether from want of knowledge of smelting, or from the poverty of the ore, does not appear.

MASSACHUSETTS. — Copper pyrites and variegated copper

* Report, p. 151.

ore occur in small quantity in veins which have been worked to some extent for lead, in Northampton and Southampton. There is no reason to suppose that it exists in sufficient abundance to be profitably mined.

The other localities of these ores in the state, although somewhat numerous, are not proved to be of any importance.

CONNECTICUT.—One of the most extensive copper mines in the country exists at Bristol, where workings were commenced about 1836. According to Prof. C. U. Shepard,* the rocks are micaceous gneiss, talcose slate, and a decomposing granite, having a general northeast and southwest direction. The ore at the surface was mostly variegated copper. From Prof. Shepard's description, there does not appear to have been at that time any evidence of a fissure-vein, and this appears also to have been demonstrated by the result of the workings, which, although extensive, have not, on the whole, proved profitable.

From the Superintendent, H. H. Sheldon, Esq., I learn that this mine has produced over $200,000 worth of copper since it was first opened, and that it has yielded to its present proprietors, since January 1848, 1600 tons of ore, averaging 32 per cent. of copper, and valued at $172,810 58. The extent of ground opened by shafts, levels, and winzes, is about 930 fathoms. With the present arrangements for crushing and washing, 26 tons of 30 per cent. ore are now shipped monthly, but as soon as the new machinery is completed, a large increase upon this amount is expected. There are 52 persons employed in and about the mine.

A deposit of malachite, in the town of Manchester, has been worked at intervals, unsuccessfully, since the middle of the last century. It occurs in gneiss, in which specular iron and iron pyrites are also disseminated.† Copper pyrites occurs in numerous localities. According to Prof. Shepard's Report, the most favorable indications of this ore are those found at the topaz-vein in Trumbull, which he thinks may become, at some future day, a mining district of considerable importance. Other localities where pyritiferous ores have been found, but in small quantity, are, in Orange, at Lambert's Mine, where it was worked to some little extent, and

* Report on Geol. Survey of Conn., p. 46. † Shepard's Rep., p. 49.

with apparent improvement in depth; at Litchfield, Plymouth, Mine Hill in Roxbury, Chaplin, Westfield in Killingly, and in Griswold. It is also found at the Middletown Silver-Lead Mine, but not yet in sufficient quantity to be of any considerable importance. Prof. Shepard predicts the ultimate discovery of rich copper mines in Connecticut, and refers especially to the topaz and fluor veins of Trumbull, the mine at Bristol, the Roxbury Mine-Hill Veins, as well as to the lead mines generally, in the primitive (metamorphic palæozoic), all of which contain copper.

Within a short time, several companies have been formed to work copper mines in the neighborhood of Bristol, Litchfield, and Plymouth; a reasonable sum, judiciously applied to the development of some of the metalliferous deposits of this district, may lead to valuable discoveries; but it will be unfortunate for the adventurers if they are misled into the belief that any of these localities are likely to prove profitable at great depths, when there is nothing like a good vein within a reasonable distance from the surface.

New York.—The number of different localities in this state where traces of copper ore have been found, is quite considerable, but there is no one which is now mined to any extent.

At Sparta, a lode six inches wide was worked, some years ago, to the depth of 40 or 50 feet, by a company chartered in New York; but the results were unfavorable.

At the Ancram Lead Mine, traces of copper pyrites are found; also at Northeast, associated with lead ore, in some quantity.* Other localities are, New Canaan, in Columbia County, and Austerlitz.

At the Ulster Lead Mine, copper pyrites occurs in considerable quantity with the galena, and forms an item of some consequence in the produce of the mine. A sale of 50 tons, of 24·3 per cent. ore, is reported in 1853.

Pennsylvania.—There are numerous localities where copper ores have been found, as well in the metamorphic rocks as in the New Red Sandstone formation. None of those in the former position have been worked with any success.

* Mather's Report, p. 500.

The Gap Mine, in Lancaster County, was first opened in 1732, and afterwards taken up by another company in 1797, and again much more recently, by a company which made large expenditures; but it has never been worked with success or profit.

At the Elizabeth Mine, near Pottstown, a shaft has been sunk to the depth of 180 feet, in a bed of coarsely crystallized calc. spar, containing a little blende, sulphuret of copper, and specular iron; but the appearances are not promising.

The only copper mines of any importance now wrought in the state, are those of Chester and Montgomery Counties, which will be noticed farther on.

MARYLAND.—Several mines have been opened in this state, and more or less extensively worked for copper. Those in the vicinity of Liberty and New London, Frederick County, are the best known. Of these, Dr. Ducatel, State Geologist, remarks as follows:*

"Copper ores were also extensively raised in the neighborhood of Liberty Town, but the old works are, at present, abandoned. New operations have since been commenced at two other places in this section of the country, which deserve especial mention. The most extensive are those near New London, where extensive operations are carried on. The ore is a sulphuret of copper, occasionally mixed with the green carbonate, and is embedded in a mixed rock of talcose slate and limestone. The perpendicular shaft sunk in pursuit of these, is 114 feet deep; but the ore is now worked in two drifts, one to the east and the other to the west, the former to the extent of 17, the latter of 50 feet. The character of the ore, associated with porous quartz, and the continuance of the veins of nearly uniform width, being interrupted only occasionally by masses of limestone, are believed to give a sufficient promise that satisfactory results may be expected hereafter."

At that time, 22 persons were employed at the New London Mine, which was worked by Isaac Tyson, Jr. It is now abandoned.

*Dolly-Hide Mine.*—Dr. Ducatel refers to another mine, now the Dolly-Hide, in the following words: "The other copper mine referred to as worthy of investigation, is near Liberty Town, on the farm of Capt. Richard Coale. Shafts have been sunk 20 to 40 feet. The ore is a mixture of oxide of iron, a little manganese, copper-black, and carbonate of copper, the last forming two-thirds of the whole, in weight. It is soft and friable."

This mine was worked at intervals, up to 1846, when it was leased to Isaac Tyson, Jr., and is now worked by a company incorporated in Maryland, and

* Geological Report, for 1839, p. 22.

called the "Dolly-Hide Copper Company," with a capital of $600,000. The published reports give the whole quantity of ore raised from 1842 up to May 1853, as 191,933 lbs., average $22\frac{13}{32}$ per cent., and 127 tons of "black dirt," average $10\frac{5}{8}$ per cent. of copper.

From various reports, and especially a manuscript one by O. Dieffenbach, kindly placed at my disposal by Dr. F. A. Genth, it is evident that there is no vein here, but a broad band of crystalline limestone, which, where best developed, is 100 feet thick, and contains numerous segregated parallel layers of ore, mixed with quartzose matter, colored brown by iron, manganese, and copper. The rock on each side of the limestone belt is an argillaceous and talcose slate. The annexed section (Fig. 32), from the printed prospectus of the company, will serve to give an idea of the position of the ore, which, according to Mr. Dieffenbach, is mainly the variegated, with some copper pyrites. Argentiferous galena, containing 45 to 50 oz. of silver to the ton, also occurs, in quantity increasing in depth. The "black dirt" is a product of decomposition of these ores. From the printed report of this company, it appears that the mine produced, from 1842 to 1853, about 33 tons of metallic copper.

Fig. 32.

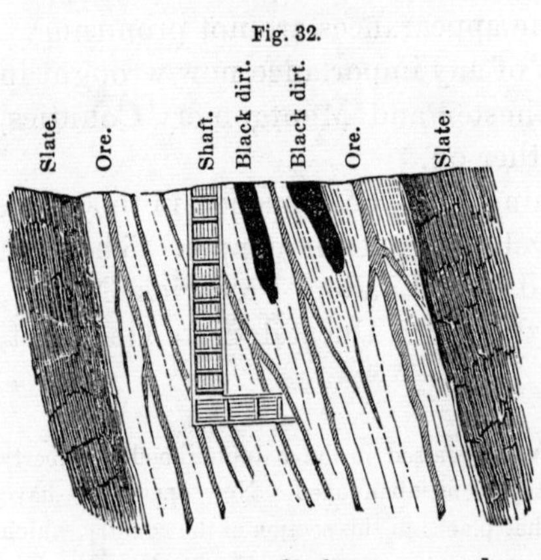


Section of the Dolly-Hide Cupriferous Bed.

The working of this mine had, until recently, been exceedingly faulty, owing partly to the abundance of water, and the ore which had been raised had been taken from small pits sunk only to the water-level. Under the present management, the mine appears likely to be thoroughly tested; but as there is evidently no vein here, too much caution cannot be used in speaking of its probable success; so long as ore can be found in sufficient quantity to pay for mining, the work may be prosecuted, but it would be attended with loss should it be carried on solely in the expectation of a richer and more abundant yield at a great depth.

The Dolly-Hide, Old Liberty, and New London Mines, seem to be all identical in character.

There is another metalliferous belt in the vicinity of Sykesville, which includes the Springfield, Carroll, Mineral Hill, and Patapsco Mines, and is of considerable interest. These mines lie, in the order named, in a line running from the southwest towards the northeast, and about ten miles in length. The rocks are talcose, chloritic, and hornblende slates, and the veins belong to the segregated class, being parallel with the formation.

*Springfield Copper Mine.* This mine, according to a manuscript report of Mr. O. Dieffenbach, is situated in Carroll County, about one mile north of the Sykesville Station of the Baltimore and Ohio Railroad. At this point the vein runs north 25° east; it underlays 13° to the southeast for the first 60 feet, then descends nearly vertically for 40 feet, then underlays 8° to northwest for a depth of 100 feet, and from that point, so far as it has been opened, a depth in all of 240 feet, it descends perpendicularly. The vein is 20 to 24 feet wide at the surface, and, in depth, never falls below 6 feet. At the surface, and for a depth of 60 feet, it consists of quartzose, magnetic, and specular oxides of iron. Here copper ores begin to be found disseminated in the vein, and the quantity increases with the depth. Copper pyrites predominates, with some variegated ore: iron pyrites is also invariably found in the vein, and is sometimes auriferous. There are also traces of cobalt and nickel, and some branches of the vein carry ores containing 4 to 5 per cent. of these metals. The workings are extended to a depth of 240 feet, the water being pumped by a steam-engine up to the adit-level, which drains the mine at a depth of 90 feet, and is 500 feet in length. The accessible workings consist of three levels driven at the depths of 90, 110, and 170 feet respectively, and which are connected by winzes.

This vein was first worked by Mr. Tyson as an iron mine, to supply the Elba furnace with ore, and it was in sinking the main shaft, for the purpose of finding better iron ore, that the copper was found sparsely disseminated through the vein. The cupriferous ores grew richer as the work descended; at a depth of from 90 to 100 feet, they yielded from 10 to 13 per cent. of metal; at 150 to 160 feet, from 12 to 15 per cent., and at the greatest depth yet reached they have increased in richness to 18 or 20 per cent. Up to this time 150 tons of ore have been raised, in addition to that obtained in sinking and driving; and, according to Mr. Dieffenbach, the operations are now conducted with energy, and a good prospect of success.

*Mineral Hill Mine.* This mine is six miles northeast of Sykesville. Old workings exist here, and the occurrence of ore at this place has been known for more than a century. There are four veins, of which three are now worked; they are parallel with each other, and run north 15° east. The rock in which they occur is a talcose and chloritic slate. One of them appears to be a fahlband of slate, impregnated with pyritous copper and small bunches of cobalt ore. The three others carry, at their outcrops, magnetic and specular ores of iron, in which traces of gold have been found; and as the depth to which they are worked increases, they become more and more intermixed with copper pyrites and variegated ore.

The mine-work thus far consists of three shafts, the deepest of which is sunk 250, and the others 160 and 90 feet, between which levels are driven at 100 and 160 feet. Out of the stopes above these drifts 100 tons of 15 to 20 per cent. ore have been taken, in addition to 15 or 20 tons raised in sinking and driving.

VIRGINIA.—Copper pyrites occurs in many localities, in connection with the auriferous iron pyrites of the gold mines, but in none of them has it yet been found in sufficient quantity to be of any importance, so far as I have been able to learn. One or two mines were formerly worked to some extent for copper in this state, but the Manassas Gap Mine, in Fauquier County, about 70 miles from Alexandria, appears to be the only one which is at present of any importance.

*Manassas Copper Mining Company.* From the report of Prof. B. Silliman, Jr., made April 18th, 1853, it appears that there are three veins which have been opened on to some extent. The map accompanying the printed report indicates two groups of veins, in one of which are three lodes parallel with each other, and with the formation, which are marked as carrying pyritous copper; of the other group are two, which are called red oxide veins; one has a course of north 30° east, parallel with the strike of the slates in which it is enclosed, and the other runs north 70° east, intersecting it at an angle of 30° or 40°. This cross-vein has been cut in a trial-shaft, at a small depth (not given), and found to be from 10 to 12 feet thick and dipping about 62°, while the slates are said by Prof. Silliman to dip nearly vertically.

These veins are well situated for drainage by a deep adit-level, and if it is the object of the company to test them thoroughly, such a work should be executed. The ore is said by Prof. S. to be both abundant and rich, the best specimens yielding 75 per cent. of pure copper.

NORTH CAROLINA.—Within the last year a very considerable amount of capital has been invested in the mines of this state, with the view of working them for copper as well as for gold. At the present stage of their development, it would be hazardous to pronounce positively that they will prove remunerative, but there can be no doubt that copper ores are widely distributed through the state, and that a considerable increase of our production of this metal may be looked for from the development of its mines, provided they are worked with skill and economy.

*North Carolina Copper Company.* This is the principal copper mine thus far opened in the state, and has produced considerable ore. It is situated about nine miles from Greensboro, in Guilford County, and was formerly worked for gold, and known as the Fentress or Stith's Mine. In 1852, it was purchased by a New York company, and has been wrought by them for copper only. The cupriferous deposit has a direction parallel with that of the slates in which it is enclosed, nearly north 30° east; its dip, which at the surface is only 15°, gradually increases, and is, at 70 feet in perpendicular depth, about 45°. The ore is almost solely pyritous copper, associated with some sulphuret of iron. It is said, on good authority, that there is a large quantity of ore exposed in this mine, but the workings have thus far been conducted with an entire want of judgment, the only aim seeming to be to raise as much ore as possible immediately, without regard to the future of the mine. It is stated by the parties interested in this concern, that they are now raising 100 tons a month of 25 per cent. ore, but no official statements of the produce of the mine have been published.

In almost all the old gold mines, more or less copper pyrites is found, and the quantity of this valuable ore which has been worked for gold, and then rejected, is said to have been, at some localities, very considerable. Some of these have been noticed in a former part of this work.* At the time I visited the region most of the mines were in such a state that no opinion could be formed of the amount of copper present in the veins, or whether they would become of permanent value for this metal at some depth. Almost all those who have visited the region more recently, concur in the opinion that it is destined to be of great value for copper as well as for gold. Some of the mines which are specified as promising well for copper, are, the Conrad Hill Mine in Davidson; the Vanderburg, Phœnix, Long and Muse's, and the Fink Mines, in Cabarrus; the M'Cullock Mine, in Guildford; and the Rhea and Cathay Mines in Mecklenburgh County.

Fig. 33.

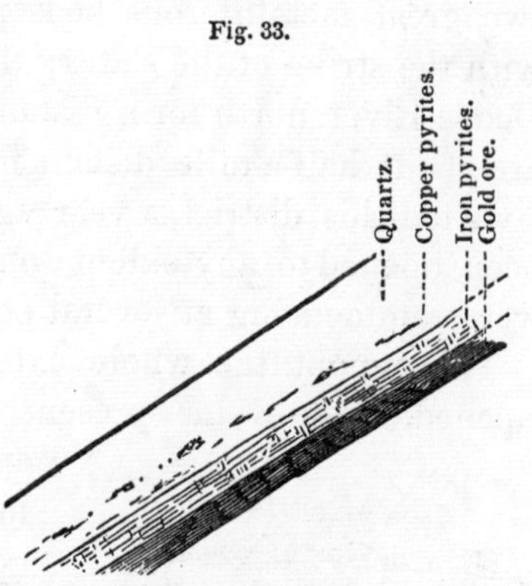


Section of the vein at the M'Cullock Mine.

The position of the copper ore in the M'Cullock Mine may be illustrated by the annexed section of the vein (Fig. 33). The pyrites is disseminated in white quartz, which overlies a thick band of almost solid iron pyrites; beneath this is a bed, of irregular thickness, of decomposed ferruginous matter, rich in gold.

* See pages 129 to 133.

At Pioneer Mills, in Cabarrus County, the new copper ore discovered by Dr. Genth, and called by him Barnhardtite, was first found. It seems to occur not unfrequently in the copper-bearing veins of North Carolina.

Tennessee.—The cupriferous deposits of Tennessee, which have recently excited so much interest, are situated in the extreme southeastern corner of the State, in Polk County, near the Ocoee River, and about thirty-five miles from the nearest station of the East Tennessee Railroad. The ores are contained in micaceous and talcose slates of Lower Silurian age, but which are so metamorphosed as not to be referable to any subdivision of the system. They dip at a high angle to the southeast, and have a course of north 20° east. There are two great metalliferous beds parallel with each other and with the strike of the slates; they have been traced from the Ocoee River north for a distance of three or four miles, and are about half-a-mile distant from each other. At the time I visited this district, a year ago, the western vein only had been opened to any extent; on the eastern one explorations were commencing at several points.

Throughout the whole extent of their course, wherever opened, these veins present a remarkable uniformity of appearance. The surface is marked by a heavy outcrop of gossan, which is particularly conspicuous along the summits of the ridges, where the ground often appears covered with masses of ferruginous material, over a width, in some places, of at least a hundred feet. On penetrating beneath the surface, the section represented in the annexed figure (Fig. 34), is obtained. Beneath the gossan is found a bed or mass of black cupriferous ore, of variable thickness and width. This, as well as the gossan, is the result of the de-

Fig. 34.

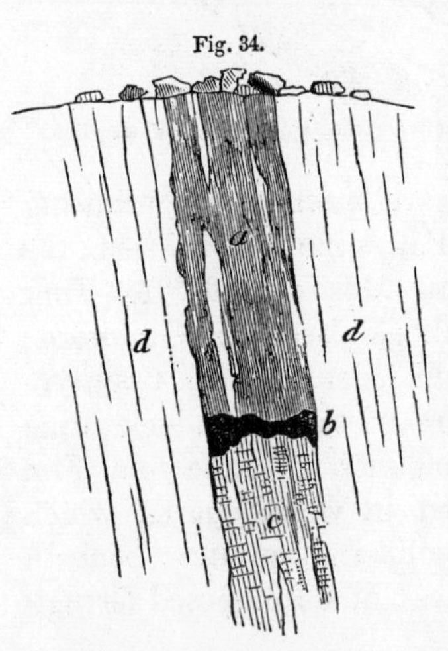


Section of East Tennessee copper vein: *a*, gossan; *b*, black ore; *c*, undecomposed portion of the vein; *d*, slates.

composition of an ore consisting originally of a mixture of the sulphurets of iron and copper, which were associated with a quartzose gangue. The place of the bed of copper ore marks the limit of the decomposition of the vein; beneath it, the ore exists in its original condition. The depth to which decomposition has extended is variable, and identical nearly with the level at which water is found. On the ridges it varies from 80 to 90 feet, in the valleys it is considerably less, probably not more than 20 to 30 feet on an average. The black ore is analogous to the "copper smut," or "copper black," which occurs so commonly in cupriferous veins, but I have never seen it in so extensive deposits in any other part of the world. It is a mixture of black oxide of copper with the sulphuret and some silicious or earthy matter. There is also considerable sulphuret of iron, in small crystals and fragments, scattered through it, as well as some sulphate of copper, and perhaps a little manganese. Its yield, as prepared for shipment, is from 20 to 25 per cent., although it varies very much in its composition in different parts of the same mine.

The thickness of the veins, from the decomposition of whose ores this black mass has originated, is in some places enormous. At one place in the Hiwassee Mine, the body of black ore was stated to be 45 feet in width, and the veins are said in some places to expand to much greater dimensions. The thickness of the black ore is equally irregular with its width. In some places it is accumulated in conical masses from which many hundred tons of ore, nearly pure, are taken. When the veins were first opened, I estimated its average width on the whole extent of the vein at 10 feet, and its thickness at 2, remarking that it would probably not fall below those dimensions. What the results of farther explorations along the line of the veins have been, within the last year, I have not been informed.

The facility with which the black ore may be mined is great, as little blasting is required. Shafts may be sunk in the gossan without being timbered, and the ore may be taken out with a pick and shovel, and is so wide that several men can work abreast in the levels driven on it. The position of the ridges affords great facilities for driving in levels transversely to the veins, through which the ore can be brought to the surface.

There has been hardly any attempt, as yet, to develop the veins below the line of the black ore. At the Hiwassee and Tennessee Mines, they had, a year ago, sunk a few feet in a hard quartzose rock, carrying a good deal of iron pyrites and some copper ore; but not enough of the latter to constitute a workable ore. There is good reason to believe, however, that farther explorations will develop valuable bunches of copper pyrites, and to this the attention of holders of property in this region should be directed; since, although the quantity of black ore which may be taken out is very large, yet the time will come when it will be exhausted, and the mines must depend for their permanent value on the development of the pyritiferous ore in the solid portion of the veins.

Several companies have commenced work in this region, within the last year, besides the Hiwassee and Tennessee Companies, which have been mining since 1852. The latter has made one or more dividends from the proceeds of its sales. The Hiwassee Company sold, in 1853, 380 tons of ore, averaging about 25 per cent. of copper. From 18 to 20 laborers of the vicinity are employed, about 8 of whom are engaged in mining. It is their intention, the new road through the valley of the Ocoee to Cleaveland, Tenn., having been completed, to forward 300 tons of ore per month.

The present great disadvantages under which these companies labor are, the cost of forwarding the ore, and the difficulty of procuring the necessary teams to transport it to the railroad or the Hiwassee River. It is understood that a furnace is to be erected at the mines, by which the ores will be reduced, and the product shipped in the form of a regulus containing 70 or 80 per cent. of copper.

Several English companies are said to have purchased property on the line of these veins; but I am unable to give any particulars of their movements.

### COPPER ORES IN THE NEW RED SANDSTONE.

Having thus noticed the principal localities of copper ores

in the metamorphic rocks of the Appalachian chain, we will proceed to describe, briefly, the deposits in the sandstones and associated trappean rocks, which are known in the Atlantic States as the "New Red Sandstone." This formation is principally developed in the Connecticut River Valley, and in the State of New Jersey, where it occupies a belt of about thirty miles in width. It extends south, along the flanks of the Appalachian chain, gradually thinning out, then appearing only in patches, and finally disappearing altogether in North Carolina.

Throughout this belt the ores of copper have been frequently found, and always under quite similar circumstances; they were the first copper mines worked in the country, and for a long time were supposed likely to prove of great value. The ores worked were mostly the red oxide and carbonate, with some native copper, and seem all to have been contact-deposits, at or near the junction of the sandstone and trap. Their mode of occurrence is illustrated in the annexed section (Fig. 35). The ores are found in irregular bunches at the junction of the two rocks, or in the sandstone, at a little distance from the trap, in deposits parallel with the stratification.

Fig. 35.

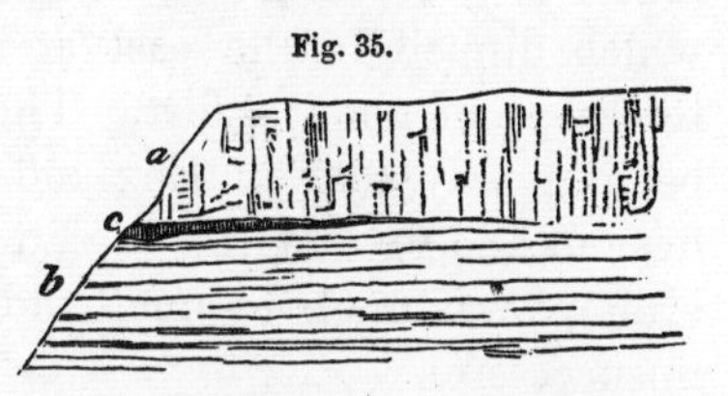


Occurrence of copper ores in the New Red Sandstone. *a*, Trap; *b*, Sandstone; *c*, Copper ores.

None of them are at present wrought, but their working forms an interesting chapter in the mining history of this country. Traces of cupriferous ores are found in the New Red Sandstone of Massachusetts, but have never been mined to any extent. The principal locality in this state is in Greenfield, at Turner's Falls, where the trap and sandstone rocks are somewhat stained by cupriferous ores; but I have never seen any well-developed veins here, which appeared worthy of becoming the object of mining enterprise.

Connecticut.—In this state, at the "Simsbury Copper Mines," in the eastern part of the town of Granby, are deposits of ore which were discovered and worked in the early part of the last century. The company was chartered in 1709, and appears to have been the first incorporated mining company in the country. According to Professor C. U. Shepard,* the ore is contained in a fine-grained, yellowish-gray sandstone, and occurs in beds of greater or less extent, as well as in nodules and strings. The principal ore was vitreous copper, with the variegated ore and malachite in small quantity. About the middle of the last century, after considerable ore had been extracted, at different times, during a period of forty years, the mines were abandoned, but were afterwards, with unheard-of barbarity, purchased by the state, and used as a prison for sixty years. Later, in 1830, the property came into the hands of a company, and was worked for a few years, and then, in all probability, finally and for ever abandoned.

At the time they were examined by Prof. Shepard, the workings were carried on at a depth of about 50 feet, in a bed which dipped to the east at an angle of 25°, and had a thickness of about 2 feet. The ores, which appear to have been mostly sold in England, were poor, a shipment sent over in 1830 averaging in value only $33 60 per ton. An attempt was made to smelt them at the mine, which proved unsuccessful.

New Jersey.—The ores of copper, especially the carbonate, red oxide, and sulphuret, exist in many places in this state, in the red sandstone region, and in every case in connection with the trappean rocks. The principal points where mining operations have been carried on are near Belleville, Griggstown, Brunswick, Woodbridge, Greenbrook, Somerville, and Flemington.

A few historical notices of the attempts which have been made to mine in this district will not be without interest. They have been principally collected from Morse's Gazetteer and Gordon's Gazetteer of the State of New Jersey.

About the years 1748, 1749, and 1750, several lumps of virgin copper, from 5 to 30 lbs. in weight (in the whole upwards of 200 lbs.), were ploughed up in a field belonging to Philip French, near New Brunswick. A lease was

* Report of the Geol. Survey of Conn., p. 42.

secured of the property, a company formed, and operations commenced in 1751. A shaft was sunk on a spot "where a neighbor, passing it in the dark, had observed a flame rising from the ground, nearly as large as the body of a man." The account goes on to state that a sheet of copper was struck, "somewhat thicker than gold-leaf," between walls of loose sandstone. Lumps of copper were found of from 5 to 30 lbs. in weight. The company followed the vein for 30 feet, when the accumulation of water exceeded their means of removing it, and the work was abandoned, some tons of copper having been obtained, and shipped to England. A stamping-mill was erected. It is said that sheets of copper, "of the thickness of two pennies, and three feet square," were taken from between the rocks, within four feet of the surface.

After the first company had abandoned this work, several efforts were made at various periods to renew the operations, and extensive excavations were made, but always without success. A shaft was sunk to a great depth, and an adit driven several hundred yards.

The Schuyler Mine, near Belleville, on the left bank of the Passaic, seven miles from Jersey City, was discovered about the year 1719, by Arent Schuyler. The ore, cropping out on the side of a hill, was easily raised, and was probably pretty abundant near the surface. At any rate, it appeared from the books of the discoverer that 1386 tons of ore had been shipped to England before the year 1731; what percentage it yielded is not known. The son of the discoverer continued the workings with an increased force, but the returns of ore raised were lost during the war. In 1761, the mine was leased to a company, a steam-engine erected, and works are said to have been carried on profitably for four years, when the engine-house was set on fire and burnt by a discharged workman, and the mine was then abandoned. A company of English capitalists afterwards obtained permission to erect smelting works at this mine, and offered Mr. Schuyler £100,000 for the estate, which was refused, and the whole lay dormant until after the revolutionary war. Several companies have since been organized, and have expended large sums of money on this mine, but have all proved entire failures.

The body of the ore appears to be imbedded in a stratum of sandstone 20 to 30 feet in thickness, and to dip at an angle of about 12°. The excavations have been carried to a depth of 212 feet. The ores occur mixed with the sandstone, and there are no indications of a true vein.

The Franklin Mine, near Griggstown, in Somerset County, is said to have been worked to a depth of 100 feet, and drained by a long adit. The ore was found in the shale, near its junction with the trap; it consisted chiefly of carbonate and red oxide. A considerable sum of money was expended here, without any return.

The Flemington Copper Mine was the only one of the mines wrought at the time of Prof. Rogers's survey (1836). He describes the ore as being intimately blended and incorporated with an indurated and altered sandstone, which constitutes a metalliferous belt of variable breadth, sometimes 20 or 30 feet wide, and preserving a nearly north and south direction for several hundred feet. The ore is a mixture of gray sulphuret and carbonate, and is generally in small particles disseminated through the rock.

The Bridgewater Mine, near Somerville, at the base of a trap ridge, was worked before the Revolution to a considerable extent, and with much loss of capital. A smelting furnace is said to have been erected here, by German workmen, and two masses of copper are mentioned as having been found in 1754, weighing 1900 lbs. This mine was again opened and worked in 1824, and was abandoned with loss.

It would seem as if these failures might have been sufficient to warn capitalists from wasting any more money in these mines. The State Geologist, in his report, remarks that there are no true veins in this formation, and warns against farther expenditures, unless made with the greatest caution.

Notwithstanding all this, New Jersey had, in 1846 and 1847, its little copper fever as well as Lake Superior. In 1847, there were six mining companies organized in this district, with 69,500 shares, and their market value exceeded $1,000,000. The Raritan Mine, 3 miles southwest of New Brunswick, was purchased at a high price, and large expenditures were made under the advice of Dr. C. T. Jackson and J. H. Blake, Esq. The Passaic Mining Company erected a steam-engine and expended a large sum of money, near the old Schuyler Mine. The Nechanic Mine, near Flemington, which had been worked before the Revolution, was reopened at a considerable expense. The Washington Mine, near the old Bridgewater Mine, at Somerville, was another of these unfortunate concerns, in which the future profits *per acre* were calculated to a fraction of a dollar.

All these mines were abandoned, after heavy expenditures, with almost total loss of the whole amount invested; and it is to be hoped that no more money will be sunk in them.

### CUPRIFEROUS VEINS AT THE JUNCTION OF GNEISS AND NEW RED SANDSTONE.

There is in Montgomery and Chester Counties, in Pennsylvania, a metalliferous district of peculiar interest, which is now developing to a considerable extent, and which has been favorably reported on by those mining engineers and geologists who have examined it. According to H. D. Rogers, State Geologist, who has made this region the subject of a special report, the metalliferous zone ranges in a general east and west direction across the Schuylkill River, occupying a belt of country six to seven miles long, in the vicinity of Perkiomen and Pickering Creeks, not far from the boundary of the gneiss, or metamorphic rocks, and the new red sandstone. Within this space are some ten or twelve lodes, some of which are said to be confined to one formation and some to the other, while others traverse both. Prof. Rogers states that, as a general fact, those veins which are confined entirely or chiefly to the gneiss, bear lead as their principal metal, whereas those which are included solely within the red shale are characterized by containing ores of copper. The Perkiomen and Ecton Lode, the United Mine Lode, the Shannonville South Lode, and a few others, are mentioned as true copper lodes, and all in the red shale. The phenomena of the lodes, the different systems, and other interesting facts in connection with this district, will be noticed in the chapter devoted to lead, as this region promises to be of more importance for this metal, and the silver which it contains, than for its copper. There seems to be but one company engaged in working for copper only; their works will be noticed in this connection:—

*Perkiomen Consolidated Mining Company.* From the Reports of this company it appears that it was organized in March 1851, by the consolidation of the Ecton and Perkiomen Mines, both of which were on the same lode, and their engine-shafts at a distance of about 1800 feet apart.

From the Report of the Manager, C. M. Wheatley, Esq., April 1, 1852, it appears that at that time, the engine-shaft in the Perkiomen Mine was sinking below the 50-fathom level, and that the lode at that depth was from 4 to 9 feet in width, made up of quartz, gossan, and sulphate of baryta, with green carbonate of copper and copper pyrites in places. It had decidedly improved from the 40-fathom level down. At the Ecton Mine, the 54-fathom level was driving from the engine-shaft west, on a lode varying from 2 to 5 feet in width, with good stones of copper pyrites, but not worth stoping.

On the 1st of May, 1853, the engine-shaft in the Perkiomen Mine was 62 fathoms deep. In the 50-fathom level the lode is represented as not productive, but expected to be at farther depth. In the Ecton Mine, the engine-shaft was down 66 fathoms, and a level driven each way a few fathoms, on a lode from 3 to 8 feet in width, but poor in ore.

From August 1851 to April 1852, 524 tons of ore were sold by this company, which realized $30,573, and varied from 7 to 23 per cent. of copper, yielding about 100 tons of pure copper. In the year ending April 1853, 143 tons of ore were sold, for $9,989.

In September 1853, the then manager, Mr. Rogers, remarks that as all the ore had been stoped out, it would require fully eight months to open the ground before farther sales could be made.

Before closing the description of the copper mines of this country, those of New Mexico should be alluded to, although I know nothing of their geological position, and have therefore been unable to arrange them in any particular group, in reference to the other mines of the United States. The region of the head-waters of the Gila is spoken of by travellers as rich in copper ores, and were they nearer a market they might become of importance; at present they must be looked on as of little value. The most celebrated mine, according to Dr. Wislizenus,* is that of Santa Rita de Cobre; it was opened in 1804, and worked from 1828 to 1835 by a Frenchman named Courcier, who is reported to have cleared half a million of dollars from it. This statement is corroborated by Mr. Bartlett, U. S. Commissioner,† who remarks, that indications of copper are abundant on the surface throughout the vicinity. The ore is chiefly the red oxide, and it is contained in a feldspathic rock, but of what age or character I am not able to state.

The copper raised is said to have been coined in the Mint of Chihuahua, by which state the mines were at that time claimed as belonging to its territory. Supplies were necessarily brought a distance of 400 miles by wagon; and, in 1838, a train was attacked by the Apaches, and taken possession of, and all farther access to the mines cut off. They remained entirely abandoned from that time up to 1851, when the place was occupied by the Boundary Commission. Dr. Wislizenus remarks, "that the whole range is intersected with veins of copper, and placers of gold."

In the preceding pages have been passed in review the principal cupriferous regions of the world, and the results will now be presented in a tabular form. And in the table

* Memoir of a Tour to Northern Mexico, p. 47.

† Personal Narrative of Explorations and Incidents in Texas, New Mexico, &c., i. 228.

immediately following, the produce of copper, in tons, of the different states of Europe, together with the most reliable estimates in regard to the other countries of the Eastern Hemisphere, are given for every fifth year, from 1820 to 1845, and from that date to the present time for each successive year. As in the other tables of this kind, the blanks are only filled up when tolerably reliable data were attainable, and a 0 in the column indicates that, previous to that period, no copper had been produced by that country.

| | Russia. | Sweden. | Norway. | Great Britain. | Prussia. | Harz. | Saxony. | Austria. |
|---|---|---|---|---|---|---|---|---|
| 1820, . . | | | | 8,127 | | | | |
| 1825, . . | | | | 10.358 | | | | 1,776 |
| 1830, . . | 3,800 | | | 13,232 | 838 av. | 196 | 19 | 2,165 |
| 1835, . . | | 849 | | 14,470 | 796 " | 220 | 21 | 2.465 |
| 1840, . . | 4,600 av. | 816 | | 13,020 | 991 " | | 27 | 2,657 |
| 1845, . . | | | 642 av. | 14,900 | 955 " | | | 2,695 |
| 1846, . . | | | | 14.950 | 1,211 | | | 2,801 |
| 1847, . . | | | | 13,780 | 1,245 | | | 3,310 |
| 1848, . . | 4,700 | | | 14.720 | 1,179 | | 11 | 3,309 |
| 1849, . . | 5,550 | | | 13,600 | 897 | | 38 | |
| 1850, . . | 6,450 | 1,423 | 567 av. | 14,700 | 1,181 | 150 | | |
| 1851, . . | | | | 14,300 | 1,450 | | 51 | |
| 1852, . . | | | | 14,300 | | | | |

| | France. | Spain. | Italy. | Turkey. | Algiers. | Asia. | Australia. | New Zealand. |
|---|---|---|---|---|---|---|---|---|
| 1820, . . | | | | | | | | |
| 1825, . . | 137 | | | | | | | |
| 1830, . . | 229 | | | | | | | |
| 1835, . . | 79 | | | | | | | |
| 1840, . . | 107 | | | | 0 | | 0 | 0 |
| 1845, . . | 142 | 285 | | | | | 450 | 30 |
| 1846, . . | | | | | | | 850 | 60 |
| 1847, . . | | 380 | 250 | 2,000 | 500 | 2,400 | 1,600 | 125 |
| 1848, . . | | | | | | | 1,500 | 50 |
| 1849, . . | | 450 | | | | | 2,500 | 75 |
| 1850, . . | | | | | | 3,000 | 3,300 | 200 |
| 1851, . . | | | | | | | 3,500 | 375 |
| 1852, . . | | | | 400 | | | 3.250 | |
| 1853, . . | | | | | | | 2,650 | |

For most of the European countries, the statements are derived from official sources, and may be relied on as being very near approximations to the truth. For Spain, Italy, and Turkey, the data are very imperfect. The estimates for Australia have been compiled from the returns of ore sold at Swansea, and the statements of exports from that colony; those for New Zealand from the former source alone. In regard to that part of Asia which is not included in the Russian Empire, but little information can be obtained. China produces a considerable quantity of this metal; but as it does not affect the commerce of the world, being all consumed within its own borders, and as no estimate whatever can be made of its amount, it is not included in the tables.

The next table gives the production of the New World, of which almost the whole quantity is furnished by Cuba and Chili, as it is only quite recently that the United States have begun to produce a notable amount of copper. The production of Cuba is made up chiefly from the sales of ores in England; that of Chili from various returns of exports from that country, some of which were furnished by the English and French consuls, and others published by the government, collated with the official statements of imports from that country into England, France, and the United States. The other countries of South America furnish a small amount, of which, at best, but a very imperfect statement can be given.

| | Chili. | South America. | Cuba. | U. S. & Can. |
|---|---|---|---|---|
| 1830, . . . . . . | | | 0 | |
| 1835, . . . . . . | | | 700 | |
| 1840, . . . . . . | 9,000 | | 4,500 | |
| 1845, . . . . . . | 13.270 | | 6,800 | 100 |
| 1846, . . . . . . | 13,800 | | 5,150 | 150 |
| 1847, . . . . . . | 11,850 | | 4.000 | 300 |
| 1848, . . . . . . | 12.275 | | 4,000 | 500 |
| 1849, . . . . . . | 12,450 | | 3,600 | 700 |
| 1850, . . . . . . | | 1,200 | 3,400 | 650 |
| 1851, . . . . . . | | | 3,400 | 900 |
| 1852, . . . . . . | | | 2,600 | 1100 |
| 1853, . . . . . . | | 1,300 | 2,500 | 2000 |

In order to exhibit at one view the gradual increase in the production of the world during the last twenty years, as well as the relative amount which each country has contributed to the grand total, a third table is appended. In this the produce of each country is given in tons, and opposite to it, in the next column, in percentage of the whole amount.

In this statement, as in others of the same kind which have been given, where reliable data were wanting, the places have been filled by the best estimates which could be formed. The whole amount given for each year is not, probably, far removed from the truth.

| | 1830. | | 1840. | | 1850. | | 1853. | |
|---|---|---|---|---|---|---|---|---|
| Russian Empire, | 3,800 | 14·9 | 4,600 | 11·2 | 6,450 | 11·8 | 6,500 | 11·7 |
| Scandinavia, | 1,500 | 5·9 | 1,500 | 3·7 | 2,000 | 3·7 | 2,000 | 3·6 |
| Great Britain, | 13,200 | 51·7 | 13,000 | 31·7 | 14,700 | 26·8 | 14,500 | 26· |
| German States, | 1,100 | 4·3 | 1.250 | 3·1 | 1,400 | 2·6 | 1,450 | 2·6 |
| Austrian Empire, | 2,150 | 8·4 | 2,650 | 6·5 | 3,300 | 6· | 3,300 | 5·9 |
| Rest of Europe, | 500 | 2· | 700 | 1·7 | 1,000 | 1·8 | 1,000 | 1·8 |
| Turkey, | 500 | 2· | 1,000 | 2·4 | 1,000 | 1·8 | 600 | 1·1 |
| Africa, | | | 100 | ·2 | 600 | 1·1 | 600 | 1·1 |
| Asia, | 2,500 | 9·8 | 2,500 | 6·1 | 3,000 | 5·4 | 3,000 | 5·4 |
| Australia and New Zealand, | | | | | 3,500 | 6·4 | 2,950 | 5·3 |
| Chili, | 200 | ·8 | 9,000 | 22· | 12,500 | 22·9 | 14,000 | 25·1 |
| Rest of South America, | | | 100 | ·2 | 1.200 | 2·2 | 1,300 | 2·3 |
| Cuba, | | | 4,500 | 11· | 3,400 | 6·3 | 2,500 | 4·5 |
| United States and Canada, | 50 | ·2 | 100 | ·2 | 650 | 1·2 | 2,000 | 3·6 |
| | 25,500 | | 41,000 | | 54,700 | | 55,700 | |

From this statement, it will be seen that the amount of copper produced in the world has more than doubled within the last twenty-five years, and that this increase has been due mainly to the opening of new cupriferous regions in Cuba, Chili, Australia, and, in some degree, in the United States. Our own production has risen in that time from a few tons, raised at irregular intervals, up to 2000, and it bids fair to go on rapidly increasing.

It is still, however, far from supplying our demand for this metal, since our importations have tended constantly to increase rather than to decrease, since our own mines began to produce a noticeable quantity. This will be seen by examining the appended statement of the value of the copper, of various descriptions, imported and retained for home consumption, compiled from official sources, and given for each fiscal year from 1840 to 1852.

| | Pigs, Bars, and Old. | Plates suited to Sheathing. | Manufactured Art. & Sunds. | Ore. | Total. |
|---|---|---|---|---|---|
| 1840, . . | $1,130,727 | $373,915 | $69,799 | | $1,574,441 |
| 1841, . . | 1,089,229 | 465,165 | 98.294 | | 1,652.688 |
| 1842, . . | 866,498 | 325,117 | 80,729 | | 1,272.344 |
| 1843, . . | 271,288 | 164,598 | 25,831 | $64,148 | 525,865 |
| 1844, . . | 568,786 | 588,565 | 135,061 | 56,485 | 1,348,897 |
| 1845, . . | 1,176,494 | 676,161 | 114,689 | 48,807 | 2,016,151 |
| 1846, . . | 1,053.294 | 824,915 | 147,111 | 98,156 | 2,123.476 |
| 1847, . . | 1,491,209 | 999,026 | 140,640 | | 2,630.875 |
| 1848, . . | 301,476 | 813,202 | 153,074 | 158,302 | 1,426,054 |
| 1849, . . | 925.202 | 984,909 | 232 296 | 177,111 | 2.319.518 |
| 1850, . . | 877,343 | 682,765 | 338,652 | 188,632 | 2.087,392 |
| 1851, . . | 1,422,206 | 710,892 | 366,527 | 65,266 | 2,564,891 |
| 1852, . . | 1,499,467 | 610,755 | | 257,357 | |

As, however, a considerable amount of the same metal is brought into the country in the modified forms of brass and sheathing-metal, it is necessary to give a corresponding statement for these also; and we have farther added, to complete the view of the whole commerce in copper of the country, an account of the "copper and brass, and copper manufactured," of domestic production, exported during the same period.

| | Brass. | | | Exported. |
|---|---|---|---|---|
| | Pigs, Bars, Old, Sheet, etc. | Manufactured Articles. | Sheathing Metal. | Copper and Brass, etc. |
| 1840, . . . . . | $1,675 | $243.695 | | 86,954 |
| 1841, . . . . . | 2,423 | 222,534 | | 72.932 |
| 1842, . . . . . | 4,683 | 160,234 | | 97,021 |
| 1843, . . . . . | 135,124 | 38.189 | $276 | 79,234 |
| 1844, . . . . . | 49,241 | 83.158 | 8,971 | 91.446 |
| 1845, . . . . . | 13,702 | 141.723 | 5,646 | 94,736 |
| 1846, . . . . . | 2,673 | 146,420 | 11,341 | 62.088 |
| 1847, . . . . . | 59.437 | 145,822 | 4.127 | 64.980 |
| 1848, . . . . . | 36,055 | 163.535 | 225,348 | 61,468 |
| 1849, . . . . . | 20,434 | 153 773 | 220 034 | 66,203 |
| 1850, . . . . . | 18,901 | 160,225 | 469.798 | 105,060 |
| 1851, . . . . . | 7,918 | 161,484 | 286.357 | 91,871 |
| 1852, . . . . . | | | 604,809 | |

By far the largest portion of our imports of pig and bar copper have been, within the last few years, from Chili.* Our sheathing and sheathing metal come almost exclusively from England. The ores which we import are chiefly Cuban and Chilian.

The metallurgic treatment of the ores of copper is a process requiring much skill and a large amount of capital. Moreover, it is usually advantageous to mix together ores of different qualities, so as to be able to smelt the poorer and more refractory kinds by adding to them the richer. These circumstances led naturally to the concentration of the smelting business at Swansea, in Wales, a point which presented unrivalled advantages in the proximity of coal and convenience of access. Hence, the ores of foreign mines, and especially of South America, Cuba, and Australia, found their way thither almost exclusively, for a long time; capital, skill, or fuel, being wanting at home. Until recently, more than half the copper of the world was smelted there. The annexed table of the sales at Swansea, from 1828 up to the present

* See Table of Imports from Chili, page 243.

time, will show the immense extent of this business, which has usually been in the hands of seven or eight different smelting companies. It presents the amount of ore sold, in tons of 21 cwts., specifying the country in which the ores were raised. The column headed "sundries," includes small items from a variety of different sources, and especially from the states bordering on the Mediterranean and from New Zealand, in which the quantities were too small and irregular in amount to be thought worthy of a separate mention.

| | England and Wales. | Ireland. | Norway. | Chili. | Cuba. | Australia. | U. States and Canada. | Sundries. | Total. |
|---|---|---|---|---|---|---|---|---|---|
| 1828, . . . . | 3,875 | 8.510 | 199 | | | | | | 12,584 |
| 1829, . . . . | 6,796 | 7,044 | 456 | 187 | | | 25 | | 14,508 |
| 1830, . . . . | 2.203 | 9,115 | 733 | 201 | | | | | 12.252 |
| 1831, . . . . | 1,982 | 9,707 | 674 | 244 | | | | 57 | 12,664 |
| 1832, . . . . | 3,830 | 11,399 | 531 | 33 | | | 15 | 62 | 15,870 |
| 1833, . . . . | 2,147 | 11,293 | 624 | 435 | | | | | 14,499 |
| 1834, . . . . | 3,713 | 17,280 | 453 | 1,107 | 517 | | | | 23,070 |
| 1835, . . . . | 4.038 | 22,123 | 329 | 2,342 | 4.087 | | | | 32,919 |
| 1836, . . . . | 2,233 | 21,013 | 1,099 | 4,402 | 3,106 | | 20 | 419 | 32,292 |
| 1837, . . . . | 2,395 | 22.306 | 1.277 | 6,825 | 6.405 | | 14 | | 39,222 |
| 1838, . . . . | 4,374 | 22,161 | 1,023 | 10,924 | 7,725 | | | 196 | 46,403 |
| 1839, . . . . | 4,449 | 23,613 | 479 | 8,436 | 15,148 | | | 29 | 52,154 |
| 1840, . . . . | 2,277 | 20,166 | 55 | 10,325 | 24,831 | | 140 | 3 | 57,797 |
| 1841, . . . . | 1,885 | 14.321 | 38 | 10,395 | 30,864 | | | 67 | 57,570 |
| 1842, . . . . | 2,767 | 15,253 | 36 | 9,475 | 34,562 | | 69 | 250 | 62,412 |
| 1843, . . . . | 1.889 | 17,600 | | 11,550 | 28,071 | | 61 | 1,057 | 60,228 |
| 1844, . . . . | 1,130 | 20.063 | | 11,857 | 33.331 | 61 | 10 | 232 | 66,684 |
| 1845, . . . . | 2,536 | 19,647 | | 4.755 | 39,270 | 1,635 | 395 | 588 | 68,826 |
| 1846, . . . . | 1,584 | 17,553 | | 7,721 | 27,279 | 3,232 | 675 | 441 | 58,485 |
| 1847, . . . . | 746 | 14.373 | | 5,795 | 21,918 | 6,321 | 407 | 1,259 | 50,819 |
| 1848, . . . . | 774 | 12,633 | | 4,163 | 25,778 | 5,891 | | 121 | 49,360 |
| 1849, . . . . | 1,677 | 9.852 | | 923 | 23.282 | 7,552 | | 307 | 43.593 |
| 1850, . . . . | 1,574 | 10,478 | | 1,537 | 21,591 | 4.561 | | 1,972 | 41,713 |
| 1851, . . . . | 592 | 11.678 | | 827 | 21,692 | 2,328 | 219 | 2,502 | 39.838 |
| 1852, . . . . | 1,504 | 10,104 | 89 | 892 | 16.177 | 1,356 | 513 | 1,019 | 31,654 |
| 1853, . . . . | 2,174 | 11,367 | | 1,203 | 14,058 | 1,040 | 1,046 | 2,086 | 32,974 |
| | 65,144 | 390,652 | 8,095 | 116,554 | 399,692 | 33,977 | 3,609 | 12 667 | 1,030,390 |

The above table of course does not include the Cornish ores, which are sold in Cornwall itself, before being brought to Swansea. The foreign ores sold at the latter place have usually averaged about 20 per cent. of pure copper. Previous to 1827, a heavy duty was imposed on foreign ores brought into England to be smelted; from that time until 1842, their free importation was permitted, but the smelters were obliged to send the whole of the copper produced from such ores out of the country to be sold, or else to pay a heavy duty. As this system was found to operate unfavorably to the English mines, by making it for the interest of the smelters to keep the price of copper lower in the foreign than in the home market, it was done away with in 1842, and a small differential duty was imposed on the ore of from 3 to 6 shillings per ton, according to the percentage of metal it contained. Under this arrangement, the quantity of copper smelted at Swansea increased rapidly, and amounted in 1844 to 27,515 tons, of which 12,674 tons were of foreign origin. Since that time, there has been a gradual falling off in the sales of foreign ores in England, although in 1848 an Act of Parliament was passed admitting them at a nominal duty of 1 shilling per ton, without regard to their percentage of copper. This may be accounted for by the fact, that smelting works have been established in Australia, Chili, and other countries, which formerly sent large quantities of ore to Swansea, and the Cuba mines have fallen off considerably; thus cutting off, in part, the supply from three of the most abundant sources.

In this country there are several establishments for smelting copper ores. The native copper of Lake Superior is separated from the small portion of gangue which accompanies it, by a single melting in large reverberatory furnaces. There are two establishments where this is principally done, one at Detroit and the other at Pittsburgh. Besides these, there are smelting works in which imported ores and those furnished by the mines of the Atlantic States, hitherto in comparatively small quantity, are treated. These are near Boston, New Haven, New York, and Baltimore.

# CHAPTER VII.

## ZINC.

## SECTION I.

### MINERALOGICAL OCCURRENCE AND GEOLOGICAL POSITION OF THE ORES OF ZINC.

MINERALOGICAL OCCURRENCE.—The ores of zinc are quite numerous, but all the metal of commerce is obtained from a few of them, and, quite contrary to what has been seen to be the case with most of the metals thus far described, the sulphuret is not one of the number. It is a curious fact, that although the ores of zinc have been employed in making brass for a great length of time, the metal itself is quite of modern use. The following are the principal ores of zinc, the metal being never found in a native state:—

#### I. COMBINED WITH SULPHUR.

*Blende,* Sulphuret of Zinc, Black-jack of the Cornish miners. This mineral is composed of one atom each of zinc and sulphur, or 33·10 of sulphur, and 66·90 of zinc. It occurs both massive and crystallized, the purest crystals being of a fine honey-yellow color. It almost always, however, contains more or less iron, forming the well-known "black-jack" of the miners. It is almost invariably associated with the ores of lead, and frequently with those of copper and tin. In order to free this ore from its sulphur, before reducing the metal, a long and careful roasting is necessary; hence the process is more expensive than that required for the oxide and carbonate, and blende is accordingly but little used for the manufacture of zinc, although so universally disseminated. Associated with blende, and mixed with it in small quantity, the sulphuret of cadmium is occasionally found.

*Rionite* is a seleniuret of zinc and mercury, a very rare mineral.

*Voltzite.* A combination of the sulphuret and oxide of zinc; of no importance as an ore.

### II. COMBINED WITH OXYGEN.

*Red Zinc Ore*, Red Oxide of Zinc. An ore of zinc which is only found in New Jersey. It consists mainly of the oxide, which, when pure, contains 19·74 of oxygen, and 80·26 of zinc, but is mixed with a small percentage of oxide of manganese, which gives it, probably, its fine red color, the artificial oxide being of a pure white, when uncontaminated by foreign impurities.

This ore was first noticed and described by Dr. Bruce, one of the earliest cultivators of the natural sciences in this country. It is a very curious fact that it should be found only in this one district, and there so abundantly. The red zinc ore is mechanically mixed with Franklinite, and associated with calcareous spar, at Franklin and Stirling. A mass of it was on exhibition at the Crystal Palace in London, which weighed 16,400 lbs.

### III. SILICATES, CARBONATES, SULPHATES, AND ARSENIATES.

*Electric Calamine*, Silicious Oxide of Zinc. This is a silicate of the oxide of zinc, with water. It contains, when pure, 25·48 of silica, 67·07 of oxide of zinc, and 7·45 of water. Some varieties, however, contain two or three per cent. more water. This is an abundant and valuable ore, and is now beginning to be worked in this country.

*Mancinite.* A silicate of zinc, containing, probably, one atom of oxide of zinc, without water; it is found only at one locality, near Leghorn.

*Willemite.* A silicate of zinc, with three atoms of oxide of zinc to one of silica, or silica 27·53, and oxide of zinc 72·47. This substance is of no importance as an ore; it occurs at the New Jersey zinc mines.

*Hopeite.* This is a rare mineral, not yet fully described, but supposed to be a phosphate of zinc.

*Calamine,* Carbonate of Zinc. This is the most important ore of zinc, and the one from which the principal portion of the metal furnished to commerce is derived. It almost invariably occurs associated with electric calamine, and the two ores are worked together at the great zinc works of Belgium and Prussia. When pure, it is a simple carbonate, with one atom of carbonic acid, and one of oxide of zinc, or 35·19 of carbonic acid, and 64·81 of oxide of zinc; but it almost always contains alumina and oxide of iron. It is rarely found crystallized, but usually in reniform, botryoidal, and stalactitic shapes, or sometimes, when quite impure, in earthy and friable masses.

*Zinc Bloom.* A hydrated carbonate of zinc, containing 71·28 per cent. of oxide of zinc. It occurs as an incrustation on other zinc ores, and is not of economical importance.

*Aurichalcite,* Green Calamine. A combination of the carbonates of copper and zinc with water; containing 29·17 oxide of copper, 44·71 oxide of zinc, 16·19 carbonic acid, and 9·93 water. It is a rare mineral; it might, with propriety, be called "native brass ore," were it sufficiently abundant to become an ore.

*Buratite.* A zinc malachite, or a carbonate of copper with a part of the copper replaced by zinc and lime. It is closely related to the last-named mineral, and, like that, not of economical importance.

*Köttigite,* an arseniate of zinc, is a rare mineral, which contains cobalt and nickel, and occurs with the ores of these metals.

GEOLOGICAL POSITION.—The ores of zinc are distributed with the greatest profusion in every country, and in almost every geological formation; but the supply of the metal furnished to commerce is drawn from a very few districts, in which abundance and good quality of ores, cheapness of labor, and facility of access are found combined.

The ores of zinc may be divided into two classes, in reference to their geological position and mode of occurrence.

The first group comprises the zinc minerals occurring in

the regular veins of the great metalliferous formations. In the older rocks, zinc is almost an unfailing associate of the more valuable metals, especially of silver and copper. The only ore found in any quantity in this position, however, is the sulphuret, and that usually more or less mixed with iron, so that it cannot be considered as of any particular value. In the lead-bearing rocks, associated with the veins and irregular deposits of that metal, zinc ores are found in considerable abundance, and are worked to some extent in England in connection with lead, but would not be were they the only object of exploitation.

The second group of zinc ores, and the only one of real importance, comprises those occurring in calcareous and dolomitic rocks, which are frequently either a part of, or closely connected with, the carboniferous system. These deposits occur either in beds intercalated in the strata, or in irregular masses occupying depressions in them, and are developed on an immense scale in the great zinc-bearing districts of Silesia and Belgium, which will be particularly noticed farther on in this chapter. The ores thus found are the carbonate and silicate, which appear in many cases to have resulted from the decomposition of the sulphuret.

---

## SECTION II.

### GEOGRAPHICAL DISTRIBUTION OF THE ORES OF ZINC IN FOREIGN COUNTRIES.

Russian Empire.—The great zinc deposits of Silesia extend into Poland, and are there worked to a considerable extent, producing about 4000 tons of metal yearly. There are also zinc mines in the vicinity of Cracow.

Great Britain.—Sulphuret of zinc, blende, or black-jack, occurs abundantly in the copper and lead lodes of Cornwall; the greater portion of it, however, is highly ferruginous, a specimen from Wheal Ann having furnished, on analysis, as much as 22 per cent. of iron. At the date of De la Beche's Report (1837), no use was made of it. Pryce, however, who wrote in 1778, remarks that black-jack was the most common mineral in Cornwall next to iron pyrites, and says that several ladings of it had been shipped off for making brass.

Plowden, in 1578, mentions that calamine was then considered to be plentiful in England, and that this mineral was fused with copper to make *lattin* (brass); he does not state whether this was done in England or out of the country.

At present, the greater number of the zinc-works are situated in the neighborhood of Bristol and Birmingham, although a few are near Sheffield. Bristol and Birmingham are principally supplied with ores from the Mendip Hills, and from Flintshire; and Sheffield from the mines of Alston Moor, in Cumberland.

The zinc ores of Alston Moor are in the carboniferous limestone, and generally found in company with the galena which occurs so abundantly in that region. Some of the veins contain galena only, others the ores both of lead and zinc, and a few contain the latter metal alone.

It seems impossible to form any estimate of the quantity of zinc smelted in England, on account of its being done at so many small establishments. It cannot, however, exceed a few thousand tons, and has lately fallen off considerably.

It is a curious fact that in Cornwall the miners believe that the blende or black-jack is an indication of richer ore, hence the proverb "Black-jack rides a proud horse;" in Derbyshire just the contrary is held, and the saying is that "black-jack has eaten up the lead."

BELGIUM.—The great Belgian zinc-works are in the province of Liège.* Dumont has divided the rocks of the carboniferous system, which is the prevailing geological formation of that district, into four groups, as follows:—

1. Group of the lower quartzose slates, made up of slate, sandstone, and conglomerate.
2. Group of the lower limestone, made up of limestone and dolomite.
3. Group of the upper quartzose slates, made up of slate and fine-grained sandstone.
4. Upper limestone; group of calcareous and dolomitic beds.

* Piot and Murailhe, Ann. des Mines (4), v. 165.

Intercalated among these beds of rock are deposits of four kinds, quartzose, ferruginous, zinc-bearing, and argillaceous, which are most developed in the limestone, and especially at the junction of the calcareous groups with the quartzose slates. The most abundant metallic substance is brown hematite, frequently containing a considerable proportion of zinc. This ore lies in the midst of the limestone or dolomite, in canoe-shaped masses, some of which are $\frac{3}{4}$ of a mile in length and 400 feet broad, with an unknown depth. They also sometimes take on the character of veins, crossing the formation in all directions, and being associated with veinstones, such as heavy spar, and with ores of lead and zinc.

Next to hematite, calamine is the most abundant ore. It is found in beds, which are entirely confined to the limestones, and never occur in the schists or sandstones. The most important deposits are those of La Vieille Montagne, La Nouvelle Montagne, Corfalie, and others near Huy, Engis, and Membach.

The most important of all, the Vieille Montagne, is situated near Aix-la-Chapelle, at the village of Moresnet. The zinc-bearing deposit is in the upper calcareous group, and fills a basin-like depression in the dolomite, extending 1400 feet in length, from northeast to southwest, and 600 to 700 feet wide, with a depth which never exceeds 190 feet. The mass of ore is divided into two parts by a dolomitic stratum, and the productive portion of the deposit may be regarded as made up of a great number of different layers, irregularly arranged, and separated by different-colored clays, each smaller mass being a confused mixture of clay and ore. This ore is a carbonate of zinc, mixed with the silicate and oxide. It contains geodes of crystals of these and other zinc-bearing minerals. The ores as they come out of the mines are divided into two classes, the red and the white; the red variety contains a much larger quantity of ferruginous matter, and is less valuable, partly because it contains less zinc, and partly because it fuses more readily in the retorts or muffles. The white variety yields about 46 per cent. of metallic zinc, the other kind 33 to 34 per cent.

There are several other deposits of calamine which have been worked at various times near the Vieille Montagne.

La Nouvelle Montagne is at Verviers. The ore is in an immense pear-shaped mass, forming a sort of envelop around an interior of dolomite. The deposit at Corfalie is a few miles from Liège. Here the calamine forms an intercalated mass between the upper limestone and the coal measures, constituting a flattened layer of from 3 to 25 feet in thickness, occupying a nearly vertical position, and having the shape of a triangular prism with its base uppermost. It is an immense stockwerk, of which the different portions are united by strings of blende. The deposit consists of galena, blende, and calamine, but the two former minerals are found together, and not mixed with the calamine.

The Vieille Montagne is said to have been mined by the Spaniards 400 years ago, but there is no record of operations carried on here extending back farther than 1640. The locality was owned by the state up to 1806, but is now conceded to a company, whose works are of great extent. The deposit of ore has been worked since 1817, by a series of concentric steps extending around it and open to the day. The establishments of the company are situated at Moresnet, Saint Léonard, Angleur, and Tilfth, and they produced in 1851 from their own ores 11,500 tons of zinc,* which is 78 per cent. of the whole production of Belgium in that year. A much larger quantity than that stated above, amounting in 1851 to 14,025 tons, is sold by the company, the additional amount being obtained by purchases of crude zinc from Silesia, which is worked up in their establishments. Of this quantity 6000 tons were sold in the form of rolled zinc; the remainder was mostly sent to France in the crude state, as the company has extensive rolling mills in that country.

There are several other companies in the province of Liège engaged in manufacturing zinc; that of Corfalie producing 1800 tons of zinc, besides 500 of lead; that of La Nouvelle Montagne furnishing 1500 tons of the former, and 300 of the latter metal. Besides these there are one or two

* Herbet, Cons. Gen. de France en Belgique, Ann. des Mines (5), ii. 609.

smaller establishments, calculated to supply from 1000 to 2000 tons annually.

The production of zinc in Belgium has been developed with great rapidity in the last few years. In 1835, it hardly exceeded 2000 or 3000 tons; and in 1851 it had reached 14,750 tons, and was still on the increase. The establishments are managed with great skill.

PRUSSIA.—The zinc business is extensively carried on in this country, and especially in the province of Upper Silesia, which produces nine-tenths of the whole amount of this metal furnished by Prussia, although there are several localities in the Rhine provinces where zinc ores are obtained in some quantity.*

The metalliferous district of Silesia is situated near Tarnowitz and Beuthen, and in this region there are numerous mines of lead, iron, and coal, as well as of zinc. The ores of the latter metal are reduced in several smelting-works, situated on the coal-basin south of Beuthen, while those of lead are smelted at a furnace near Tarnowitz.

The town of Beuthen is situated on a limestone, which is referred to the Muschelkalk, or Upper New Red Sandstone, and which extends towards the east into Poland, and is surrounded on three sides by the coal measures. The zinc ores rest on strata of the Muschelkalk, which occurs in not very thick beds, regularly stratified, and nearly horizontal. It is almost always changed in character, and bleached, in the neighborhood of the metalliferous deposits. It presents a very irregular surface, being often hollowed out into cavities, of greater or less extent, in which masses of dolomite have been accumulated. These have no regular stratification, and are everywhere easily to be distinguished from the limestone. The largest of the dolomitic deposits occurs near Tarnowitz, and extends into Poland, being everywhere intimately associated with the principal beds of calamine. Another one, which lies south of Tarnowitz, contains deposits of galena. These metalliferous deposits are found in depressions in the surface of the limestone, either near the

* See Rivot, Ann. des Mines (4), xiii. 271, and v. Huene, Zeitschrift der Deutschen Geol. Gesells., iv. 571.

dolomitic beds, or directly under them. The ore thus occurring is calamine, and there are two forms of it, the so-called "white deposits," and "red deposits." The former are interstratified with white clay, the entire mass of alternating layers of clay and zinc ore being generally from 3 to 6 feet thick, rarely as much as 12 feet. The whole appearance of the layers indicates that they were originally deposited, in a succession of thin strata, in the tranquil waters of small basins in the limestone. The yield of this variety of ore is from 20 to 30 per cent. of metallic zinc. Sometimes the calamine is associated with iron ore, or is colored by a large percentage of oxide of iron. Such deposits are confined to the dolomite, and are destitute of stratification, and seem to have had a different origin from that indicated above. Occasionally there are two such beds of ore, one above the other.

The mine of Scharley is worked in one of the largest of these ore beds. In this locality, the white deposit, which is comparatively thin, is separated from the red by a seam of clay. It is worked, in part, by open pits. The upper portion of the metalliferous stratum contained a large quantity of galena, which was formerly extensively mined. The depth already attained by the workings is a little over 100 feet. At the surface, the open pit has a length of 1500 feet, and a width of 750.

At the Lydognia furnace, a considerable quantity of cadmium is obtained from the zinc, the former metal being separated from the latter by distillation. The first portion of the oxide of zinc collected from the apparatus is much the richest in cadmium, and this is reserved, and afterwards intimately mixed with charcoal and subjected to a very slow distillation, at as low a temperature as possible; this process is repeated several times, until a sufficient degree of purity is attained.

In Westphalia, there are zinc works near Iserlohn, but they have produced only a small quantity of metal.

In the Rhenish mining district there are, near Stolberg, some deposits of calamine, but they are not to be depended on for any large supply. Attempts have been made to work zinc blende at Mühlheim, Düssel, and Bohrbeck, without much success, owing to the low price of zinc.

In 1852, an English company commenced mining and smelting zinc at Bergisch Gladbach, nine miles east of Cologne: their stock stands at a small premium. The geological features of the deposit of ore are interesting.* The calamine lies in tunnel-shaped depressions in dolomite of palæozoic age, which is covered by a shale containing brown coal. Blende is associated with the calamine in such a way as to show that the latter is the result of the decomposition of the former. This is well illustrated at the Frühling Mine, near Altenbrück, where there is a powerful vein of blende, which, at the surface, is converted into calamine, and in which large masses of this ore are found with undecomposed blende in the centre of them. At the Gladbach Mine, the ores are found in fragments lying surrounded by clay, as if they had been washed together into the depressions of the dolomite, at the time of the deposition of the overlying shales.

The production of zinc in Prussia is very large, being fully three-fifths of all that is furnished to commerce in the world. Since 1830, it has increased from 8600 tons to over 30,000. A large quantity of this goes to England, and is partly consumed there, and partly exported to India. There is also a great demand for home consumption, as the metal is constantly being introduced into use in new forms; and as no other part of Germany furnishes any considerable quantity of zinc, their supply is drawn chiefly from the Silesian mines.

Austria. — The production of zinc in Austria is but trifling, compared with that of Belgium or Prussia. It amounted in 1847 to 1395 tons, and has probably increased somewhat since that time. The mines are chiefly in the provinces of Tyrol and Illyria.

* V. Huene, Zeitsch. d. Deutsch. Geol. Gesells. iv. 571.

## SECTION III.

### DISTRIBUTION OF THE ORES OF ZINC IN THE UNITED STATES.

THE ores of zinc are distributed over the United States in great abundance, but have, as yet, hardly begun to be worked. Some of the more important localities will be noticed.

NEW HAMPSHIRE.—Eaton is the most important locality of this metal, but it occurs only as blende. The vein will be noticed more particularly in the chapter on lead. The same ore is found at the Shelburne Mine, and, in fact, in greater or less quantities in almost every metalliferous vein in the state. There is, at Warren, a heavy bed of black blende mixed with copper pyrites and galena. The zinc itself is of no value at present.

NEW YORK.—The only locality in this state which has been worked to any extent for zinc, is that near Wurtsboro, in Sullivan County, formerly called the Shawangunk Mine, more recently known as the Montgomery Zinc Mine. According to Prof. Mather,* this deposit of ore occurs in a bed parallel with the strata of the Shawangunk Mountain, about two miles northeast of Wurtsboro. He considers it a true vein, which, from his own description of it, it can hardly be. It forms a segregated mass, varying from 2 to 5 feet in thickness, and the larger portion of which is made up of a silicious rock, like that forming the roof and floor, and containing particles of greenish and blackish slate. The metalliferous contents of the vein are blende, galena, copper pyrites, and iron pyrites, which are associated with crystallized quartz, the two former minerals largely predominating; the leader of solid ore varies from a mere seam to three feet in thickness.

In 1837, and for some time afterwards, considerable work was done here, and about 100 tons of lead ore obtained: the zinc, however, greatly exceeded the lead in quantity. Traces of cobalt and silver are said to be contained in the ore.

* Report on Geology of N. Y., p. 359.

The New York and Montgomery Mining Company took up this mine, which had been abandoned for several years, and in 1851 and 1852 undertook to work the ore in the *humid way*, or, in other words, to separate the various metals it contains, in the form of oxide of zinc, red lead, blue vitriol, oxide of cobalt, and silver, by the processes followed in the laboratory of the analytical chemist. This attempt, as might have been foreseen by any one acquainted with metallurgic processes, failed entirely.

The quantity of the ore at this locality is said to be large, but as the zinc greatly predominates, the mine cannot be considered at present as of any value.

There are many other places in this state where blende occurs in small quantity. It is found in some of the lead veins of St. Lawrence County, but not to any extent, and does not appear to become more abundant in depth. In the geodes of the Niagara Limestone, in the vicinity of Rochester, Lockport, and Niagara Falls, it is found in small crystals. Massive zinc blende occurs at the Ancram Lead Mine.

New Jersey.—The zinc ores of this state are of very considerable interest and importance, and are now quite extensively worked, being, until recently, the only mines of this metal on the American Continent.

The zinc deposits of New Jersey are in Sussex County, on a range of hills which commences near Sparta, and extends in a southerly direction through Sterling to Franklin, in which latter places are the only beds of ore at present known. They are found in connection with a white crystalline limestone, which can be traced from Orange County, in New York, to several miles beyond Stirling. This limestone is regarded by Prof. H. D. Rogers as a Lower Silurian rock, altered by the heating agency of igneous injections. It is associated with a quartz and feldspar rock, which appears to have the character of an igneous intrusive mass, and to have rendered the limestone crystalline, and given it its present position and appearance. The intrusive rock seems to form dykes in the limestone, which is tilted up at a high angle. At the Stirling Hill, the ore lies in a position which may be represented by the annexed section (Fig. 36). The bed of zinc ore rests, with a steep southeast dip, against a bed of Franklinite, and both coincide in dip with the limestone in

Fig. 36.

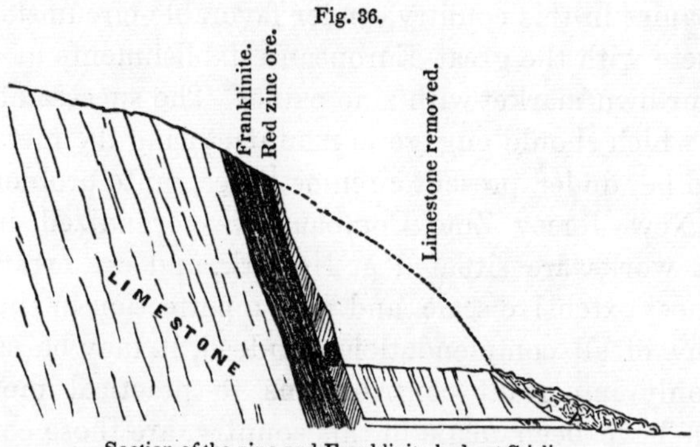

Section of the zinc and Franklinite beds at Stirling.

which they are contained. A portion of the rocks covering the ore-bed on the face of the hill has been removed, to a depth of about 70 feet, as indicated on the section by the dotted line. The outer bed, thus exposed, is a mixture of the red oxide of zinc with Franklinite; its width at the surface was about 3 feet, but it widens out to 8½ feet in descending. Next to this is the bed of Franklinite, which is from 20 to 30 feet in width. Beyond is the crystalline limestone, dipping 70° to 80° to the southeast, like the intercalated metalliferous beds, and succeeded by the before-mentioned quartz and feldspar rock.

At Mine Hill, in Franklin, the same succession of limestone and metalliferous beds may be observed, the intrusive rock being there a kind of sienite, and the blue limestone having been converted into a white crystalline mass along the line of contact of the two formations; the zinc ore and Franklinite forming intercalated beds within it.

The existence of zinc ores at Stirling Hill has been known for many years, and numerous attempts have been made to work it at different times, all of which, up to the time when the present company commenced operations, had proved unsuccessful, more from want of knowledge and of a demand for zinc, than from any inherent difficulty of reducing the ore. The substitution of the oxide of zinc for white lead as a pigment, one of the most important improvements in the application of chemistry to the arts, as a measure both of sanitary and of economical value, rendered it possible for companies in this country, under favorable circumstances, to compete with the great European establishments in furnishing our own market with zinc paint. The success of a company which should engage in manufacturing the metal itself, would be, under present circumstances, quite problematical. The New Jersey Zinc Company was organized in 1848. Their works are situated at Newark, and are arranged on the most extensive scale, and with a perfection in the details worthy of all commendation. Indeed, it may be said that the only important contributions to practical metallurgy which have been made in this country, are those connected with the working of zinc ores. The white oxide is manufactured directly from the ore, which is a mixture of the red oxide and Franklinite; the metallic vapors, as they are given off by the reduction of the heated mixture of ore and coal, being burned into oxide by the introduction of atmospheric air into the flue through which they are passing. The oxide thus formed is drawn, by the suction of a fanblast, after depositing the heavier particles of mechanically-intermixed coal and other impurities, into a cotton cylinder of great length, from which depend bags of the same material, in which the oxide is collected. It is the intention of the company to grind most of the oxide in oil, before disposing of it, in order to prevent adulteration.

The amount of oxide of zinc manufactured by this company has been as follows:—

| | Year ending November, 1852. | 1853. |
|---|---|---|
| No. 1, . . . | 2,041,736 | 3,832,036 |
| " 2, . . . | 282,890 | 157,000 |
| " 3, . . . | 100,880 | 54,379 |
| Total, . . . | 2,425,506 = 1083 tons. | 4,043,415 = 1805 tons. |
| Or metallic zinc, . . . | 860 tons, | 1440 tons. |

The estimates of the production for the current year are 3570 tons of oxide, equal to 2850 of metal, as the furnaces have recently been doubled in number, and the arrangements in every way rendered more perfect.

During the year 1853, the company netted a profit of $90,592 16, and paid $42,944 50 in dividends.

PENNSYLVANIA.—Among the numerous deposits of zinc ore in this state, the only ones which have become of any importance are those of the Saucon Valley, near Friedensville, Lehigh County. These have been known for several years, and quite extensive explorations have been made at various times to ascertain the quantity and quality of the ores. Within the last year, furnaces for manufacturing the white oxide have been established at Bethlehem, about four miles from Friedensville, and mining commenced on an extensive scale.

The ore is almost entirely the silicate of zinc, of a very good quality, being remarkably free from intermixture with lead or iron. The deposits are in the form of included beds, in a blue compact limestone, which belongs to the Lower Silurian system, and is apparently the equivalent of the Calciferous Sandstone of the New York geologists. Numerous openings have been made at various points in the vicinity of Friedensville, which have demonstrated the existence of an immense body of ore of excellent quality. The principal excavations have been made on the Stadiger and Ueberoth estates. On the latter, the main opening, a year ago, was 70 feet in length, and it exposed the zinc ore in one place to a depth of 56 feet, and over a width of more than twenty feet. The direction of the body of ore is about north 70° east, which is that of the strata of limestone adjacent; its whole width has not been ascertained, but it is very considerable, as there are several parallel bands, which together must have a width exceeding fifty feet, so that the quantity may be considered as amply sufficient to encourage the erection of the most extensive works.

The whole of this property is now under the control of the Pennsylvania and Lehigh Zinc Company, which was organized in 1853. The smelting furnaces are at Bethlehem, and were erected under the direction of Samuel Wetherill, Esq., for the purpose of manufacturing the oxide directly from the ore, by a patented process of his own invention. The works are calculated to produce ten tons of the oxide per day. The ore yields in the furnace from 40 to 60 per cent., and costs to mine and deliver at the furnace, as estimated by the company, $1 50 per ton; the expense of manufacturing, which is done by contract, averages $50 per ton of oxide produced.

Operations had hardly been commenced at the close of 1853.

WESTERN LEAD REGION.—The ores of zinc are plentifully distributed through the lead mines of the Mississippi Valley, but nowhere, so far as I have been able to ascertain, in sufficient abundance, or so advantageously situated, as to render them of any value for working.

In Wisconsin, the silicate of zinc occurs frequently, associated with galena, and is usually called "dry-bone"* by the miners; black-jack, or the ferruginous sulphuret, is still more common.

The same ores of zinc are also found, in small quantities, at several of the Missouri lead mines. Their mode of occurrence is similar in every respect to that of the Wisconsin ores.

No one acquainted with the manufacture of zinc ores into metal or oxide, would recommend the establishment of works for this purpose in the western lead region, as the business cannot be made profitable, against the competition of the Belgian and Prussian manufactories, except under the most favorable circumstances of situation, and an abundant supply of ore which can be obtained without any considerable mining cost. The zinc deposits of the West do not satisfy these conditions, either as regards quantity or quality of the ore, or the proximity of fuel.

There are no farther deposits of zinc ore in this country worthy of notice at present.

The statistics of the production of zinc throughout the world, so far as they could be procured, are given in the following table, in tons. Full statistics of Belgium could

* The dry-bone of the Missouri miners is an impure earthy carbonate of lead.

not be obtained, but the figures given will indicate the rapidly-increasing production of that country, and its present approximate amount.

| | Russia. | Sweden. | Great Britain. | Belgium. | Prussia. | Harz. | Austria. | United States. |
|---|---|---|---|---|---|---|---|---|
| 1825, . . . . | | | | | | | 104 | |
| 1830, . . . . | | | | | 8,590 av. | 4 | 30 | |
| 1835, . . . . | | | | 2,500 | 7,200 " | 6 | | |
| 1840, . . . . | | | | | 10,980 " | | 105 | |
| 1845, . . . . | | | | | 17,100 " | | 388 | |
| 1846, . . . . | | | | | 22,260 | | 413 | |
| 1847, . . . . | | | | | 22,400 | | 352 | |
| 1848, . . . . | 4,000 | | 1,000 | 6,500 | 20,190 | | 1,395 | |
| 1849, . . . . | | | | | 26,270 | | | |
| 1850, . . . . | | 36 | | | 28,670 | 3 | | 0 |
| 1851, . . . . | 4,000 | | | 14,750 | 30,620 | | | |
| 1852, . . . . | | | | | | | | 860 |
| 1853, . . . . | | | | | | | | 1,450 |

The production of zinc for 1853 was nearly as follows:—

| | Tons. | Relative Amount. |
|---|---|---|
| Russia (Poland), . . . . . . . | 4,000 | 7·3 |
| Great Britain, . . . . . . . . | 1,000 | 1·8 |
| Belgium, . . . . . . . . . | 15,000 | 27·3 |
| Prussia, . . . . . . . . . . | 32,000 | 58·2 |
| Austria, . . . . . . . . . . | 1,500 | 2·7 |
| United States, . . . . . . . . | 1,500 | 2·7 |
| | 55,000 | 100·0 |

According to the estimates of the two companies now producing the oxide in this country, we may calculate on a production for the present year equivalent to between 5000 and 6000 tons of the metal.

The universal use of the oxide of zinc, in preference to white lead, is most earnestly to be desired, not so much on account of its superiority as a pigment, as because it is perfectly free from the poisonous qualities which are so ruinous to the health of those who use paints in which lead is an ingredient.

# CHAPTER VIII.

## LEAD, AND SILVER IN PART.

## SECTION I.

### MINERALOGICAL OCCURRENCE AND GEOLOGICAL POSITION OF THE ORES OF LEAD.

MINERALOGICAL OCCURRENCE.—The variety of forms under which lead occurs in nature is very great, as will be seen from the following list of the plumbiferous minerals. Almost the whole quantity furnished to commerce, however, is derived from one ore, the sulphuret.

#### NATIVE METAL.

*Native Lead.* The metal lead combines so readily with oxygen that it is hardly probable that it would be found in its native state, except in minute quantities, as an accidental product of the decomposition of its ores, and not in a permanent form.

#### COMBINATIONS WITH SULPHUR, SELENIUM, TELLURIUM, ANTIMONY.

*Galena,* Sulphuret of Lead; containing one atom of each: in percentage, 13·34 of sulphur, and 86·66 of lead. Hardly any other ore occurs in sufficient quantity to be an object of much commercial importance. Galena is almost invariably argentiferous; it may be safely stated that no galena is destitute of at least a trace of silver. In many cases this metal is present in sufficient quantity to be worth separating, and a considerable portion of the silver of commerce is obtained from this source. The silver of the American Continent is mostly derived from the ores proper, but all this metal produced in England, and a very large part of that of the Continent, is separated from lead.

The percentage of silver in galena is very variable, even in the same locality; but, as previously stated, it is almost impossible to find a specimen of this ore which would not show, by delicate analysis, at least a trace of it. In the European lead ores, the amount of silver present varies from 0·03 to 7·0 per cent. In England, the average quantity contained in the lead which is worked for silver is 7 or 8 ounces per ton. The galena of the lead region of the Mississippi Valley hardly contains any silver. But there are, in some of the eastern mines, ores yielding as high as 70 or 80 ounces to the ton.

*Cuproplumbite*, Sulphuret of Lead and Copper. A rare mineral found in Chili.

*Clausthalite*, Seleniuret of Lead; occurs in the Harz mines; it resembles galena in appearance.

*Altaite*, Telluret of Lead; found in the Altai Mountains.

*Zinkenite*, Sulphuret of Lead and Antimony; occurs in the Harz.

*Dufrenoysite*, *Heteromorphite*, *Boulangerite*, *Jamesonite*, *Geocronite*, *Plagionite*. These are combinations of sulphur with lead and antimony, or lead and arsenic; interesting mineralogically, but not important in a commercial point of view.

*Kobellite*, Sulphuret of Lead, Antimony, and Bismuth; occurs in the cobalt mines of Sweden.

### COMBINATIONS WITH OXYGEN.

There are three oxides of lead which are found native, but all in small quantity.

*Plumbic Ochre*, Lead Ochre, Litharge. An oxide of lead, with one atom of each constituent; in percentage, 7·17 of oxygen, and 92·83 of lead. This is the yellow oxide, which was formerly used as a paint, under the name of massicot; but it is now superseded by the chromate of lead. Litharge is the same, after having undergone fusion, when it is converted into delicate scales with a high lustre. It is very rare as a native ore.

*Minium*, Red Lead. An oxide containing three atoms of lead to four of oxygen, or 9·34 of oxygen to 90·66 of metal. As it occurs in nature, it is generally mixed with plumbic ochre. It is found, in this country, at Austin's Mine, in Virginia.

*Plattnerite*, Superoxide of Lead; contains one atom of lead to two of oxygen. A rare ore.

### COMBINATIONS WITH CHLORINE.

*Mendipite*, Chloride of Lead. A combination of the oxide and chloride of lead, containing chloride of lead 38·4, and oxide of lead 61·6 per cent. It is a rare substance.

*Corneous Lead*, Horn Lead. A rare ore, containing one atom of chloride and one of carbonate of lead. Occurs at Matlock, in England.

### OXYGEN SALTS OF LEAD.

These are very numerous, and some of them are of some importance as ores.

*Cerusite*, White Lead Ore. A carbonate of lead, containing carbonic acid 16·4, and oxide of lead 83·6 per cent., or 77·7 per cent. of metallic lead. The finest specimens in this country were from the Washington Mine, in North Carolina, the Phœnixville Mines, Pa., and Mine La Motte, in Missouri. It is a valuable ore, and is not unfrequently found in some quantity in the lead-bearing veins, near the surface, as one of the products of decomposition of the sulphuret.

*Anglesite*, Sulphate of Lead, Lead Vitriol. The sulphate contains 26·4 of sulphuric acid to 73·6 of oxide of lead. It is another product of the decomposition of the sulphuret. The principal locality in this country is the Wheatley and Perkiomen Mines, in Pa. It also occurs in small quantity in the Missouri mines.

*Linarite*. A combination of sulphate of lead with hydrated oxide of copper; occurs at Leadhills, England, and Linares, in Spain.

*Leadhillite*. A combination of the sulphate and carbonate of lead. Its principal locality is Leadhills, in England.

*Lanarkite*. Another ore analogous to the last mentioned, with one atom of the carbonate and one of the sulphate.

*Caledonite.* A sulphate of lead combined with carbonate of lead and carbonate of copper.

*Pyromorphite*, Phosphate of Lead. A complex combination of the phosphate of the oxide of lead with chloride and fluoride of the same metal. It contains about 90 per cent. of phosphate of lead. Fine crystallizations were formerly obtained at the Washington Mine, North Carolina, and it is so abundant at the Chester County Mines, Pennsylvania, as to have been worked as an ore. It is found in small quantity at most of the lead mines of Europe.

*Mimetene*, Green Lead Ore. A mineral resembling pyromorphite in composition, except that it contains arsenic acid instead of phosphoric. It is found in small quantity in some of the Cornish and Cumberland (England) lead mines.

*Vanadinite*, Vanadate of Lead. A rare mineral, first discovered in Mexico.

*Crocoisite*, Chromate of Lead. A beautiful, but rare ore, of which the finest specimens come from the Ural Mountains.

*Melanochroite*, is another rare chromate of lead.

*Vauquelinite*, is a chromate of lead and copper.

*Bleinierite*, is an antimoniate of lead.

*Plumbo-resinite*, an ore containing oxide of lead, alumina, and water.

GEOLOGICAL POSITION.—The ores of lead are abundantly distributed through the geological series, but their greatest concentration is in the Lower Silurian and carboniferous groups. In this position they do not usually form true veins, but occur in irregular bunches and gash-veins, often very rich in a certain stratum of rock, and giving out entirely on entering another. Deposits of lead ore are not generally rich in silver, unless the formation in which they occur has been metamorphosed and rendered crystalline. The argentiferous ores are mostly in the older rocks, and although not so abundant near the surface as the deposits in the less crystalline rocks, they are more persistent in depth, and often make up by their richness in silver for their smaller yield of lead.

The ores of lead frequently undergo decomposition at the surface, the normal ore, galena, being converted into a great variety of oxidized combinations, of which the carbonate, sulphate, and phosphate are the most common. If the lead ore is rich in silver, this metal is often found near the surface in its native state, in dendritic bunches and spongy masses, having been evidently separated and concentrated from the galena during the process of its decomposition. Lead and zinc are metals whose ores occur intimately associated with each other. The great veins of argentiferous galena which have been so extensively worked in various parts of Europe, usually carry a considerable amount of blende, which sometimes exceeds in quantity all the other ores. The zinc ores do not generally contain so much silver as the accompanying galena; frequently they are destitute of it altogether, although there are instances in which the blende associated with lead ores has the same richness in the precious metals which they have. In many instances, in Europe and in this country, it has been observed that the zinc gains upon the lead in depth. Hardly an instance has been noticed where there was a decrease in the quantity of the zinc ores in descending. In the argentiferous lead mines it is frequently found that the tenor of silver in the ore becomes less as the depth increases, and the same is true, with some exceptions, where silver ores proper are worked.

In general, it may be said of the ores of lead, with more positiveness even than of those of the other metals, that the nearer the approach of the deposit to a true vein in character, the more persistent are its metallic contents in depth, and the greater the chance of its being profitably wrought.

Hitherto, the larger part of the lead furnished to commerce has been obtained from irregular deposits, and segregated and gash-veins, of extraordinary richness near the surface, but not holding out in depth; as these are gradually worked out, the number of mines in the older rocks is increasing, and many which had been previously abandoned are being resumed; indicating that hereafter a larger portion of this metal than heretofore is to be the produce of workings in true veins, in the older, crystalline rocks.

A sufficient number of facts illustrating this will be found in the succeeding section, and the changes which are going on in the production of lead will not fail to be noticed.

## SECTION II.

### DISTRIBUTION OF THE ORES OF LEAD IN FOREIGN COUNTRIES.

Russian Empire.—The silver-lead ores of Siberia are of very considerable importance. According to the Russian mining engineers, the principal mines are situated in the Altai and Nertschinsk districts; but there are others, the exploration of which has but recently commenced, in the Caucasus, in the country of the Kirghises, and beyond the Irtysch. There are also lead and silver veins in the Ural Mountains, and in the country bordering on the Don.

In the Altai, as well as the Ural, there are abundant traces of mining at a very early period, of which no records have preserved any account, and which is usually ascribed to the Fins; their excavations, however, were confined to the surface, for want of suitable tools to excavate the solid rock. The Altai mines, in the valley of the Ob, have been worked since the beginning of the last century. The total yield of the mines of this district, up to 1835, had been over 3,000,000 lbs. of silver, which contained nearly 100,000 lbs. of gold. The annual production at present is about 44,000 lbs. of silver. Of eleven mines, now working, those of Zerinofsk and Krioakofsk possess the richest ores: their average yield of silver is 0·1 per cent.

The mining region of Nertschinsk is situated in the south-eastern part of the Government of Irkutsk, between 50° and 53° north latitude, and 131° and 138° east longitude.* The existence of ores was made known to the world in 1691, by Admiral Golownin, and they soon after began to be worked, and seem to have attained their maximum production in 1771, when they yielded about 27,600 lbs. troy of silver; from this time forward the produce fell away, till 1827, when a systematic scientific attempt was made to place these mines on a more permanent footing. They are now subdivided into five sub-districts.

* Von Wersilow, Verhandlungen der Russ. Kais. Min. Gesellschaft, Jahrgang 1848–1849, p. 44.

The general character of these deposits seems to be that of stockwerk mines, and of lenticular masses of ore arranged in linear succession in the limestone, and connected by thin thread-like veins of mineral. They all give out at a comparatively small depth.

The Wodwishkenski Mine consists of a series of nests of ore, connected by sometimes hardly-perceptible threads. It has been worked to a depth of 65 fathoms. The gangue is ochrey iron (gossan), and talc. The Ivanow Mine is in a stockwerk, 14 fathoms long, about the same in breadth, and some 20 fathoms deep.

The produce in lead of all the Nertschinsk Mines was, in 1847, 214 tons, and of silver 7940 lbs. troy.

The total amount of lead furnished by the Russian Empire hardly comes up to 1000 tons; that of silver, which is exclusively derived from the working of argentiferous galena and from the native gold, has been given in the table of silver statistics at about 60,000 lbs.

SWEDEN.—The number of active silver-lead mines in Sweden was, in 1849, only five.* The principal ones are at Sala, which I visited in 1843. They are in the azoic limestone, which forms a band in the gneiss. The ore is not obtained from regular veins, but from parallel bands running with the stratification. It is chiefly galena, containing silver in the proportion of from 0·0015 to 0·0125, or 49 to 408 ounces to the ton; but it reached its maximum richness at a depth of from 500 to 650 feet, and the mine has, since that time, been falling off. The last returns give the yield of lead, in 1850, as only 196 tons, and of silver obtained from it, 3418 lbs. The former metal is not sufficiently abundant to be profitably worked, except for the sake of the silver it contains.

GREAT BRITAIN.—On examining the tables, it will be seen that England maintains a superiority in the production of lead, as well as of tin, copper, and iron. The ores of lead are found in that favored country in a variety of positions, and under a variety of circumstances; and the production of this metal, as well as of silver, with which it is intimately connected, is increasing from year to year with regularity, if not

* Durocher, Ann. des Mines (4), xiv. 339.

with rapidity. The introduction of new processes has rendered it possible to work ores which, a few years since, were neglected, and great changes have consequently taken place in respect to the locality of the production. The lead-producing districts are scattered over England, Wales, Scotland, and Ireland.

*North of England.* The lead mining district of the North of England is the most important from the quantity of ore raised, and is also of great geological interest. It lies chiefly in the vicinity of Alston Moor, where the three counties of Northumberland, Durham, and Cumberland come together. The geological formation to which it belongs is the carboniferous limestone, lying beneath the coal measures; the name of "mountain limestone" is also frequently given to it by the English geologists. Directly above this limestone lies a coarse-grained sandstone, called the "millstone grit," which forms the lower limit of the productive coal-beds. The carboniferous limestone is made up of beds of limestone, with intercalated strata of marl, the stratification being nearly horizontal, and quite regular. There are about twenty beds of limestone, each one of which is known to the miners, and generally designated by a name. Two of the thickest are called the "great limestone" and the "scar limestone;" of these, the former is about 60 feet thick, the latter 130 feet. There is also a horizontal mass of trap, of variable thickness, sometimes amounting to 60 feet, which is intercalated between the beds of limestone, in an irregular manner. It is known to the miners as the "whin-sill," and is generally considered to be an igneous rock, injected laterally between the strata. The principal exploitations are on the so-called "rake-veins," or true transverse veins; but the miners distinguish two other classes of deposits, called by them also veins, namely *pipe-veins* and *flat-veins.* The rake-veins present some extraordinary anomalies, although in general they exhibit the characteristics of true veins; in some instances they do not descend through the various strata in an uninterrupted manner, but are arranged in zigzags, one portion of a vein not being in a vertical line with another part of the same vein, above or below, but connecting with it by a horizontal prolongation. This flat portion of the vein is generally contained in a rock of a different character from that in which the vertical part is, the latter being in the limestone, while the transition from one vertical part of the vein to another, by a horizontal or oblique line, is in the slaty or marly beds. The veins, also, are much narrower in the marl or sandstone than in the limestone, and, at the same time, less rich in ore. The character of the stratum through which they pass has a most marked influence on their productiveness, as well as their width, and there are certain beds which are peculiarly metalliferous. The "great limestone" is the one in which the veins are widest and richest, and in that belt they furnish more ore than in all the others together. Generally, the veins are not worked below the "four-fathom limestone," which is 153 fathoms below the millstone grit, and their productive portion is limited to a thickness of 100 fathoms. The workings have in some instances, however, been carried to a depth sufficient to reach the whin-sill, and it is said that veins have been traced in the scar limestone, although not worked there. The pipe and flat veins are masses and sheets of ore, subordinate to the principal veins; but in some instances they are of sufficient extent to be quite productive.

The lead region of Derbyshire is, in some respects, quite similar to that just described, but is more complicated in its details. The whole district, which is some 25 miles long, has been broken up by faults, which have deranged the continuity of the strata. Besides this, instead of one intercalated bed of trap, as in Cumberland, the whin-sill, there are three distinct belts of igneous rock associated with the limestone, the whole series of beds of both rocks having a thickness of over 1500 feet. There are four principal strata of limestone, the two uppermost being about 150 feet in thickness each, the others 200 and 350 feet; the three belts of trap which separate them are generally of an amygdaloidal structure, the amygdules being filled with calc. spar and quartz; the local name of the rock is "toad-stone."

The same names are applied to the same varieties of form of metalliferous deposit as in Cumberland, but the rake, or transverse veins, are the only ones sufficiently developed to be of any practical importance. The disposition of these veins in regard to the strata in which they are contained is still more remarkable than in Cumberland. They appear to be entirely interrupted in the toad-stone, the vein disappearing as soon as the workings enter the rock; this is not true in all cases, but in much the larger number. It is also said that, in some instances, veins thus cut off by the toad-stone are found again on the other side, on reaching the limestone below.

The ordinary gangues of these veins are fluor-spar and calc. spar, and some heavy spar, or sulphate of baryta. Specimens of the fine crystallizations of fluor-spar occurring here are to be found in all mineralogical cabinets. At the Great Exhibition in 1851, beautiful models and sections were shown, illustrating the position of the lead mines of a considerable tract of mining ground near Alston Moor in Cumberland, belonging to W. B. Beaumont, Esq., from which about one-fourth of the whole quantity of lead produced in England is raised.

*Cornwall and Devon.* The Cornish lead-bearing veins have been noticed in speaking of copper and tin. These mines were wrought, some of them for two or three hundred years, but they had been almost abandoned for many years previous to the introduction of the Pattinson process for desilverizing lead, which effected a great change in the working of argentiferous lead ores. At the time Borlase wrote, in 1758, only one lead mine was worked in Cornwall. In 1839, according to De la Beche, the whole produce of Cornwall amounted to scarcely 180 tons, while in the years from 1845 to 1850 over 10,000 tons of ore were raised annually. One mine alone, of extraordinary richness, East Wheal Rose, produced from 1845 to 1849 from 3000 to 5000 tons of metallic lead annually, and although this mine has somewhat fallen off, others have more than made up the deficiency.

In Devonshire, the Combe Martin and Beer Alston Mines have long been celebrated for their argentiferous lead ores. Large amounts of silver were raised in Devon as far back as 1293. The Combe Martin Mine was worked at various times, and according to De la Beche, who had in 1837 an opportunity of examining the old workings, very unskilfully. The lodes at Beer Alston are very rich in silver, containing often from 80 oz. to 120 oz. of silver to the ton; 140 oz. per ton is said to be the yield of the richest ore.

The lead raised in Cornwall in 1852 averaged 35 oz. of silver to the ton, that of Devon 40 oz., the highest of any mining district in England.

*East Wheal Rose Mine.* As an illustration of what lead mining is, where the lode is large and rich, this, which is probably the most productive lead mine in the world, may be described.

This mine, which is situated near Newlyn in Cornwall, was originally worked on east and west lodes, and was exceedingly poor, so that in 1830 it was about to be abandoned; but on making up the accounts for the purpose of closing the concern, it being found that there was a balance of £200 in hand, it was resolved to expend it in driving a cross-cut, at a venture, and in a few fathoms distance a rich vein was discovered running north and south.

Above the 130-fathom level the mine is now worked out, but in the bottom of this level, for 200 fathoms in length, the lode is said to be still very rich, worth £16 per fathom, although it will be some time before it can be taken down. The returns in 1853 were equal in value to about £3000 per month. The yield of the mine in metallic lead was as follows:—

| | Tons. | | Tons. |
|---|---|---|---|
| 1845, | 5191 | 1848, | 2856 |
| 1846, | 3114 | 1849, | 3191 |
| 1847, | 3854 | | |

At this mine there are ten steam-engines, two of them having 85-inch cylinders, and one 36-foot water-wheel. In addition to this, there is another new 70-inch cylinder engine at the South Cargoll part of the mine, the whole being valued at £17,460.

There are 128 shares, on which £50 per share has been paid in, and they have stood as high as £1500 each. £2,245 per share has been divided, the last dividend, of £10, having been paid in March 1852, and the present price of the shares is about £140.

It is worthy of notice that the silver-lead ores of Cornwall and Devon, unlike those of copper and tin, occur at a distance from granite and elvans, the conditions under which they were formed being quite different from those favorable to the development of the last-named metals. In general, here as almost everywhere else, the lead-bearing lodes are in calcareous rocks, or in near proximity to them. The lodes of Beer Alston cut through slates which are in part calcareous;

those of Combe Martin occur in beds which alternate with limestone.

*Cardiganshire and Montgomeryshire.* This metalliferous district is about 40 miles in length, by from 5 to 22 miles in width, and consists of beds of Lower Silurian age.* The usual strike of the lodes is east-northeast and west-southwest, and their dip is most frequently to the south. The gangue consists mostly of fragments of slate, cemented together by quartz and a little calcareous spar. The ore is mainly galena, sometimes containing as much as 75 or 80 oz. of silver to the ton: it is much mixed with zinc blende, which forms the principal mass of the poorer lodes. The brecciated character of these lodes is the most interesting feature about them. The mines have been worked at intervals for many centuries, and the uncertainty of mining enterprises is well illustrated by the history of one or two of them, which are now in successful operation.

"The mine of Logaulas had long been worked by shallow shafts with various success, till the adventurer, resolving to make a bold push at a greater depth, commenced, towards the close of the last century, to drive an adit-level from the north, which after a course of nearly half a mile should reach the lode at a depth of 60 fathoms from the surface. The rock was hard and the progress slow, but for upwards of 30 years did the miners persevere, till at length, after piercing about 360 fathoms, a lode was cut; but so miserable was the aspect it presented, that after driving right and left upon it for a few feet, the disappointed speculator gave up all his cherished hopes and abandoned the undertaking. After a short interval, some Cornish adventurers were led to believe that something yet remained to be done, and having set a party of men to push forward the same level, in the course *of a few feet* cut the true lode, in the midst of a vast deposit of ore which yielded rich returns for several years. This company, however, in their turn, fell into a similar error, and losing the true lode, mistook for it a small vein on the south, dispirited with whose poverty they surrendered the mine. The present holders, after making an accurate survey, were satisfied that they must be too far southward, drove a "cross-cut" towards the north, and very shortly discovered not only the lode, but a rich bunch of ore, parallel to which their predecessors had been toiling for many a fathom through barren rock at the distance of only a few feet. The mine has ever since been yielding thousands of pounds profit per annum."†

The Goginan Mine presents another curious example. After being abandoned as worthless, by an adventurer who

* Mem. Geol. Survey, Great Britain, ii. 655. † Ibid., p. 671.

declared that he "would carry on his back to Aberystwyth" all the ore that could be got out of it, it was taken up by other parties, and was recently producing 1500 tons of silver-lead ore per annum.

The miners of this part of Wales have little confidence in the lead ore continuing in depth below 40 or 50 fathoms; and several instances are adduced of lodes, several fathoms wide at the surface, dwindling down to a few feet at the depth of forty fathoms. This, however, is not the case with the lodes before noticed, which, at the depths of 105 and 110 fathoms, are as large as anywhere above. As many as 158 lodes are described by the geologists of the survey as existing in this district, although they are not all productive.

The yield of the United Kingdom, of lead and silver, has been steadily increasing since the beginning of the present century. In 1810, the production of lead in the country was estimated, by Heron de Villefosse, at 12,500 tons, a quantity exceeding that furnished by all the rest of Europe. In 1835, according to Mr. Taylor's estimates, it had increased to 46,112 tons, of which Northumberland, Durham, and Cumberland furnished 19,626 tons. In 1852, the yield of lead of the United Kingdom was, according to the most authentic returns, as collected by Mr. R. Hunt, 64,960 tons.

The amount of metallic lead furnished by the different lead-producing districts, at different times, may be seen from the following table:*

| | 1845. Tons. | 1846. Tons. | 1847. Tons. | 1852. Tons. |
|---|---|---|---|---|
| Cornwall and Devon, | } | } | 8,330½ | } |
| Cumberland, Durham, and Northumberland, | 38,401 | 36,718 | 18,615 | 43,812⅓ |
| Derbyshire, | } | } | 4,570 | } |
| Shropshire, | } | } | 2,769 | } |
| Yorkshire, | } | } | 5,223 | } |
| Wales, | 11,014½ | 10,027½ | 12,294 | 13,708 |
| Scotland, | 901 | 942 | 822½ | 2,381⅓ |
| Ireland, | 855½ | 811 | 1,380 | 3,223 |
| Isle of Man, | 1,523 | 1,663 | 1,699 | 1,835⅓ |
| | 52,695 | 50,161½ | 55,703 | 64,960 |

* Compiled from the returns of R. Hunt and J. Y. Watson.

Mr. Hunt estimates the produce of lead of the United Kingdom, for the five years from 1848 to 1852, at 308,108 tons, which gives an annual average of 61,621 tons.

The following is a statement, by the same authority, of the yield of silver from the silver-lead mines of Great Britain and Ireland, in 1852:—

| | Ounces Silver in a Ton of Lead. | Amount in Ounces. | Value. |
|---|---|---|---|
| Cornwall, . . . . . | 35 | 250,008 | £62,502 |
| Devon, . . . . . | 40 | 91,340 | 22,835 |
| Cumberland, . . . . | 9 | 52,893 | 13,223 |
| Durham, Northumberland, and Westmoreland, . . . | 12 | 191,736 | 47,934 |
| Cardigan, Caernarvon, and Caermarthenshire, . . . | 15 | 91,680 | 22,920 |
| Flintshire and Derbyshire, . . | 7 | 47,138 | 11,784 |
| Montgomery and Merionethshire, | 6 | 5,562 | 1,390 |
| Ireland, . . . . . . | 10 | 32,220 | 8.055 |
| Scotland, . . . . . | 8 | 19,048 | 4,762 |
| Isle of Man, . . . . | 20 | 36,700 | 9.675 |
| | | 818,325 | £205,080 |
| | | (68,194 lbs.) | |

Belgium.—The sulphuret of lead is found, in considerable quantity, associated with the very interesting deposits of zinc, which have been already noticed in the preceding chapter. The ores occur in funnel-shaped cavities, which are generally situated above, and in a line with, some great fissure, or dislocation of the rocky strata.* Together with the oxidized ores, there are almost always found the sulphurets of zinc, lead, cadmium, and iron, sometimes mixed with sulphur, and always associated with black clay. The sulphurets occur in the lower part of the deposit, or in the fissure beneath it, in which position they apparently form regular veins. The oxidized ores are evidently a secondary product of the decomposition of the sulphurets, since a nucleus of the latter is frequently found within nodules and masses of the carbonates and silicates. Of the combinations thus formed from the sulphurets, the carbonate appears to be the oldest, and lies

* Delanoue, Ann. des Mines (4), xviii. 455.

the deepest; above it are the silicates, which have a cellular or stalactitic form, and usually contain organic matter, and are often fossiliferous.

The actual amount of lead produced in Belgium appears to be but small. The Nouvelle Montagne Company smelted in 1851, 168, and in 1852, 600 tons of this metal.

PRUSSIA.—The production of lead in this country more than tripled itself between 1847 and 1851, the last year for which returns have been received. The localities are numerous, but are chiefly concentrated in Silesia and the Rhine provinces. In the former, near Tarnowitz, the mining of galena has been carried on since 1526. The red deposits of calamine are almost all plumbiferous in their upper portions, the ores being argentiferous galena and carbonate of lead. The larger portion of the metal at present produced is from a mass of ore lying between the Muschelkalk and the dolomite, south of Tarnowitz. This deposit has been explored over a length of more than five, and a width of over one mile. It rarely exceeds 12 feet in thickness, and the galena occupies only 2 or 3 inches of this.* This deposit was producing, in 1843 and 1844, about 450 tons of lead per annum.

In the vicinity of Siegen, not far from Coblentz, there are lead mines of some importance. Near Cologne are also interesting deposits of lead and zinc ores, which are rapidly increasing in importance. Mines have also been recently opened at Stolberg, near Aix-la-Chapelle. An English Company, called the Wildberg Great Consolidated Mining Company, was formed in 1853, to work some mines of argentiferous galena and copper ore at Wildberg, 10 miles from Olpe, in Westphalia. The locality was examined and favorably reported on by J. A. Phillips. There are 8 principal lodes, running east and west, which have been worked, besides many branches. The lead obtained yields about 80 ounces of silver to the ton.

The whole yield of lead in Prussia amounted, in 1851, to 7195 tons; that of silver, all of which was obtained from lead or copper ores, to 26,493 lbs. troy.

* Ann. des Mines (4), xiii. 303.

THE HARZ. — This mining district is one of the most interesting in the world, not so much from the great amount of its metallic produce, as on account of the high degree of skill in mining engineering and metallurgy which the works above and below ground exhibit, and the great extent and complexity of the systems of veins, and the variety of the metalliferous combinations which are there wrought. The ores are not rich, and it is only the most skilfully-contrived machinery, combined with systematic and economical management, which allows of the workings being carried on.

The mines of the Harz belong to four different states. Those of the Upper Harz, which are the most important, since they include the veins in the vicinity of Clausthal and Andreasberg, belong to Hanover; the Rammelsberg mines, near Goslar, are four-sevenths the property of Hanover, and three-sevenths of Brunswick. Those of the Eastern Harz are in the territory of Anhalt-Bernburg, but are of minor interest.

The Harz Mountains form an island of azoic and palæozoic rocks, which may be said to rise out of a sea of the North German secondary strata.* The longest axis stretches from west-northwest to east-southeast, a distance of 60 miles. Its width is from 18 to 20. The celebrated mountain of the Brocken forms its culminating point, being elevated about 3500 feet above the sea. This and the neighboring heights constitute a central granitic mass, around which the palæozoic rocks are folded, the metamorphic strata having but little development.

The veins of the Upper Harz are concentrated into two groups, that of Clausthal and that of Andreasberg.

In the vicinity of Clausthal and Zellerfeld, the region in which the veins are most fully developed and have the greatest extent and width, there are six principal lines of fracture (German, Züge), all the fractures belonging to each one of which are of the same age, and may be considered as the result of a force of tension acting along a

* For various important papers on the Harz, and the processes adopted there, see De Hennezel in Ann. des Mines (4), iv. 330; Rivot, same journal (4), xix. 465; also, Karsten and Dechen's Archiv, x. 3.

certain line, but only producing a visible effect in certain parts of that line. These lines of fracture have a general east and west direction, and are nearly parallel with each other. The mass of the veinstone is made up of a breccia of the "country," held together by calc. spar, brown spar, spathic iron, quartz, and heavy spar. The principal ore is argentiferous galena, with some copper pyrites and blende. The most productive portion of these veins is where they are most split up, and it is a very curious fact that where the lode is undivided and compact, it is almost invariably poor; where, on the other hand, it is separated into numerous branches, so as to form a stockwerk, there the ore is most abundant. For instance, the Rosenhöfer Zug, a little west of Clausthal, widens out into a stockwerk 300 feet wide, which is nothing more than a collection of narrow veins and branches, and there rich returns of ore are obtained.

The Andreasberg system of veins is much more limited than that of Clausthal, and differs from it in character. The space which it occupies is not much over a mile long, and two-thirds of a mile broad. Its formation is exclusively argillaceous and silicious slate. The ores proper of silver, as well as argentiferous galena, occur here. Among them are the pyrargyrite or dark-red silver ore, antimonial sulphuret of silver, and light-red silver ore.

The lead-bearing veins of the Upper Harz are by far the most important, as the ores of this metal are always argentiferous. Their principal metalliferous contents are galena, containing from 13 to 123 ounces of silver per ton, blende, a little copper pyrites, iron pyrites, and gray copper ore; the gangues are principally calc. spar, quartz, heavy spar, and spathic iron. The ores are mixed together in such a way as to require the most delicate processes of washing before they can be economically smelted; and as they are very similar to those which occur in the silver-lead veins of the Atlantic States, the admirably-contrived machinery now in use in the Harz should be studied by those interested in the mining of argentiferous ores in this country.

Besides the mines of the Andreasberg and Clausthal districts, noticed above, there are the Rammelsberg mines, which are situated in the neighborhood of Goslar, and are next in importance. The nature of this deposit of ore has been alluded to in a preceding chapter.*

The other metalliferous deposits of the Harz are: 1st. The cupriferous veins of Lauterberg; these carry almost exclusively pyritous ores of copper, with a gangue of quartz and fluor spar. 2d. Ores of antimony, wrought to some extent in the veins of the neighborhood of Wolfsberg, near Stollberg. These veins have a gangue consisting principally of fluor-spar, and they are similar in their position to those of Andreasberg. 3d. Manganese, worked at Ilfeld. 4th. Cobalt, at Hasserode, now abandoned. 5th. Iron, extensively worked between Lerbach and Altenau, forming contact deposits between the slates and igneous hornblende rocks.

In the Andreasberg district, the great Samson vein has been worked to over 420 fathoms in depth, being the deepest mine now wrought in the world; the workings have been extended downwards 130 fathoms in the last 30 years. The rich ores in this vein usually occur in courses, occupying an extent of about 100 feet in each direction. At the depth of 360 fathoms, one of the finest accumulations of ore ever met with was struck, and the works have been carried down to their present depth without any considerable change in the richness of the mine.

The Harz has been for a great length of time the scene of extended mining operations. The Rammelsberg has been uninterruptedly worked since 1449. In the Clausthal and Zellerfeld district, mines were opened in the sixteenth century, and in the former there were 45 at work in 1591. Their produce has been remarkably steady for many years. In 1836, the "Silberbergwerkshaushalt," which includes the principal mines of the Upper Harz, namely those of the Clausthal, Zellerfeld, and Andreasberg districts, produced as follows: Silver, 29,375 lbs. troy; lead and litharge, 4080 tons; copper, 30 tons. The Rammelsberg mines produced, in the same year: Gold, 7 lbs.; silver, 2509 lbs.; lead and

* See page 47.

litharge, 567 tons; zinc, 6 tons; copper, 227 tons. The present production of the whole Harz may be estimated as follows: Gold, 5 lbs.; silver, 30,000 to 35,000 lbs.; lead, 5000 to 6000 tons; copper, 150 tons; zinc, an amount hardly worthy of notice; iron, 5000 tons.

SAXONY.—The mines of Saxony have already been described with sufficient detail, in the chapter relating to silver.* The production of lead and silver from argentiferous galena is of minor importance; still, the whole kingdom furnishes nearly 2000 tons of the former metal annually.

NASSAU.—In proportion to its size, Nassau is one of the first mining states in Europe. At Holzappel, silver-lead ores have been mined since 1158; and at Ranzenbach, copper has been obtained since 1465. The whole Duchy is only 82 square German miles in extent, but there are several hundred mines in operation in it, the larger number of which are worked for iron and manganese.

The mines of Holzappel, Obernhof, Marienfels, Welmich, and Werlau, resemble each other most strongly. Their direction is nearly east-northeast and west-southwest, and they dip at an angle of from 50° to 80°.† The veinstone is generally quartz; the ores, argentiferous galena, blende, copper pyrites, and spathic iron. Near the surface, the above-mentioned ores have been converted into carbonate and sulphate of lead, malachite, azurite, gossan, &c. The lodes generally consist of several branches, running parallel with each other, and their whole width is about 3 or 4 feet.

This "Zug," or group of veins, stretches from Holzappel on the Lahn, to Welmich and Werlau on the Rhine, and is enclosed in argillaceous slate and grauwacke, of the Silurian system; its whole length is between 30 and 40 miles, although there are portions of the ground included in this range which have not yet been worked, and the veins are not actually known to be continuous for the whole distance; yet they are so much alike in their principal features, that they must be regarded as belonging to the same system. The metalliferous deposits, wherever opened, are characterized by a

* See page 163.

† Bauer, Karsten and Dechen's Archiv, xv. 137.

near coincidence in strike and dip with the strata in which they are enclosed; and they are cut through by cross veins, running east and west, which heave them from their regular course. There are two of these cross-courses which are particularly conspicuous, and they are known as the eastern and western cross-courses. They divide the lodes into three portions, each one of which exhibits certain peculiarities of structure. The veins have regular, smooth, and polished selvages. Frequently the mass of the lode is traversed by fissures, running across its whole width, at right angles to it, whose sides are lined generally with fine crystals; they are in some cases empty, and in others filled with flucan. The productive, or ore-bearing portions of the lodes, are confined in their extent within certain limits, marked by planes dipping east, at an angle of 14° to 20°, which divide the veins into a number of alternate rich and poor sections; the rock adjacent to the unproductive portion being usually softer and more decomposed than where the lode is rich in ore.

About 30 mines are working on these veins, producing together 600 tons of lead, and 2500 lbs. of silver, besides a small quantity of copper.

An English company commenced operations on a large scale, at Obernhof, in 1853. They have their principal mines near Obernhof and Vinden, the veins producing argentiferous galena, blende, and copper. The yield of the former is said to be 70 to 80 ounces of silver to the ton.

BADEN AND WÜRTEMBERG.—Silver and lead mines were formerly extensively worked in the Schwarzwald and Odenwald, but had been mostly abandoned until quite recently. Within the last few years, several of the old mines have been taken up, with what success I have not learned. An English company commenced operations in Baden during the past year.

AUSTRIAN EMPIRE.—The greater part of the production of this country is from the celebrated mines of Bleiberg and Raibl, in Carinthia.

The village of Bleiberg is situated near Villach, in the Carinthian Alps. The mines are stretched along the valley of the Nötsch, for a distance of five miles, from Bleiberg to

Kreuth. The metalliferous rock is a light-gray limestone, with veins and geodes of calc. spar.* Its geological age is probably that of the Muschelkalk, although this question appears not to be sufficiently determined. The ore is chiefly galena, with some carbonate of lead, calamine, and blende, and it occurs in deposits which do not seem to have much resemblance to true veins. In the Ramser Mine, the workings had, in 1845, been carried to a depth of over 1200 feet. The ore is raised from the mine by means of cars, on a tram-road, laid down in an inclined shaft of large dimensions, which dips with the lode, at an angle of 52°. The cars are moved by water power; the descending one, being filled with water, raises by its weight a load of ore in the ascending one. The whole arrangement is simple and elegant, but, of course, only practicable in mines peculiarly situated with reference to water power.

The lead mines of Przibram, in Bohemia, are next in importance to those of Bleiberg. The veins consist principally of diorite, and the galena is accumulated on one of the selvages, forming a kind of contact deposit. The gangue is principally quartz, with some heavy spar and brown-spar. The metalliferous portion of the lodes contains argentiferous galena, antimonial galena, blende, and iron pyrites, with some silver ores and gray copper. The greater portion of them are too poor to pay for working, until the depth of 300 feet has been attained;† from this point downwards, the richness of the galena in silver increases. There are 13 veins, worked in two different mines. They are contained in sandstones and conglomerates, which are apparently of Lower Silurian age.

The average annual production of the different provinces of the Empire, during the ten years from 1838 to 1847, was as follows:—

* Ann. des Mines (4), viii. 239.

† De Hennezel, Ann. des Mines (4), i. 27.

| | Lead. | Litharge. | Sold as Ore. |
|---|---|---|---|
| Illyria, . . . . | 3,253 tons. | | |
| Bohemia, . . . . | 47 | 851 tons. | 1,073 tons. |
| Hungary, . . . . | 246 | 572 | |
| Tyrol, . . . . | 125 | 8 | |
| Gallicia, . . . . | 12 | 28 | |
| Venice, . . . . | 16 | | |
| Military Frontier, . | 138 | 85 | |
| Styria, . . . . | | 17 | |
| | 3,837 | 1,561 | 1,073 |

The whole production of the Empire was, in 1848, equal to 6737 tons of metallic lead.

SPAIN.—This is the country where, next to England, the mining of argentiferous galena and of simple lead ore is of the greatest importance. The produce of these metals, however, has undergone very great fluctuations within the last few years, and the same has been the case with all the other mineral productions of this extraordinarily rich country, from the earliest times. It is only quite recently that we have been able to obtain anything like a connected view of the geology of Spain, Messrs. De Verneuil and Collomb having made a geological reconnaissance of that country in 1852.* For our knowledge of its mines, we are principally indebted to the French mining engineers, Le Play, Paillette, and Pernolet.

Strabo, Diodorus Siculus, and other writers assure us that Spain, in the earliest times, was a country pre-eminently rich in mines and metals. Pliny describes the mines with minuteness, and that they were of immense extent is evident from the ancient excavations now to be seen in the mining districts, especially along the borders of Spain and Portugal, where gold-washings were also carried on extensively. The same evidences of former mining-work may be seen on the southern slope of the Sierra Nevada, where the Romans worked numerous mines of copper, lead, silver, and iron. Under the Moorish dominion, the art of mining is said to have been in a flourishing condition, but at their expulsion it fell into complete decay. Mines have been

* Bull. Soc. Geol. de France (2), x. 61.

reopened and worked within the last few years, which had been abandoned for from 500 to 1500 years, and in which the old Roman and Moorish lamps and tools were found. Enormous piles of slag, of Roman origin, are seen around the city of Carthagena, which still retains its ancient name. The discovery of America and its rich mineral treasures, occurring near the time of the expulsion of the Moors, caused the mines of Spain to be entirely neglected, with the exception of those of mercury, at Almaden, which were worked in order to furnish the necessary supply of this important agent in obtaining the silver from the American ores. Thus it happened, singularly enough, that Spain was possessed of the richest silver mines and the most extensive deposits of mercury in the world at the same time. The loss of the American colonies, and the subsequent pillage of Spain by Napoleon's troops, reduced that country to a state in which it was necessary for her to make some effort to open her own mines, and prosper on her own abundant resources. For this purpose, the royal decree of Ferdinand VII. was issued in 1825, which laid mining open to all, whether natives or foreigners, under certain tolerably liberal conditions. This law was supplanted by a new one issued in 1849, which, however, did not differ very materially from the old one, except in increasing the size of the "pertinencias," or setts, which were extended from the previously small dimensions of 200 varas by 100, to 300 by 200. Previous to 1820, the only mine worked in Spain was that of Almaden, unless we except a few unimportant iron mines in the Biscayan provinces; and now this country stands among the first in point of metallic production, especially in silver, lead, and mercury.

One of the first points to which mining enterprise directed itself, after the political changes of 1820 and the promulgation of the ordinance of 1825, was the lead mines of the Sierra de Gador, in Andalusia, in the mountain district of Alpujarras; in 1826, more than 3000 mines had been opened in the Sierras of Gador and Lujar. The production of this district was almost fabulous in quantity, and the price of lead fell throughout Europe with great rapidity, ruining many mines in England and Germany. In 1823, these mines yielded 25,000 tons, and in 1827 the production had reached the enormous amount of 42,000 tons, so reducing the price of lead that the miners were obliged to enter into a mutual agreement to work the mines only half the year.

These deposits of lead ore present the greatest analogy with those of the Mississippi Valley. Like them, they are in the Lower Silurian formation, and in the calcareous beds, which, however, are more metamorphosed than at the West. The deposits are not veins, but are compared by Le Play to an immense amygdaloid, in which the paste is limestone and the amygdules galena. The Loma del Sueño was said to look like a hillside burrowed .in by gigantic moles, and more than 4000 shafts had been sunk in the Sierra de Gador alone. From the very nature of these deposits, it will be evident that so enormous a production could not be continued for many years; and, in fact, from 1827, when it attained its highest point, there was a rapid falling off, and these deposits are now comparatively exhausted.

The silver mines of the Sierra de Almagrera were discovered in 1839, and the excitement which the discoveries there made produced was extraordinary. In 1845, 8000 miners were employed in that district, in 826 mines, and there were 38 smelting-works in operation, which produced in that year 108,230 lbs. (troy) of silver, and 8350 tons of lead. The ore which furnished this great quantity of lead and silver was argentiferous galena, of great richness in silver, yielding from 130 to 180 oz. to the ton. The rock is a micaceous slate, belonging to the metamorphic Lower Silurian formation, and containing intercalated beds of trap and porphyry. The deposits of ore do not seem to be of the nature of true veins, but rather bunches lying in the direction of the stratification, and not holding their richness in depth. The vein called the "Jaroso" is the principal one of the district, and had seven working mines on it in 1847. Near the surface, the silver was abundant in the decomposed part of the lode, with sulphate of lead and hydrated oxide of iron. On working downwards, the silver became more rare, the sulphate of lead was replaced by galena, and the oxide by carbonate of iron. The returns of the mine show a gradual impoverishing in silver, as the work descended in the veins.

More recently, attention has been turned to the carbonate of lead, which occurs in abundance from Carthagena to Almeria, although of low percentage, and refractory, being mixed with blende and pyrites; but as the ores lie near the surface, they can be raised very cheaply, and the recent improvements of the smelting furnaces in Spain render it possible to work them with profit. The Pattinson system of separating lead from silver has been recently introduced, and found to assist greatly in the development of these mines. It is an interesting fact, that the immense piles of ancient Roman slags are now re-worked with very considerable profit.

The following remarks on the nature of the mines of Andalusia, are communicated to the English Mining Journal* by a correspondent signing himself "An English Miner," but who is evidently well acquainted with that district, and capable of giving an opinion.

"In a mining point of view, I have repeatedly had occasion to remark—

1. The comparative paucity of mineral veins or lodes, and that, when they do occur, they are generally short, of a bunchy nature, and do not appear to make ore very deep.

2. The great abundance of mineral deposits in beds, masses, nodules, isolated bunches, and other irregular forms, some of which are very remarkable, especially those of lead and iron.

3. The fact that the lead found in slaty rocks, and in north and south veins, is very argentiferous, while that derived from limestone rocks is, without exception, poor in silver.

4. That mineral deposits are most developed in the vicinity of the eruptive rocks, and there richest in silver."

As a consequence of these facts it is inferred that these deposits, notwithstanding the general richness of the ores, require extraordinary caution and great local experience in

* No. 900, Nov. 20, 1852.

their working, and that, as compared with true veins, they are little to be depended on.

The mining region of Linares, to which attention has lately been much directed, is situated a few miles northwest of Linares, in the province of Jaen. There are numerous excavations, extending over a surface of ten square miles, which show that the ancients were well aware of the existence of these lodes. At the commencement of the present century they began to be worked on an extensive scale. There are several lodes known, which are parallel with each other, and run northwest and southeast.

Three or more English companies have recently commenced work in this region, and have been quite successful in their operations.

The principal one, which is called the "Linares Lead Mining Association," purchased mining property at Linares and commenced operations there in 1849. The lodes are represented as being large, and well developed, the adjacent rock being a decomposed granite. The ore seems to lie in bunches, which are occasionally 2 to 3 fathoms wide; other portions of the lode are from 1 to 4 feet in width, carrying gossan, calc. spar, quartz, decomposed feldspar, and galena, with small bunches of copper ore.

They are now raising about 300 tons of ore per month, the lode producing in the 75 fathom level in some places 4 tons per fathom. This company has commenced paying dividends, having cleared about £16,000 in the year ending June 30, 1853.

The New Linares Company is opening three mines in the same vicinity, with favorable prospects, and the San Fernando Company has also valuable silver-lead mines, on whose stock a small dividend has been paid.

Of the total produce of lead in Spain it has been difficult to procure reliable statistics. It has been seen how much it has fluctuated within the last thirty years. For the last years whose yield has been ascertained, 1847 and 1849, it is given at about 30,000 tons. Of this a considerable quantity comes to this country, and a large part of the remainder goes to supply France and the countries bordering on the Mediterranean.

ITALY.—*Sardinian States.* The mines of Pesey and Macot, in the Piedmontese Alps, have been worked for many years, by different companies. These mines and that of Saint Jean de Maurienne produced, from 1745 to 1842, about 25,000 tons of lead, and 150,000 lbs. of silver. Their yield is given by Burat, in 1846, as about 250 tons of lead, and 1600 lbs. of silver, yearly.

*Sicily.* There are irregular deposits of lead, ore in Sicily, which are worked to some extent, but I have no information of the amount of metal produced.

FRANCE.—In this country the production of lead, like that of all the other metals except iron, is quite insignificant. The principal silver-lead mines, indeed the only ones of any importance, are at Pontgibaud, and are now worked by an English company.

At these mines workings have been carried on for a great length of time. The official documents go back to the 16th century.* There are three concessions belonging to the present company, containing three mines, which have been repeatedly taken up and abandoned at various times. Their geological position is in the granitic, gneissoidal, and schistose series of Central France, and the general direction of the veins is north 15° to 45° east.

They have, with one exception, a feldspathic gangue, little different from the rock itself. Sulphate of baryta also occurs in them near the surface. The ore is highly argentiferous, the lead generally containing from 0·3 to 0·5 per cent. of silver. The portions of the veins which carry ores are arranged in a series of vertical columns or masses, quite limited in length on the course of the veins, but holding their richness in depth. The production in silver of the Pontgibaud mines had steadily increased, from 2026 lbs. in 1842, to 4168 lbs. in 1849. That of lead, in the same period, rose from 90 tons to 1260.

The mines of Poullaouen and Huelgoet were formerly

* Rivot and Zeppenfeld, Ann. des Mines (4), xviii. 137.

quite important, but their richness diminished as they were worked downwards, the veins appearing to give out entirely. There are three principal veins at Poullaouen, of which the most important, which had been traced for nearly a mile, has a direction of north 22° west. The ore is galena, with 0·3 to 0·5 per cent. of silver, in strings which are sometimes 8 inches in width. Work was commenced here in 1729, and over 1000 men were employed in 1760.

At Huelgoet, the vein varies at the surface from 2 to 80 feet in width; it is contained in the older crystalline slates, and has been traced for a mile in length. It contains galena with 0·1 per cent. of silver. At the surface rich silver ores were found, analogous to the *colorados* of the Mexican mines. This vein was worked before 1578.

There appear to be no mines of lead of any importance except in Europe and the United States, at least I have been able to procure no information with regard to any such.

In Australia, a small amount of galena had been mined previously to the gold-discoveries, and that ore is said to exist in numerous localities. There are undoubtedly abundant supplies of this metal within the metalliferous districts of South America; but thus far they have been but little worked; indeed, where such rich deposits of silver lie almost neglected, it could hardly be expected that those of a metal so much inferior in value would be noticed. Without doubt any considerable increase in the price of lead would lead to the opening of new regions producing that metal.

---

## SECTION III.

### GEOGRAPHICAL DISTRIBUTION OF THE ORES OF LEAD IN THE UNITED STATES.

We now turn to this country, whose mines of lead are abundantly scattered over its surface, and which has produced a larger amount in value of this metal than of any other, with the exception of iron and gold. The productive mines are chiefly concentrated within a district of comparatively small extent, known as the Upper Mississippi Lead Region; but there are numerous lead-bearing veins in the Atlantic States, in various geological positions, some of which promise at a future day to become of importance.

For convenience of description, the lead-bearing veins and deposits of this country will be arranged according to the following scheme:—

### I. MINES OF THE ATLANTIC STATES.

*a.* In the azoic formation; mines of St. Lawrence County, New York; transverse veins; ore, galena, free from zinc and iron.

*b.* In the metamorphic palæozoic rocks; ore, galena principally, generally more or less argentiferous; almost always associated with blende, copper pyrites, and iron pyrites; veins usually parallel with the formation; localities numerous, especially in New England.

*c.* In the unaltered Lower Silurian rocks; the only localities in the Atlantic States, so far as known, are in New York; mode of occurrence apparently irregular; not extensive; recently worked, but not largely.

### II. MINES OF THE MISSISSIPPI VALLEY.

*a.* Lead Region of the Upper Mississippi; ore galena, not argentiferous; deposits, irregular and gash veins, in limestone of Lower Silurian age; worked in numerous localities, occupying a space principally in Wisconsin, but extending into the adjacent States of Iowa and Illinois.

*b.* Lead Region of Missouri; localities numerous, and ores similar in character and position to group *a;* formerly more extensively worked than at present; not now of much importance.

### MINES OF ST. LAWRENCE COUNTY, NEW YORK.

There are numerous well-developed lead veins in the azoic of Northern New York, which have been worked to some extent, and which promise hereafter to become of considerable value. The most important of these, so far as can be judged from their present development, are those of Rossie, in St.

Lawrence County. The veins of this vicinity, which have attained a wide celebrity among mineralogists for the splendid crystallizations of galena and calc. spar which they have furnished, have been worked, at intervals, since 1835; but thus far, not with that degree of success which ought to have attended the opening of so rich veins.

The rocks in this region are mostly gneiss, interlaminated with hornblende and mica slates, of which the stratification is not always clearly to be made out. They belong to the azoic period, and are overlaid on the north and west at a short distance by the strata of the Potsdam Sandstone.

The attention of the public was first turned in this direction in the winter of 1835–6, when the remarkable vein generally known as the "Coal Hill Vein," was first opened. This vein is one of a group of four or more, exhibiting the same characters, and similar in position, while differing in their width and the quantity of ore which they carry; and it has been more extensively worked than any other of them. Dr. L. C. Beck, of the State Geological Survey, gives the following description of it, as it appeared soon after it was opened, in August 1836.* "The vein of galena and white decomposed ore was distinctly visible for some distance, passing down a precipitous ledge of primitive rock, about fifty feet in height. The average width of the vein was two feet, and it cut the rock in a nearly perpendicular direction; at the lower part, however, inclining slightly to the north. On ascending the ledge, the course and extent of the vein could be easily determined by the excavations which had been made, and by the appearance of the surface in those parts which had not been opened. Its course was found to be about south-southeast and north-northwest; and its length, as exposed at that time, was about 450 feet; and throughout this whole extent, the vein seemed to be so distinctly characterized as to excite surprise that it had not long before been noticed." Farther on, he remarks that the average width of the calcareous gangue is about four feet, and that the proportion of the ore in the vein was quite variable.

* Mineralogy of New York, p. 48.

Prof. Emmons, State Geologist of the district in which these mines are situated, remarks,* that the vein is three or four feet wide, and filled with calcareous spar and galena; the latter having a width of *only* from two or three inches to eighteen; the average width of the solid ore he estimated at ten inches.

*Coal Hill Mine.* This mine was worked in 1837 and 1838 with great activity, but with an entire ignorance of the first principles of mining. The length of the vein which was opened was about 400 feet, and this was divided into two sections, and worked by two distinct companies, one of which was called the Rossie Lead Mining Company, and the other the Rossie Galena Company. The two sections were divided by a bar of ground left standing between the Engine and Rodda's Shafts (see Section, Fig. 37), which has since been taken down. Thus, of course, a double expense was required for freeing the mine from water; and in every other respect the workings were as ill-arranged as possible, the whole excavation being little more than an open cut, without even an adit-level for carrying off the surface-water. The ore was smelted on contract, at $25 per ton of lead produced. According to C. L. Lum, Esq., who had charge of the books of Messrs. Moss and Knapp, the smelters, the amount received for lead sold from this vein, up to the time it was abandoned in 1839, was about $241,000; 3,250,691 lbs. of lead were smelted, according to the record in the company's books, together with another lot of 151 tons, and other smaller amounts of which no record was kept.

As would naturally have been expected, where so entire an ignorance of the method of opening a mine prevailed as here, the Coal Hill Vein was abandoned, after the richest portion of it had been worked out down to a depth where the water began to be unmanageable without well-contrived machinery. At this time the great commercial depression throughout the country drew all attention away from mining, and the whole of the St. Lawrence County region, in regard to which there had been so much excitement, was deserted.

After lying dormant for some years, the Coal Hill Mine was taken up again by a company formed in New York, called the Great Northern Lead Company, and operations commenced in the winter of 1852.

The annexed section (Fig. 37) will show the present state of the excavations on this vein, and from it an idea may be obtained of the amount of work which had been done at the time the Great Northern Company commenced their operations. There was an excavation about 440 feet in length, and 180 feet deep for a part of that distance; there was no adit-level, a block of ground having been left standing in its course; from the surface down to the present adit, a depth of 67 feet, the mine was an open cut, and had become choked up with every kind of rubbish and filth, which it was a tedious and expensive job to clear out. This was done, however, and a steam-engine of 60 horse

* Geological Report, p. 355.

power was erected. But little was done during the winter of 1853 in the way of mining; and as the erection of the necessary machinery, furnaces, and buildings, together with the clearing out of the old mine and making good the levels, had consumed the funds of the company before the mine was opened so that returns of lead could be made, and as the stockholders had not sufficient confidence in the mine to contribute the necessary amount for continuing operations, the work was stopped, almost before it was commenced.

Fig. 37.

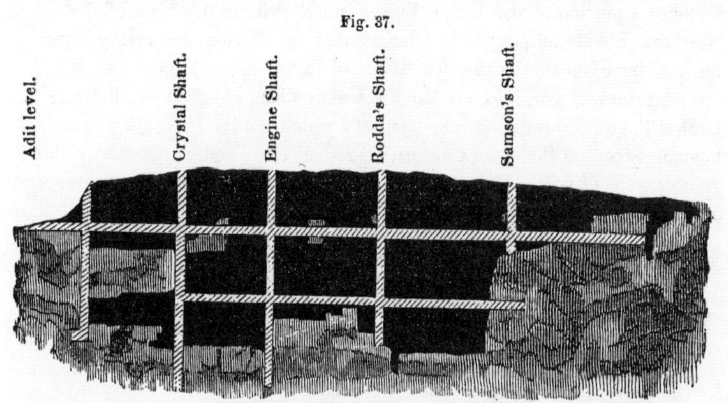


Section of workings on the Coal Hill Vein.

At this time two shafts had been sunk 40 and 45 feet below the old workings, as shown on the section, and a winze had been carried down 18 feet in the western part of the mine. In the bottom of all these excavations, the lode appeared large, regular, and well-defined. In the Crystal Shaft, the solid galena occupied a width of from 6 to 18 inches for the last 4 or 5 fathoms which were sunk. In the winze, the lode appeared to be equally rich, carrying large masses of pure lead ore.

The vein at this depth, about 200 feet, is apparently of the same width as at the surface, varying from 2 to 4 feet. The gangue is almost pure calc. spar, frequently forming vugs in which splendid crystals occur. Some of the finest crystallizations of this mineral ever found have been obtained from this mine. One gigantic crystal, nearly transparent, in the cabinet of Yale College, weighs 165 pounds. In the same cavities the galena itself is often finely crystallized. Groups of crystals weighing over a hundred pounds are said to have been obtained near the surface. The ore is remarkably free from any association with iron, copper pyrites, or blende, differing in this respect from all the lead-bearing veins in the metamorphic palæozoic rocks of New England and the Atlantic States. It contains a trace of silver, but not enough to be worth separating.

About 2000 fathoms of the vein have been removed, and if we adopt Mr. Lum's statement of the value of the lead taken from this vein, at a time when its price was considerably lower than at present, adding $5000 for the ore sold and taken out by the Great Northern Company, it appears that the average value of the lode per fathom is over $120.

It is understood that this mine has been quite recently taken up by a company called the Ontario Mining Company, and, under judicious management, there seems to be no reason why it should not be made profitable.

In the same sett, there is a parallel vein to the Coal Hill, called the Indian River Vein. It is not opened to any extent, but it appears on the surface to have the same character as the one just described.

*Victoria Lead Company.* The sett of this company, containing 640 acres, includes a portion of the Union Vein, formerly within the lease of the Great Northern Lead Company, and a large tract in addition, embracing a part of the western extension of the Coal Hill and Indian River Veins.

The principal workings are on the Union Vein, which is parallel with the Coal Hill, and distant about one-third of a mile from it. This vein was wrought to some extent at the same time with the Coal Hill; and, if possible, with still greater want of skill. The work was done by tributers, and the openings were of the most irregular description. The annexed section (Fig. 38), will show

Fig. 38.

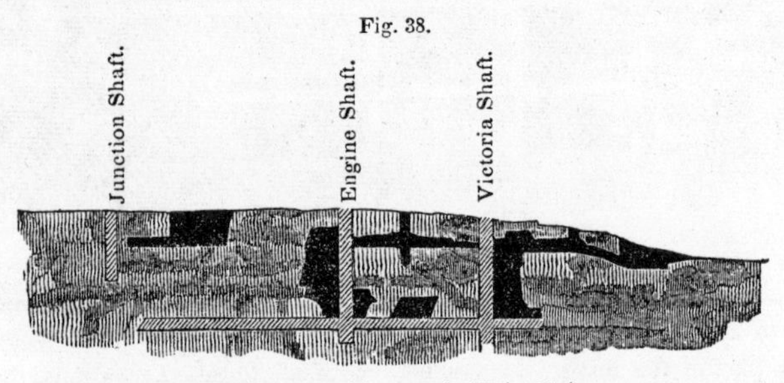


Section of workings on the Union Vein.

the present state of the work, the shafts and level having been excavated since work was resumed on the mine. According to Mr. D. W. Baldwin, the amount of metallic lead smelted from the ore of this mine in 1836 and 1837, was 213½ tons. The same vein was worked about half a mile farther east, and was there known as the Victoria Mine; and the two mines together produced 524 tons of lead before they were abandoned.

While the Union Mine was in the hands of the Great Northern Company, but little was done towards opening it, as the shafts could not be sunk below the first level, for want of a steam-engine for drainage. The level, 96 feet from the surface, was extended a short distance west of the principal shaft, and some explorations were made on the line of the vein.

The Union Lode is very regular, and has lead disseminated through it in considerable abundance. In general, it is from 3 to 4 feet wide; and its gangue is like that of the Coal Hill Mine, except that it is less crystalline. The walls are well-defined, and generally pretty smooth; but there are no slickensides, or flucan selvages, and the spar does not separate readily from them. The whiter, more transparent, and more coarsely-crystalline the gangue, the more abundant is the ore; and where it is most plentifully distributed, the lode is vuggy, and the cavities are lined with crystals of calc. spar and galena. In this vein, as in the Coal Hill, there is hardly a trace of any other ore than galena.

The present company commenced working the mine quite recently, and have erected a steam-engine, and propose sinking immediately to a depth of 10 fathoms below the present level. A considerable part of the lode is worth fully $100 per fathom, and it seems highly probable that it may be worked with profit, if opened properly.

There are two lodes here, which seem to come together near the Junction Shaft; one of these is probably the same vein which was worked in the Old Victoria Mine, but they have not been traced into each other, owing to the low ground to the east of the Coal Hill and Union Mines.

*St. Lawrence Mining Company.* This company has until recently been working what was formerly called the Macomb Mine, about nine miles from the Coal Hill Mine, which it is said very much to resemble. The vein is thought to be a valuable one, but capital and confidence appear to have been wanting, and the works have been stopped, temporarily only, it is hoped. A steam-engine, furnace, and other machinery have been erected here, and probably about $50,000 expended in preparations for mining. The company owns 1420 acres of land in fee, and it is to be hoped that they will not allow so important a property to lie dormant.

At Mineral Point, on Black Lake, six miles below the village of Rossie, lead-bearing veins similar to those described are mentioned as occurring, and as promising favorably for mining. Indeed, numerous lodes have been discovered in various parts of the adjacent country, but nothing is known with regard to their value.

### LEAD-BEARING VEINS OF THE METAMORPHIC PALÆOZOIC.

In this group are included a great number of plumbiferous veins, occurring in the belt of metamorphic rocks which extends through the Atlantic States, along the southeastern flank of the Appalachian chain. They are especially numerous in the New England States, where they have been repeatedly the object of mining enterprises, and unfortunately have thus far rarely been profitably worked or thoroughly developed. They appear, in almost all cases, to belong to the segregated class of veins, although often developed on a large scale, forming powerful and well-marked lodes. The reasons why they have failed to make profitable mines seem to be, first, because the valuable ores which they contain are too much mixed with other metalliferous substances of no value, from which they must be separated by expensive machinery, requiring skill in its construction and management; second, because the ores are not sufficiently concentrated in the veins, the rich bunches being too much scattered through a mass of barren veinstone; and lastly, though in a less degree, because capital and patience have been wanting for the development of the mines.

I propose to commence with the New England States, and to notice most of the prominent mines of lead and argentiferous galena which have been wrought at different times, in the hope of throwing some light on the prospects of this branch of mining in the United States; trusting that what may be said may have some influence in directing the application of capital, or at least, in some instances, of checking unnecessary and foolish expenditures upon mines which give no promise of profitable working.

MAINE.—The Lubec lead mines are the only ones mentioned by the State Geologist as worthy of being worked. These are situated on the estate of John Ramsdell, four miles west of Lubec. The rock in which they occur is an argillaceous limestone, and the veins are found at the points where it is traversed by dykes of trap. The veins were wrought for a few months after their discovery, in 1832. There are several of them, all exhibiting the same character. They are apparently contact deposits between the trap and the limestone. The principal one is said to be 2½ feet wide, with a gangue of quartz and compact feldspar. It appears on the face of an abrupt precipice of limestone 100 feet in height, into which a drift has been carried for 155 feet. There appears to have been no sufficient encouragement to carry on the works, which were stopped many years since, and have not been resumed.

A small vein of lead and zinc ore has been discovered at Parsonsfield, but it has not been considered worthy of being worked.

VERMONT.—The only localities in this state where galena has been found, so far as I know, are at Chittenden, Thetford, and Morristown. At none of these places are the indications sufficiently promising to warrant working.

NEW HAMPSHIRE.—The localities where lead ores have been found in this state are numerous, and the veins wide and well defined; but the valuable metalliferous substances are too sparsely scattered through the veins, and too much mixed with blende and iron pyrites, to be, at present at least, worthy of being worked.

Among the more interesting mines which have been opened and worked to some extent, the following may be mentioned.

*Eaton Lead Mine.* A powerful vein occurs here, which was discovered about 1826, and several times wrought for a little while and then abandoned. Previous to 1840, a shaft had been sunk on it to the depth of 40 feet, and 15 barrels of picked lead ore sent to Baltimore to be smelted. The vein is stated to be six feet in width,* and to consist mostly of yellow blende, including masses of galena. Its course is north 21° east, and it dips west 60° to 65°. It is favorably situated for drainage. The lead ore contains, according to Dr. Jackson's assay, 2 lbs. of silver per 2000 lbs. Dr. Jackson remarks: "It is hoped that persons interested in mines will attend to this valuable vein, for it is one of the largest and richest in New England;" although, as he states, the lead ore is the smallest part of the vein, and does not *average more than eight inches in width.*

On the strength of these recommendations, parties were induced to resume the working of this mine a few years since, but after a large amount of money had been expended, with no returns, it was abandoned, simply because there was no ore in the vein, as I am informed by those engaged in working it.

*Shelburne Lead Mine.* A considerable amount of mining has been done at Shelburne, on a powerful vein. The vein is thus described by J. T. Hodge, Esq.† "The position of the vein is nearly vertical, its inclination being 71° to the north. At the surface it appears to consist almost wholly of quartz, but by close inspection a small seam of metallic ore is discovered near the northern or upper side of the vein. In some places this is an inch or two wide, and then narrows away to a small fraction of an inch; never, however, being completely lost. The ores at the surface were found to be the sulphurets of lead, zinc, copper, and iron; and the first named, the prevailing ore, proved to be highly argentiferous, containing, according to Dr. Jackson's analysis, three pounds of silver to the ton of ore." Mr. Hodge advised that in consideration of the richness of the lead in silver, and the appearance of the vein, openings should be made here to some depth, in the hope that the metalliferous portion would increase in width, or that valuable bunches of ore would be found.

The mine was worked from 1846 to 1849, and considerable money expended, with the following results, as I am informed by C. L. Lum, Esq., who was connected with the work. There are three shafts, one of which is 275 feet deep. In this shaft, a rich bunch of ore was struck, at the depth of 30 feet, and a drift was run off to the northwest about 40 feet on it; here the vein was 2 feet

* Final Report on Geology of New Hampshire, p. 83.

† Mining Magazine, i. 29.

wide, and from about 30 fathoms of the vein removed by underhand stoping at this point, 20 tons of good ore were obtained. A short distance farther down, another bunch of ore was struck, of less importance, but from that point, nearly to the bottom of the shaft, the lode was very poor, although from six to ten feet wide. In the best part of the vein, it contained six inches of solid galena. A shipment of the ore to England, of a little over 5 tons, netted there about £16 per ton. The richest portion gave 84 oz. of silver to the ton of lead. Another shaft, only 90 feet to the southeast of the main shaft, was sunk to the depth of 80 feet, and the vein found to be tolerably well filled with ore. The works were abandoned in 1849, and have not since been resumed.

It appears possible, from an examination of the works and the nature of the ground, that the vein may not have been explored sufficiently in longitudinal extent. A single shaft might be sunk where it would, perhaps, pass through a poor part of the lode for the whole distance, while, at the same time, there might be rich bunches of ore within a short distance on each side of it. The region appears to be a metalliferous one, and should be farther explored.

Galena has been found in many other localities in New Hampshire, but in none of them in veins of sufficient magnitude to require notice.

MASSACHUSETTS. — Among the localities where lead ores have been found in this state, the only ones of any importance are those of Northampton, and of the vicinity, within the boundaries of Southampton and Easthampton. The veins are large, and very conspicuous on the surface, and were among the first known and worked in the country, and have a degree of historic importance, although of no value economically.

The principal and oldest mine is in Southampton, and has been well known to mineralogists for the last half century. A company was formed and workings commenced here in 1765, by a party of Connecticut adventurers, and the old records speak of stones of ore taken out of the back of the vein, which weighed 200 lbs. Operations were suspended by the Revolutionary war, and it was not until 1809 that they were resumed. The old shaft was cleared out, and had been sunk to a depth of 60 feet at the time it was visited by Prof. Silliman, in 1810. Soon after, an adit-level was commenced, intended to intersect the lode at a depth of about 140 feet, by driving from 1100 to 1200 feet. This work was carried on by a single miner until 1828, when it was abandoned, in consequence of his death; at that time the adit had been driven about 900 feet, two-thirds of the distance being in a light-colored, coarse-grained sandstone, belonging to the New Red, and the remainder in a granitic rock, very difficult to break. Thus the works remained for many years, the adit about two-thirds completed, and an open cut excavated on the vein 46 feet deep, 40 feet long, and from 6 to 8 feet wide.

This open-cut shows a vein occupying its whole width, of which the gangue is mainly crystallized quartz, with some heavy spar, and with galena scattered through it in small quantity, but not sufficient to form a workable lode. From near the surface, a variety of the usual ores resulting from the decomposition of the sulphuret were taken, and are to be found in the principal mineralogical cabinets of the country.

After lying dormant for more than twenty years, this vein was taken up and worked for a short time on a small scale.

In the spring of 1853, I found the adit-level driven in 970 feet; the rock at the end was a coarse-grained feldspathic granite, costing probably $150 a fathom to break, as the miners had refused to take it on contract at $132, and they were proceeding at about three feet a week. In March last, I found that the work had been stopped, although the end of the level was supposed to be within a few feet of the lode.

The same vein, apparently, has also been recently opened, at a distance of about half a mile in a direction nearly north 20° east, in the town of Easthampton. Here the lode is large, and contains porous and crystallized quartz and gossan, but is destitute of ore; it is what the Cornish miners would call a "hungry lode."

Farther on, in Northampton, about two-thirds of a mile from the last-mentioned opening, another attempt at mining was made in 1852 and 1853, but the work was abandoned when I was last there. Here two shafts have been sunk at 80 feet distance from each other, and to a depth of about 50 feet. The apparent direction of the lode here is north 64° east, so that if this is the same vein which is opened at Southampton and Easthampton, it must have changed its course somewhat. The gangue resembles that found at the last-mentioned excavation, and the appearance of the lode is the same as to width, &c., but there is more blende and copper pyrites mixed with the galena. In the most northern shaft, the vein is over 10 feet wide, and has little patches of lead scattered through it.

The only place where any work is now doing (March 1854), is at the Northampton Silver Lead Mining Company's mine, next adjoining on the north to the one just described, and distant from it only 200 or 300 feet. Here a shaft has been sunk 65 feet, which is now discontinued, and an adit is driving towards the vein, and will intersect it near the bottom of the shaft. It is driven through coarse-grained granite, costing $12 50 a foot to break: it is expected that the vein will be cut by July 1st. The lode here, as in the other places where it has been opened, consists of quartz, with galena, some blende, and a few little bunches of copper pyrites.

All these openings appear to be on one great vein, which has a general direction of about north 25° east; the distance from the most northern opening in Northampton to the most southern one in Southampton being about two miles. If not the same vein, there must be two veins very similar in character and in mineral contents, and not far distant from each other.

Although this lode is so well defined and wide, I cannot recommend its being farther worked. The cost of opening it would be very great, and there is no reason to suppose that it will increase in richness in depth; at least, it has not done so in any of the openings which have been made on it, up to this time. I should recommend that the Southampton adit be continued to intersect the lode, provided the distance, when accurately determined, be not too great; and unless a considerable change shall be found to have taken place in the character of the vein, all farther attempts at working it should be abandoned. There is no reason to suppose that the patience of the adventurers will hold out until that depth is reached, where, in the opinion of some, the vein is to become rich enough to be worth working.

Quite a number of lead veins, similar to the one just described, have been noticed in this part of Hampshire County, in the towns of Westhampton, Williamsburgh, Goshen, Hatfield, and Whately. In the latter, there are three veins which are known to contain lead, but they have not been explored to any extent. According to President Hitchcock, State Geologist, there is in Russell a vein from 2 to 3 feet wide, containing galena and copper pyrites. In Leverett, the occurrence of galena in two veins, with a gangue of quartz and heavy spar, is noticed.

The State Geologist remarks,* "that the central parts of Hampshire County contain extensive deposits of lead, which may be of great value to posterity, if not to the present generation." Thus far, the result of the explorations has not been of a nature to render us very sanguine of the speedy approach of the time when these veins will be profitably worked.

CONNECTICUT.—The indications of galena are abundant in this state, but there seem to be but few localities where mining for it has been carried on to any extent. A mine was opened at Brookfield some years since, which was expected to prove rich in lead; but the deposit, which was contained in a white limestone, proved to be a mere bunch, and not a continuous bed, or vein. At the locality in Monroe, called Lane's Mine, which is so well known to mineralogists from the variety of minerals occurring there, galena has been found disseminated through a bed of quartz, but not in workable quantity; it is, however, very rich in silver, the lead produced from it containing, according to Prof. Silliman's analysis, from 2 to 3·5 per cent. of that metal.

At Plymouth, a mine has been worked to some extent, on

* Report on the Geology of Massachusetts, p. 202.

what is supposed to be a vein; I have no information as to its value.

The most important silver-lead mine in the state, is that near Middletown, about two miles south of the city, on the right bank of the Connecticut.

*Middletown Silver and Lead Mining and Manufacturing Company.* This mine is thus noticed by the State Geologist in his Report.* "It (galena) here occurs in a thin bed or seam of quartz included in mica-slate, having a thickness of from 10 to 20 inches. The strata dip west between 35° and 140°. The ore is associated with blende, iron pyrites, and, rarely, with yellow copper pyrites. These ores, however, form but a small proportion of the seam, into the composition of which a plumbaginous argillite, or mica-slate, often enters. The galena seems even less abundant than some of the other sulphurets. The excavations prove that the mines must have been wrought formerly to a considerable extent."

The annexed section (Fig. 39.) will serve to show the extent of the workings at present, and that portion of them which was executed many years since. The excavations shaded by oblique lines represent the old mine, as it was when opened in 1852, and those in vertical lines show the work done since that time by the present company.

Fig. 39.

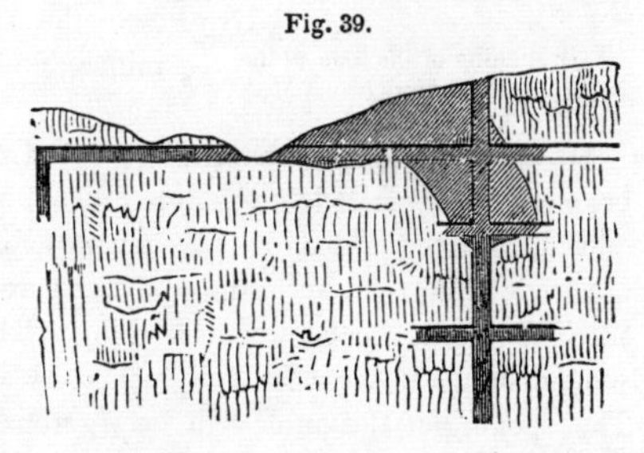

Section of the Middletown Silver-lead Mine.

By whom this mine was first wrought there remains no tradition among the inhabitants; it is certain that it must have been many years previous to the revolutionary war. The early history of Connecticut shows that Gov. John Winthrop had obtained, as early as the year 1651, a license giving him almost unlimited privileges for working any mines of "Lead, Copper, or Tin, or any minerals; as Antimony, Vitriol, Black Lead, Allum, Salt, Salt Springs, or any other the like;" and he was allowed "to enjoy forever said mines, with the lands, woods, timber, and water within two or three miles of said mines." A special grant was made to Gov. Winthrop of any mines or minerals he might discover in the neighborhood of Middletown. There seems to be no positive evidence that the Middletown mine was worked by him, but his grandson John Winthrop, F.R.S., was evidently well acquainted with the existence of deposits of lead, silver, and other metals in this state, of whose minerals he made a large collection, which he forwarded to the Royal Society.

The vein is included within the strata of a silicious slate, and at the surface appears to have the same general direction and dip as the formation itself. Its

* Report on the Geol. Surv. of Conn. p. 52.

course in the adit-level is about north 50° east, and it dips to the northwest about 45°. In the adit its width averages about 18 inches, but in the level below it expands to nearly three feet. The gangue is principally quartz, sometimes in crystallized plates or combs, with some calc. spar, and a very little sulphate of baryta and fluor spar, in fine crystals. In the lower level the vein is made up of silicious matter and a dark slate-colored argillaceous substance, in alternating bands, with strings and seams of ore of a few inches in width.

The ore is a highly argentiferous galena, some of which is fine-grained, and other portions largely crystalline. There is some blende associated with the galena in these seams, and generally occurring on the sides of the bands of that ore, and not intermixed with it. Some copper pyrites has also been found, in small strings and bunches, associated with the galena. The vein sends off flat branches, seemingly in the direction of a set of cross-cleavages of the strata, as shown in the annexed section (Fig. 40.), one of which, near the adit-level, was worked by the old miners, and found rich in ore for a limited distance.

Fig. 40.

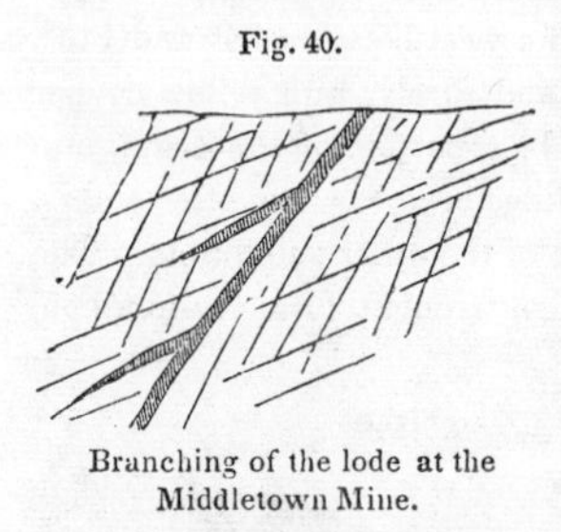

Branching of the lode at the Middletown Mine.

Since the present company commenced operations, the principal shaft has been sunk on the lode to a depth of about 120 feet below the old workings, and the 20-fathom level driven each way for a few fathoms; parts of the lode are wide and rich, but other portions are poor, and too much mixed with silicious matter to be worth working. The vein has also been traced for some distance to the west, and opened by an adit-level driven in on its course. The opening of this mine will be vigorously pushed by the company, with a view to ascertain whether it can be worked with profit. The galena contains, according to Mr. Pattinson's assays, from 25 to 75 oz. of silver to the ton (of 21 cwts.) of lead, the fine-grained variety yielding only one-third as much of this metal as the coarsely cubical ore; a fact quite contrary to what is commonly observed with regard to the richness of lead ores in silver, the finer grained almost always containing a considerably higher percentage of this metal than the coarse varieties.

There are several other veins on the property of the company, which are, however, of minor importance.

New York.—The lead deposits in the metamorphic rocks of this state are chiefly in Columbia, Dutchess, Washington, and Rensselaer Counties. According to Prof. Mather,* State Geologist for that district, the ores of lead are situated in veins traversing the strata near the junction of limestone with slate-rocks, where they had been upturned and exposed

* Report on the Geol. of New York, p. 498.

to great derangements, and more or less affected by metamorphic agency.

The mine which has been most extensively wrought is the Ancram or Livingston Mine, in Ancram, Columbia County.

It is opened in an argillaceous slate, near its junction with a sparry limestone. Prof. Mather describes the lode as being about 4 inches wide, and having a gangue of quartz, carrying galena, blende, and copper pyrites. It has a course of south 70° west, and dips nearly vertically, while the lines of bedding of the slates dip to the east from 60° to 70°. The vein, which is narrow at the surface, widens in its descent.

Prof. Beck notices this locality in his description of the mineralogy of the state, and remarks that it has not the characteristics of a true vein, but is rather a collection of strings and bunches parallel with the strata, the ore not being bounded by any regular walls, but gradually losing itself in the adjacent rock.

He adds that the mine has been extensively worked at various times, with large expenditures, but no returns; and, furthermore, that there is nothing at this locality which would warrant any additional outlay.

This mine has, however, been recently taken up again by a New York company, called the "American Silver-Lead Company." The old shafts have been cleaned out, and found to present sufficiently favorable appearances, in the opinion of A. C. Farrington, Esq., to justify a farther expenditure in opening the mine.

Lead mines have been worked in Northeast, about five miles southeast of Pineplains, in Dutchess County, on the farms of Judge Bockee and Mr. Ward Bryan. A company of German miners made some explorations here in 1740, and procured some ore, which was sent abroad to be smelted. Afterwards, during the revolutionary war, an attempt was made, under the direction of the Committee of Public Safety, to procure lead here for supplying the army. The veins are said by Prof. Mather to be too small for successful working. A mine in this county has been recently taken up by a New York company, but I am unable to state whether it is the same one which is alluded to above.

Although the lead does not seem to be abundant in this region, yet it is generally rich in silver; some specimens have been assayed with extraordinary results. It is to be hoped that the companies who have recently commenced mining here will soon ascertain satisfactorily what the real value of these argentiferous veins may be.

Pennsylvania.—The interesting metalliferous region in Montgomery and Chester Counties has already been alluded to in a former chapter.* Not having visited the mines of this district, my notice of them will be chiefly extracted from the published reports of Prof. H. D. Rogers and Dr. Genth.

These mines are situated near the junction of the New Red Sandstone and the gneiss, or metamorphic palæozoic; as has already been mentioned, the cupriferous veins of the district are chiefly confined to the sandstone and shale, and bear those ores in that formation, while the lead-bearing lodes, although in some instances entering the sandstone, are best developed in the gneiss, and are there almost solely worked.

Four lodes carrying principally lead are mentioned by Prof. Rogers as more or less explored, and there are several others less known. These are the Wheatley and Brookdale, the Chester County, the Montgomery, and the Charlestown Lodes. The two first-mentioned do, however, bear lead in the shale, which forms only a superficial cap of inconsiderable thickness upon the metamorphic strata below.

The gneiss is described as being decomposed down to a very considerable depth, so that the shallow excavations in the mines are made with great facility. It is intersected by numerous dykes of granite, greenstone-trap, and other igneous rocks, which sometimes cut the strata vertically, and sometimes are parallel with the planes of the enclosing rock.

The greater number of the lodes of this district have a course of about north 32° east, and they dip to the south-east; another set, however, runs north 52° to 54° east; but there seems to be no marked distinction in the character of the two systems, if such they may be called.

* See page 328.

The principal gangue of the veins is quartz, with some heavy spar. The ores found comprise almost every variety of lead ore usually found on the backs of plumbiferous lodes when decomposed. Specimens of extraordinary beauty from this region were exhibited in New York at the Crystal Palace, among which sulphate, carbonate, phosphate, molybdate, and chromate of lead were conspicuous for their beauty. The following are the principal mines which have been or are worked for lead and silver.

*Chester County Mining Company.* This company took possession of the mine in June 1850, at which time an adit had been driven 837 feet, cutting two veins and draining a third by a cross-cut, and some other work had been done, from which about twenty tons of lead ore and a little copper had been taken.

There are two veins, of which the principal one has a course of north 53° east, and dips 75° to the north. The other has a course of nearly north 34° east, and dips south. The principal ore obtained from the workings was pyromorphite, or phosphate of lead, mixed with galena; the sulphate and carbonate were also found in some quantity. The pyromorphite contained, according to Dr. Genth's analysis, 71·5 per cent. of lead, and 0·0054 of silver, with a considerable quantity of sesquioxide of chromium, to which substance the dark-green color of the mineral was found to be owing. A little blende and indigo copper were also found. The coarsely-granular galena gave 16·2 oz. of silver to the 2000 lbs., the radiated and finely-granular 0·0406 per cent., or 11·9 oz. to the 2000 lbs.; while the pyromorphite contains only 1·6 oz. in the same quantity.

Some difficulty was experienced in smelting the ores of this mine, on account of their being so much mixed with the phosphates. A furnace was erected by the company in 1851, and after some experimenting was found to answer the purpose, and is said to have smelted the ore with entire success. Up to Nov. 1851, 190,400 lbs. of dressed ore had been worked, producing about 47 per cent. of metallic lead; the ores smelted up to that time were almost exclusively phosphates.

Mining has not been carried on at this place for some time, but there is a furnace and establishment for desilverizing and smelting lead ores, which is the only one for this purpose in the United States.

*Wheatley Mine.* According to Prof. Rogers's Report, dated May 1st, 1853, this lode has been opened for a length of 3072 feet. The adit-level is 1279 feet long. The vein in this distance varies in width from 1 to 2½ feet, averaging about 18 inches. Its dip is about 68°. On the 1st of August last, the main shaft was down 234 feet, and is expected to cut the lode at the 50-fathom level, if its dip continues regular. Another shaft, at 194 feet distance from the engine shaft, is sunk on the lode to a depth of 174 feet, and a third is 100 feet deep, and will cut the lode at the 20-fathom level. The 20-fathom level has been driven 560 feet, and the 10-fathom 935 feet.

At the date of Prof. Rogers's Report about 360 fathoms of ground had been stoped, but the amount of ore produced is not given. He estimates that one half of the lode is too poor to pay for taking down, and that the productive portion will yield from 1¼ to 1½ tons per fathom.

A high-pressure engine of 24-inch cylinder is erected for pumping at this mine.

*Brookdale Mine.* This is on a continuation of the Wheatley Lode, the engine shafts being 2076 feet distant from each other. The lode is stated to average 2 feet in width. May 1, 1853, the engine shaft was down 75 feet. Prof. Rogers remarks that the indications of a productive vein in the lower levels of this mine seem as encouraging as in the Wheatley Mine. But little ore appears as yet to have been raised. An engine of 24-inch cylinder and 8 feet stroke is working at this mine, where the influx of water is somewhat troublesome.

*Charlestown Mine.* This mine is opened on a lode parallel with the Wheatley Lode, and distant from it about half a mile. It varies in width from 2½ to 4 feet. Near the surface it is much decomposed, and contains but little ore. It is recommended by Prof. Rogers as being worthy of the necessary expenditure to open it to a depth of from 25 to 30 fathoms. For that purpose a fine Cornish steam-engine has been erected, and the shaft is said to be down nearly 200 feet at this time.

Several other companies have worked here more or less at different times, but the above-described seem to include all which promise to be of any consequence. It remains to be seen what developments will be made in sinking on these interesting lodes to a considerable depth. Should they, as is to be hoped, be found to improve in character as they descend, they may be profitably worked.

North Carolina. — Throughout the auriferous belt of rocks extending from Pennsylvania into Georgia, galena has been occasionally found, in connection with the iron and copper pyrites, and the other ores of the gold-veins. The Lemmond Mine has already been noticed as producing a galena rich in gold and silver. The quantity thus far found has been but small; it is thought, however, to increase in some cases, as the mines are worked downwards.

The only mine in the southern gold region which has been worked to any extent, up to the present time, for lead or silver, is the Washington Mine, in Davidson County, North Carolina; it is not now in operation, having been stopped in

October 1852, but its past history is interesting to those who are watching the development of the veins of the Southern States.

*Washington Mine.* This mine, which is situated about 10 miles south of Lexington, was discovered about 1836 by Mr. Roswell King, and worked by him for several years, when it was sold to a Philadelphia company, by whom it was worked until 1852. On visiting the locality in 1853, I found the mine inaccessible, as it was filled with water; but it was carefully examined and reported on in 1845 by R. C. Taylor; and from his report, as well as from information obtained at the mine, and from Dr. F. A. Genth, who formerly had charge of the smelting department, the account here presented has been drawn. The annexed transverse section (Fig. 41), gives a view of the lodes, and of the shafts and cross-cuts by which they have been worked: it represents the state of the workings at the time the mine was stopped, when the deepest shaft had been sunk about 15 feet below the 200-foot level.

Fig. 41.

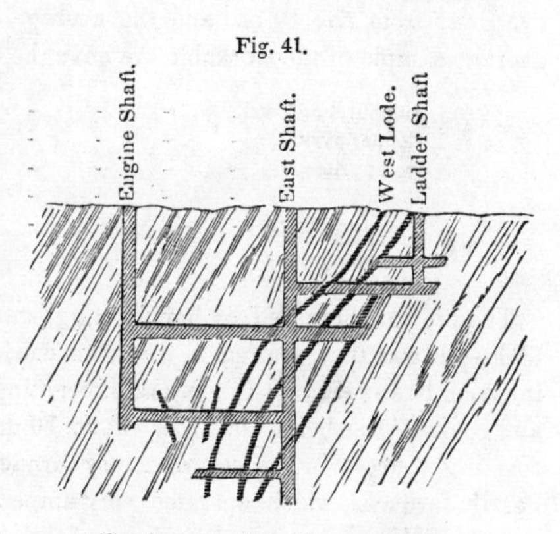


Section of the Washington Mine.

There are two principal lodes, nearly parallel with each other, and both coincident in dip and direction with the stratification of the rock in which they are enclosed, which is a silicious and talcose slate, very variable in its character, but usually much softened and decomposed in the neighborhood of the lodes. These are quite variable in size, expanding from a few inches to 10 or 12 feet. The workings upon them are confined within a space of about 150 feet, and it appears that the vein has been lost, or cut off by a heave, in each direction.

From the surface to the 200-foot level, the ground removed in the vein gradually decreases in length; and, according to Mr. Taylor, it appears that the planes which limit the lodes, although nearly vertical, are not parallel to each other, but inclined at an angle, so that they would meet if prolonged; and as the lodes dip in the direction of their intersection, the extent of ground comprised between them at each lower level is less than in the one above.

A great variety of beautiful ores of lead were found near the surface, such as the carbonate, phosphate, and sulphate. The west vein yielded, in 1844, between the 60 and 100-foot levels, very rich silver ores, and beautiful arborescent and dendritic masses of the native metal were obtained. These, however, became more and more infrequent as the mine was wrought in depth, and below the 125-foot level had pretty much disappeared. This lode was excavated along a space of 300 feet in length in the upper levels, and for 200 feet found rich in argentiferous lead.

The east lode, on the whole, was found richest in silver; at 170 feet it still exhibited arborescent native metal, and was considered by Mr. Taylor as the more valuable of the two.

The ores obtained from the mine, during the more recent years of working, consist mostly of fine-grained blende mixed with galena, and both containing a not inconsiderable portion of auriferous silver. Dr. Genth remarks* that the tenor of the sulphurets in this mine in silver is very variable. They contain from 2·5 to 195 oz. to the 2000 lbs. The usual yield of the ores, however, was from 7 to 10 oz., and the average of 200 assays gave 7·5 oz. An average sample of the workable ore gave the following as its composition:—

| | |
|---|---|
| Sulphuret of Lead, - - - - - - | 21·9 |
| Copper pyrites, - - - - - - - | 1·8 |
| Iron pyrites, - - - - - - - | 17·1 |
| Blende, - - - - - - - - | 59·2 |
| Gold and silver, - - - - - - - | 0·025 |
| | 100·025 |

The ore was prepared by hand-sorting, stamping, and washing on joggling-tables; it was then roasted in ovens measuring 12 feet by 5, and 5 feet high, in which it was arranged in layers, alternating with other layers of mixed wood and coal. After being thus roasted for 10 days, it was submitted to a second roasting process in the reverberatory furnace. It was then smelted in slag-hearth furnaces, which operated very imperfectly, returning not much more than half the lead required by assay of the ores. The argentiferous lead was then cupelled and the silver separated on a bone-ash test, and the litharge afterwards reduced to metallic lead.

The produce of this mine from the beginning cannot be given. In the printed reports of the company, that of the year 1844 is stated at $24,009 07 of silver and $7253 69 of gold, produced from 160,000 lbs. of lead, an average of 240 oz. of auriferous silver to the 2000 lbs.

From information received at the mine it appears that, in 1851, the produce of the mine was 56.896 lbs. of lead, and 7942·16 oz. auriferous silver, an average of 11·2 oz. to the ton of ore, and 279 oz. to the ton of lead produced.

It is believed by Dr. Genth that this ore may be profitably smelted, if a proper system should be adopted. The question of the quantity to be obtained from the mine would appear to be the most important one to be settled in determining the value of the property. If the lodes are confined within as narrow limits as appears from Mr. Taylor's reports, then the supply of ore must give out at a very moderate depth, and any farther idea of reopening the mine would have to be abandoned.

### LEAD MINES IN THE UNALTERED PALÆOZOIC ROCKS.

The localities included under this head are few in number in the Atlantic States, and, so far as I know, they are all in

* MS. communication.

the State of New York. Many instances of the occurrence of galena in the various groups of the New York system, are noticed by the State Geologists, but in few cases have they been thought worthy to become the object of mining enterprise.

Along the southern edge of the great azoic nucleus of this state, in the counties of Herkimer, Montgomery, and Lewis, many specimens of lead ore have been obtained from the Lower Silurian strata. In the last-named county, near the village of Martinsburgh, according to Prof. Beck, galena is found associated with pyrites, in narrow veins, traversing the Trenton Limestone. Some explorations have been made here, at an expense of a few thousand dollars, without success. The ore is interesting to mineralogists from its being crystallized in perfect octohedra. Excavations have been made in Herkimer County, near Salisbury Corners, but no mines are opened, and the locality is of no importance. The same may be said of the veins near Flat Creek, in the town of Root, in Montgomery County.

Specimens of blende and galena are not unfrequently met with in the Upper Silurian strata, in the neighborhood of Rochester, and from thence to Niagara Falls, but they are of no value, although at least one company has been formed for working such deposits; they seem to be mere isolated bunches in the rock.

The principal deposits of the ores of lead in the unmetamorphosed rocks of New York are in Sullivan and Ulster Counties, where they are now being mined to some extent. The Montgomery Zinc Mine, near Wurtsboro, which was considered a lead mine at one time, has already been mentioned. This vein, if the ore is as abundant as is claimed, may become of importance at some future day, but at present it is of no value.

In Ulster County, a mine was opened and worked, in 1837, by the "North American Coal and Mining Company," near Redbridge. According to Prof. Mather, the deposit of ore is on a line of fault, where the fractured grit-rock abuts against the broken and bent edges of the slate of the Hudson River group. The space between the slate and the grit is filled with a silicious gangue, containing quartz crystals in great numbers and of large size, more or less interspersed with masses of blende, galena, copper pyrites, and iron pyrites. The mine was opened by a shaft and levels, driven into the mountain at different heights, but, according to Prof. Beck, there was no encouragement to carry on the work.

The Ellenville Mine was opened near the village of that name, at the base of the Shawangunk Mountain; it is said by Prof. Mather to be in a transverse break of the strata, ranging north 60° east, and dipping nearly vertically. The mine was first worked about 1820, and was again taken up at a later period, with little success, as it appears.

Another mine near Ellenville has been recently opened and worked by the Ulster Company, and considerable ore obtained. From the Reports of the President of the Company, J. T. Hodge, Esq., it appears that work was commenced here in February 1853. The vein is represented as being one of a series of nearly vertical fissures which traverse the strata of the Shawangunk grit, of which the mountain is made up. This fissure is found to open in some places to a considerable width, and in others to close up entirely, there being only a mere seam or crack, destitute of veinstone or ore. Where the fissure expands, which it does sometimes to five feet, it is found more or less filled with loose fragments of sandstone, bunches of quartz crystals, and lumps of copper and lead ores, all bedded in a sticky and tough yellow clay.

The strata of the grit dip towards the valley at an angle of about 48°, and the courses of ore in the fissure appear to follow the lines of stratification pretty nearly. A shaft has been sunk 100 feet on the fissure, at the base of the mountain, and a drift extended into it for a distance of 200 feet (Jan. 1854), from near the same point. Above the drift, at a distance of a few feet, the walls of the fissure close together; but below they open out into a wide cavity, whose bottom has not been reached.

This appears to be an immense gash-vein in the sandstone, and will probably be found limited to that rock, which seems throughout this region to be congenial to the development of lead ores. Some of the most magnificent specimens of crystallized quartz, copper pyrites, and galena, ever found in the country, have been obtained here. The rock is very hard, and the cost of driving the level amounts to $130 per fathom.

Two Scotch hearths have been erected for smelting the lead ore, which averages in the furnace about 70 per cent. of metal. Up to the 2d of January, $29,915 18 had been received from sales of pig lead and of a small quantity of copper ore, and there was on hand lead to the value of $3375, and copper ore estimated at $2500.

The monthly running expenses of one furnace are $280 25, including fuel, and the product 66,768 lbs. The consumption of fuel by each furnace is 3 cords of pine wood per week, at $2 25 per cord. The running expense of the engine is about $70 86 a month. The lead produced contains, according to Mr. Hodge, about 12 oz. of silver to the ton, an amount, under present circumstances in this country, not worthy of being taken into consideration.

At the Old Ellenville Mine, before-mentioned, preparations have been made to sink a shaft at the foot of the mountain, where the indications are considered by Mr. Hodge as favorable. A third vein has also been discovered on the property of the company, about 20 rods from the present mine.

The value of these veins, of course, depends on the continuance of the fissures in length and depth, which can only be determined by actual mining. According to James Hall, the thickness of the Shawangunk grit at this point is over 500 feet, so that there is no lack of room for the development of an extensive deposit of ore.

### LEAD DEPOSITS OF THE UPPER MISSISSIPPI.

The great lead deposits of the Mississippi Valley may be considered under two heads: the Upper Mississippi and the Missouri Mines. At the West, they are commonly distinguished as the Upper and the Lower Mines. The first of these divisions comprehends the lead region lying in the southwestern portion of Wisconsin, and including a small part of the adjacent states of Illinois and Iowa. The second embraces the mines of the State of Missouri, lying principally south of the Missouri River. Although these deposits possess many features in common, yet, as they are geographically, and in some respects geologically, distinct, they will be considered separately, commencing with those of the Upper Mississippi.

The lead deposits of the Northwest were undoubtedly well known to the aboriginal inhabitants. Galena has been repeatedly found in the western mounds, but no metallic lead;* it would seem hardly possible, however, that the race which had sufficient skill in mining to procure the copper of Lake Superior from a depth of 50 feet in the solid rock, should not have had ingenuity enough to perform the simple operation of smelting the pure galena of the Mississippi mines. It is generally understood that the "Buck Lode," near Galena, was well known to the Indians.

The first excitement on the subject of mining in this region dates back to the famous expedition of Le Sueur, in 1700 and 1701. On his voyage up the Mississippi, he noticed many mines of lead along its banks; but his most wonderful discoveries were a short distance up the St. Peter's River, where he found what he supposed to be mountains of copper ore, ten leagues long. He wintered at the mouth of the Mukahto, or Blue River, as it was called by him, and in the spring he returned down the Mississippi with a cargo of this valuable article; in regard to the disposition which was made of it when it arrived in France, history is silent.

The Missouri mines had been worked for some time before any farther attention was given to the remote region of the Upper Mississippi. In 1788, however, a French miner named Julien Dubuque, who had previously settled there, obtained a grant from the Council of the Sacs and Foxes, which was afterwards confirmed by Carondelet, at that time Governor of Louisiana, of a tract of land situated upon the western bank of the river, and including the site of the now flourishing town of Dubuque. Here he remained engaged in mining until his death, which took place in 1809. The land occupied by him was relinquished to the United States by the

* Squier and Davis, Smithsonian Contributions, Vol. i. p. 208.

Indians in 1832, and although Dubuque's representatives claimed it, they were forcibly ejected.

By the Act of March 3d, 1807, all the government lands containing lead were ordered to be reserved from sale, and leases were authorized. None, however, were issued until 1822, and but a small quantity of lead was raised previous to 1826, from which time the production began to increase rapidly. For a few years the rents were paid with tolerable regularity; but, after 1834, in consequence of the immense number of illegal entries of mineral land at the Wisconsin Land Office, the smelters and miners refused to make any farther payments, and the government was entirely unable to collect them. After much trouble and expense, it was, in 1847, finally concluded that the only way was to sell the mineral lands, and do away with all reserves of lead or any other metal, since they had only been a source of embarrassment to the Department.

In 1839, a geological survey of the Lead Region of the Upper Mississippi was authorized by Congress, in order to ascertain the extent of the productive lead formation, with a view to the preparation of a plan for the sale of the lands previously reserved as mineral. This survey was intrusted to Dr. D. D. Owen, by whom, with the aid of 139 assistants, it was completed in the course of the autumn of the same year.

The extent of the lands reported by Dr. Owen as belonging to the productive lead region, is thus given by him.* They lie "chiefly in Wisconsin, including, however, a strip of about eight townships of land in Iowa, along the western bank of the Mississippi, the greatest width of which strip is on the Little Mequoketa, about twelve miles from east to west, and including also about ten townships in the northwestern corner of Illinois. The portion of this lead region in Wisconsin includes about sixty-two townships. The entire lead region, then, comprehends about eighty townships, or two thousand eight hundred and eighty square miles, being about one-third larger than the State of Dela-

* Report of a Geological Exploration of a Part of Iowa, Wisconsin, and Illinois, &c. Senate Doc., 1844.

ware. The extreme length of this lead region, from east to west, is eighty-seven miles; and its greatest width, from north to south, is fifty-four miles."

The principal mining centres are Galena, in Illinois; Mineral Point, in Wisconsin; and Dubuque, in Iowa. The Mississippi runs along the western edge of the tract, and the course of the Wisconsin River is nearly parallel with its northern line, and distant from it only a few miles. The face of the country occupied by the mining region is not broken by any mountain ranges; the highest points, namely the Blue Mounds, hardly rise more than 200 feet above the general level. The streams generally run through valleys excavated from a hundred to a hundred and fifty feet in the limestone, and bounded by steep escarpments, above which the land is only gently undulating. The soil is, in the main, of high fertility. Wood, however, is comparatively scarce.

In tracing along the members of the great Silurian formation, from the east towards the west, we find some of the groups which were most conspicuous and strikingly marked in New York, thinning out and vanishing altogether as they pass through the Lake Superior district and bend south into Wisconsin. The subject of these changes in the various subordinate groups of the Silurian system and their palæontological relations, has been ably investigated by James Hall.* Dr. D. D. Owen has given us some of the details of the groups, as they appear in the Northwest; and I have myself, in the course of two visits to the lead region, in one of which I was accompanied by I. A. Lapham, Esq., of Milwaukee, made an examination of many of the most important localities. The most striking change in the character of the Silurian rocks is the gradual increase of calcareous matter, and the passing into one another of beds previously lithologically distinct. The intercalated masses of conglomerate and grits, which in New York separate groups differing in their organic contents, are wanting at the West; and limestones and shales, with fine-grained sandstones, take their places, and there are fewer subdivisions recognizable by lithological characters. This the annexed table of equiva-

* See Foster and Whitney's Report on the Geology of Lake Superior, Part II.

lent formations in the States of New York and Wisconsin will illustrate:—

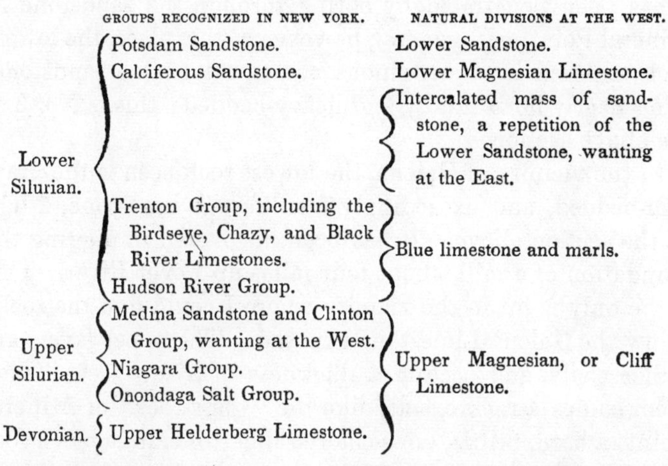

| | GROUPS RECOGNIZED IN NEW YORK. | NATURAL DIVISIONS AT THE WEST. |
|---|---|---|
| Lower Silurian. | Potsdam Sandstone. | Lower Sandstone. |
| | Calciferous Sandstone. | Lower Magnesian Limestone. |
| | | Intercalated mass of sandstone, a repetition of the Lower Sandstone, wanting at the East. |
| | Trenton Group, including the Birdseye, Chazy, and Black River Limestones. Hudson River Group. | Blue limestone and marls. |
| Upper Silurian. | Medina Sandstone and Clinton Group, wanting at the West. Niagara Group. Onondaga Salt Group. | Upper Magnesian, or Cliff Limestone. |
| Devonian. | Upper Helderberg Limestone. | |

Confusion was caused for a long time by the fact that the western geologists included rocks of so many different ages under the rubric of "Cliff Limestone;" and it was left uncertain, until the investigations of Mr. Hall settled the matter, whether the lead-bearing rock was really Lower or Upper Silurian.* Since that time, Mr. Lapham and myself have made large collections of fossils in the lead-bearing region, and have fully confirmed Mr. Hall's views, so far as the fact of the occurrence of the lead in the Lower Silurian is concerned.

We pass now to some details of the mode of occurrence of the lead in the rock. The first important fact to be observed is this, that the deposits of lead are confined almost exclusively to a certain part of the Lower Silurian formation. The general order of succession of the rocks throughout the lead region is as follows, in an ascending order:—

1. A soft, friable sandstone.
2. Limestone, Lower Magnesian, or Calciferous Sandstone.
3. Sandstone; a repetition of No. 1.
4. Blue limestone.
5. Galena Limestone, or lead-bearing rock.
6. Niagara Limestone.

---

* See F. and W.'s Report, Part II., p. 147.

No. 1 crops out in the valley of the Wisconsin River, but nowhere comes to the surface in the lead region proper. No. 2 has been penetrated by boring through the sandstone at Mineral Point. In general, however, at this place, the lowest rock exposed is a ferruginous, somewhat friable sandstone, lying nearly horizontally and heavy-bedded; this is No. 3 of the above section.

In the vicinity of Galena, the lowest rock seen is the gray, thin-bedded, and exceedingly fossiliferous limestone, No. 3 of the section above. It has been uncovered in digging the foundation of a mill, about four miles up Fever River. This is the only point in the vicinity where I could find the rocks below the Galena Limestone exposed. The upper layers are rather shaly, and occupy a thickness of a few feet; below them comes a rock exactly like the "glass-rock" of Mineral Point, a hard, brittle, dove-colored limestone, in layers a foot thick, but increasing in thickness, while the rock becomes less brittle and more arenaceous, as it goes down; a thickness of 25 feet is exposed below the Galena Limestone. The thin-bedded portion of this limestone, and the slaty, thin seams between the beds of the "glass-rock," are filled with fossils of the Trenton Limestone.

Above this rock comes the Galena Limestone, the lead-bearing rock, which forms the mass of the escarped hills around Galena, and which is about 200 feet in thickness. No lithological difference can be observed between that rock and the Upper Silurian which is seen in the summits of the highest points of land in the vicinity, for instance Pilot Knob, where Niagara fossils are found. The excavations for lead, around Galena, appear nowhere to reach the blue limestone, and certainly none of them go through it. Neither are they in the highest portions of the Galena Limestone proper, but chiefly confined to the middle; so that, in reality, the productive portion of the formation does not exceed 100 feet in thickness.

The same general facts are true with regard to the Wisconsin rocks. In the vicinity of Madison a soft yellowish rock is quarried for building purposes. It is a portion of the Calciferous Sandstone, interstratified with the upper part of the Potsdam. The two are so intermixed that it is difficult to separate one from the other. No fossils were observed in them, but portions of the limestone bands are abundantly covered with dendritic impressions. There is evidently a great expansion of the inferior or Potsdam sandstone throughout the northern part of Wisconsin. Above these alternations is seen the blue limestone, which in the neighborhood of Madison is filled with large orthoceratites. At the Blue Mounds a complete section of the rock is obtained, from the blue limestone, which is reached in the deepest excavations in the neighborhood, up to the Niagara, which forms the summit of the West Mound. About 75 feet of the upper part of this elevation is made up of a cherty rock, filled with geodes of quartz and agate and containing Niagara fossils, such as *Favosites Niagarensis*, and *Pentamerus oblongus*. There is no lithological distinction to be made between the rocks which occupy the space from the blue limestone to the Niagara, although they seem to represent both the Trenton and Hudson River groups. It is interesting to remark that as we go farther west, the lithological distinction which is so apparent at the Blue Mounds between the Upper and Lower Silurian becomes entirely lost, for, as shown above, in the Mississippi Valley the rock is homogeneous from the blue, or Trenton, up to the Upper Helderberg.

At Mineral Point the lead is recognized by the miners as occurring generally at three different stages of level, known by them as "openings," of which the middle is much the most productive. The following is the section there presented:—

8. Yellow Galena limestone, 75 feet.
*Lead.*—7. Upper opening, 3 to 8 feet.
6. Glass-rock or cap-rock; a brittle, hard limestone, 9 to 12 feet.
*Lead.*—5. Middle opening, 3 to 8 feet.
4. Soft spongy limestone, 13 to 18 feet.
*Lead.*—3. Lower opening, 4 to 8 feet.
2. White, soft, Galena limestone, 1 to 2 feet.
1. Red sandstone.

The glass-rock represents the blue limestone; thus the true limit of the lead-bearing rock, in this district, is the sandstone above the Lower Magnesian Limestone.

In the vicinity of Dubuque, the lead-bearing rocks rise to between 200 and 225 feet above the river, and attain to a greater thickness than farther east. The principal fossils here are the "lead-fossil" (*Receptaculites*), and *Lingula quadrata.* The diggings here seem to be entirely in the Galena Limestone, and have not reached to the blue. Of forty-five species of fossils collected by Mr. Lapham and myself in the lead-bearing beds of Wisconsin, thirty-two are of Trenton age, five are of the Birdseye, Black River, and Chazy groups, and the rest common to the Hudson River group and the Trenton Limestone.

MODE OF OCCURRENCE OF THE GALENA.—There are no deposits of lead in the Valley of the Mississippi which can be considered as coming under the head of true veins. They are invariably limited in depth. It has been shown above that the diggings are all in a certain geological formation, and that the productive part of this does not generally exceed a hundred feet in thickness. In the vicinity of Galena, no ores are found below the blue limestone; and throughout the Mineral Point district, the limit of the fissures and horizontal deposits is the sandstone, in which hardly a trace of metalliferous substances occurs, and where there are no fissures. This rock is exposed in the valleys of the principal streams, so that abundant opportunity is afforded to examine it; and if it contained anything of value, the explorations would long since have revealed the fact.

Below this sandstone, in the limestone commonly called the Lower Magnesian, a few instances are known in which deposits of lead have been worked; but so rare are they, and of so recent date is their discovery, that Dr. Owen stated, after his survey of the district, that no discoveries of value had been made below the blue limestone. Were the fissures found to extend through the sandstone, so that they could be followed into the limestone beneath, it might possibly be advisable to sink into the latter rock; but they are not thus continuous, and there is no reason to suppose that corresponding fissures would be found in the rock below the sandstone; sinking through the latter, therefore, in mere random explorations, with the hope of finding valuable deposits, would be a very foolish enterprise. The space over which the Lower Magnesian Limestone crops out is so small in comparison with that occupied by the productive lead rock, and the discoveries thus far made in it are so insignificant, that no material increase in the production of lead can be anticipated from that source.

The fissures from which the lead is obtained in the productive rock are very irregularly distributed. It may perhaps be said that their prevailing direction is either proximately north and south, or east and west. According to Dr. Owen, the north and south fissures to the west of the Mississippi are always very thin. My own observations did not enable me to deduce any general law respecting them.

The width of the fissures is very variable; they sometimes widen out into caverns of 30 feet or more in diameter, but generally occupy from a few inches to a couple of feet. Their inclination is usually nearly vertical, with numerous zigzags and other irregularities.

Fig. 42.

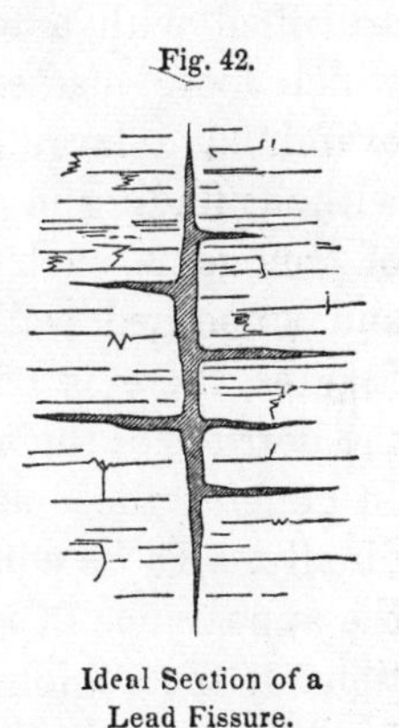

Ideal Section of a Lead Fissure.

The annexed figure (Fig. 42) will represent a frequent form of these gashes, with horizontal branches running off parallel with the stratification, and gradually losing themselves within a short distance. Their linear extent is rarely very considerable. The surface indications of such fissures are various, but not very reliable, as is evident from the immense number of "prospecting holes" in the vicinity of every productive "lead;" the whole surface is usually filled with them. Depressions in the surface soil, pieces of "float-mineral," even the growth of plants with long radicles in a linear direction, are noticed by the miner, and help to direct him in his researches. Of course, where the fissure closes up before reaching the surface of the rock, as represented in Fig. 42, or, in other words, where the formation has not been denuded so as to lay bare the productive stratum, or cut into the fissures, all signs may be said to fail.

The ore found is almost exclusively pure galena, containing hardly an appreciable trace of silver. Carbonates, phosphates, and other oxidized combinations, seem to be quite rare. Hence the lead produced is very pure and soft, and brings a higher price in the market than any other.

The varieties of the forms of deposit of the galena may be classed as follows:—

1. *Simple Alluvial Deposits.* These are masses of ore, "gravel-mineral," or "float-mineral," as they are commonly called, which are imbedded in the superficial soil. The localities where such pieces are sufficiently abundant to be collected, are called "clay diggings." The fragments of ore thus found have been washed out from the decomposing rock, and left irregularly distributed through the clay and sand.

2. *Deposits in Vertical Fissures.* These fissures appear, so far as my observations have extended, to be destitute of any proper gangue; certainly in the majority of instances, they are filled with a tough, dark-red, and ferruginous clay, in which loose masses of ore are promiscuously scattered. I examined a large number of diggings in vertical fissures, without finding in any of them any appearance characteristic of true veins, such as distinct gangues, selvages, or striated and smoothed walls. Besides the clay and ore in these fissures, there is often intermixed with them a quantity of the detritus of the adjacent rock, and fragments of veinstone. In general, these fissures have a very limited extent longitudinally. To be convinced of this, it is sufficient to examine the appearance of the diggings in any of the productive portions of the region; it will be found entirely impossible to account for their arrangement and position on any hypothesis of a system of extensive parallel fissures.

In the strata favorable to their development, the fissures expand out and become productive in lead; hence such strata are called openings. In the Mineral Point region, the miners say that there are three such openings, of which the middle one is the most productive. Often the fissures commence and terminate entirely in one opening; at other times they are continued by a mere crack or seam from one to another, across an intervening bed of unproductive rock.

Sometimes the expansions become sufficiently extensive to deserve the name of caves, and are found to contain very large quantities of ore. Dr. Owen notices one very interesting instance of this kind, discovered at the Vinegar Hill diggings in 1828. It was 35 feet in length, expanding in the centre to the width of 6 or 8 feet, and terminating in a point at each end. The walls were cased with galena for about a foot in thickness, forming a hollow shell of ore. This appears to have been the most remarkable discovery of ore ever made in the lead region.

3. *Deposits in Flat Sheets.*—This appears to me to have been the original form in which most of the lead was deposited. Here it is often observed accompanied by a gangue, and unmixed with clay and detritus. These sheets are of various dimensions, generally elongated in one direction, and thinning out on all sides from the centre. Several such flat sheets are sometimes found connected together by vertical or oblique fissures containing ore, as represented in the annexed figure (Fig. 43), thus descending from one stratum to another by zigzags.

Fig. 43.

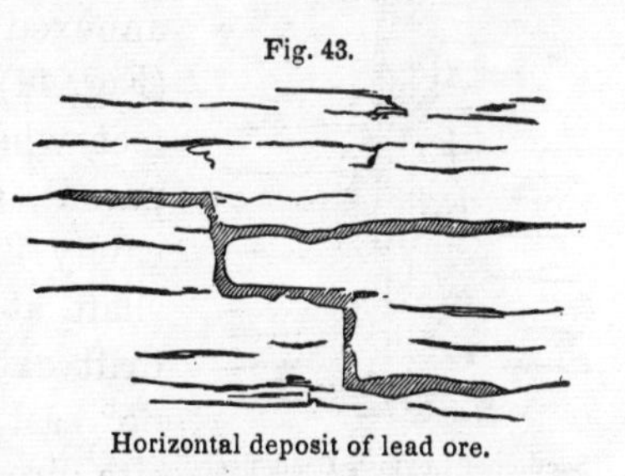

Horizontal deposit of lead ore.

The principal veinstone associated with the galena is calc. spar, or "tiff," as it is usually called by the miners; sometimes heavy spar is found. Flat sheets of these veinstones alternate with others of calamine or "dry-bone,"* black-jack, and iron pyrites. In some places the latter minerals are more abundant than the lead itself. The calamine not unfrequently is the predominating mineral.

As an illustration of the mode of occurrence of the lead of this region, a description of one of the most interesting loca-

* Both the carbonate and silicate occur with the lead ore of the West.

lities which I had an opportunity of inspecting will here be given. It must be remembered that most of the diggings are mere "prospecting holes," and that one may visit fifty of these in succession without seeing any lead actually taken out, or any indications of ore or veinstone in the rubbish at their mouth. But occasionally, a fine deposit of ore is struck, and the returns of a few months make up for years of toil and anxious expectancy. Such a one was the deposit of ore about 2½ miles northwest of Dubuque, called Levins's Lead or Cave, from which, in 1852, great quantities of ore were being taken. This deposit was first struck in prospecting, by a drift run in on the side of a gently-ascending hill, and afterwards a shaft was sunk upon it, as represented in the annexed transverse section of the cave (Fig. 44). The shaft descends for 90 feet, when the fissure begins to be apparent, and rapidly widens out to a cavity 30 feet in width. From the shaft, at a depth of about 130 feet, the drift extended 875 feet to the north 65° east, and 200 to 300 in the opposite direction. The fissure is not absolutely continuous for that distance, but nearly so. It varies from a mere inch-wide crack to a foot or more, the cave-like expansion occupying a length of perhaps 50 or 100 feet. The lead lies in the fissure irregularly; sometimes in heavy masses, at other times entirely wanting. It is without gangue, or any other accompanying minerals whatever, but is generally in bunches of large crystals, some of which measure three inches on a side. Besides the vertical deposits of ore, there are subordinate horizontal ones, sometimes carrying from 3 to 4 inches of solid galena. Numerous small horizontal strings shoot off from the main fissure; but the principal one, and that which seems to be at the bottom of the deposit, is about 110 feet below the mouth of the shaft, represented in the figure by the dark-shaded patches on each side of the cavity at the foot of the ladder.

Fig. 44.

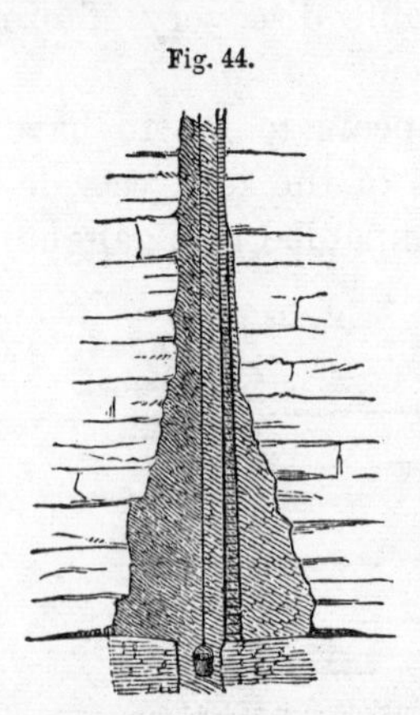

Section of Levins's Lead, near Dubuque.

This does not form one continuous horizontal deposit, but may plainly be seen to descend from one stratum to another, as represented in Fig. 43.

The action of water in wearing out this cavity and breaking up the original deposits of lead, which undoubtedly extended in flat sheets across the whole distance now excavated, is beautifully seen throughout the mine. At the bottom, a stratum of loose materials of unknown depth has been deposited. This consists of red clay and disintegrated limestone, through which is scattered an abundance of slightly rolled masses of lead ore. In many places the clay is seen to be distinctly stratified, and the walls of the fissure are smoothed and the projecting corners rounded off; while, what is still more interesting, the lower surface of the cap-rock is distinctly water-worn and grooved, as if currents of water had poured through the fissure beneath it.

From this place 2,000,000 lbs. of ore had already been obtained, and it was estimated that at least as much more remained to be taken out. The water was found very troublesome, but no other apparatus was used for removing it than a horse-whim and bucket.

At another locality near by, the lead was found by drifting from another mine, at a depth of 100 feet, the cap-rock here being 90 feet thick. 800,000 lbs. of galena had been taken out, in lumps mixed with clay in a vertical fissure.

It will not be necessary to describe in detail any more of these localities; the mode of occurrence is similar through them all. In 1852 the diggings were mostly in Wisconsin, in the neighborhood of Mineral Point, Platteville, Shullsburgh, Hazle Green, Jamestown, Potosi, &c.

The theory of the deposition of the ores occurring in the western lead region cannot here be entered into at length. It may be asserted without hesitation that they are deposits from aqueous solutions, which took place either in depressions of the surface, or in vertical fissures, of the nature of gash-veins, produced by the shrinking of the calcareous strata. Thermal springs containing salts of lead in solution, coming in contact with sulphuretted hydrogen, would deposit their lead in the form of sulphuret of lead. Casts of fossils in sulphuret of lead are not unfrequent in this region, a fact which alone is sufficient to demonstrate the aqueous origin of the ore. The occurrence of deposits of galena in the greatest abundance directly over, and almost in contact with, the most fossiliferous formation of the West, is worthy of notice. The beds of the blue limestone, which are replete with organic remains, form the base of the lead-bearing beds, and it is by no means impossible that the gases evolved in the decomposition of such an amount of animal matter as must once have existed there, may have had a material influence in determining the precipitation of the lead from the plumbiferous solutions lying upon them.

From what has been said thus far with regard to the mode of occurrence of the galena and the position of the rocks in which it is found, a few inferences of a practical nature may be drawn, which are not without importance in the guidance of expenditures in this region.

1. The deposits of lead are not in true veins, but are limited in depth. They cannot, therefore, be worked on an extensive scale; and the formation of companies with large capital and expensive machinery to develop them will in all cases be attended with loss. A simple, portable steam-engine, with a moderate outlay of money, a small number of hands, and a limited amount of time, are all that is necessary to exhaust any deposit of lead which ever has been or ever will be discovered in this region. All the sneering at the "surface-scratching" of the West is absurd, since nature has placed the mineral at or near the surface.

2. The development of deposits of ore in the Lower Magnesian Limestone, on any scale which will compare with those of the proper lead-bearing rock, is not to be expected. Did any such exist, they would long since have been found; and a few isolated instances of lead ore obtained in this formation are not sufficient evidence of its productiveness. To sink through the sandstone into this rock, in the expectation of finding the veins continuous below, will be surely followed by disappointment.

3. The production of the lead mines of the Valley of the Mississippi has now reached its maximum, and will, although

perhaps with many fluctuations, on the whole continue to decline. The quantity of lead produced in 1852 and 1853 was but little more than one-half of what it was in 1845 and 1846, and that in spite of the increase of value in the ore raised, amounting to almost the double. No doubt the attractions of California drew away many of the miners; but, under the stimulus of a greatly increased price, others would have taken their places, were the quantity sufficient to hold out inducement of reasonable profits.

The decrease in the yield of these mines, however, will be no loss to the vigorous and beautiful State of Wisconsin; for it admits of a demonstration that the same amount of labor applied to manufacturing and agricultural pursuits which is now wasted in prospecting, would more than earn the million and a half of dollars which she annually receives for her lead.

Some remarks will be added, after noticing the Missouri mines, in regard to the smelting processes followed in the Upper Mississippi region, and the statistics of the yield of both the Upper and the Lower mines will be furnished in one table a few pages farther on.

### LEAD REGION OF MISSOURI.

The first mining operations in Missouri were commenced in 1720, under the authority of the patent granted to Law's famous company. Renault came over in that year, with a large number of miners, and a mineralogist, named La Motte; and they immediately commenced their explorations, in the course of which numerous discoveries of lead mines were made, but none of the more valuable metals of which they were in search. The Mine La Motte was one of the first and most important localities opened by the person whose name it still bears. About that time some lead ore may have been smelted, but regular mining operations did not commence until 1798; until then, the ore obtained had been raised from open cuts, and smelted on log-heaps; but in that year Moses Austin erected a reverberatory furnace, and commenced sinking the first regular shaft for raising ore.* He also erected a shot tower, and made many other

* Schoolcraft's View of the Lead Mines of Missouri (1819), p. 19.

improvements, so that lead-digging began to be a regular business. According to Mr. Schoolcraft, there were forty-six mines working in 1819. From 1834 to 1837, Mine La Motte produced on an average 1,035,820 lbs. of lead per annum. The largest production of any one locality was from Mine Shibboleth, which, during the year 1811, is said to have furnished 3,125,000 lbs. of lead, from 5,000,000 lbs. of ore. Mine à Burton and Potosi Diggings together produced, in the eighteen years from 1798 to 1816, 9,630,000 lbs. of lead, or an average of half a million pounds per annum.

All these mines have now fallen off very much, and most of them are completely exhausted, so that but little information can be gathered respecting them, even by travelling through the region.

The principal mines of galena in Missouri are in Washington County, near Big River and Mineral Creek, branches of the Maramec River. There are a few others in Franklin County, on the last-named river, and one or two in Jefferson County, not many miles south of St. Louis.

The geological position of the metalliferous deposits of Missouri is nearly the same as that of the Upper Mississippi lead mines, but the detailed succession of the stratified rocks in the former region has not yet been studied with sufficient care to enable us to say with certainty into what groups the Lower Silurian of Missouri is to be divided, and in which one of them the lead deposits are principally concentrated; nor has it been possible hitherto to make out a parallelism between them and the Lower Silurian groups of New York or the Northwest. Throughout the Southwest, the great carboniferous formation exceeds all the others in its development, and is the only one which has been much studied.

In that part of Missouri which I have examined, including portions of St. Genevieve, Washington, Madison, and St. Francis Counties, I found the stratified rocks overlying the azoic nucleus to be made up of beds of limestone and sandstone, frequently alternating with each other, and never developed to any very great thickness; they are also almost entirely destitute of fossils. The beds are thin, and apparently not so favorable to the development of metalliferous ores as those of the Upper Mississippi region. There are heavy deposits of cherty and silicious rocks, in which there would seem to be little encouragement to look for anything of value.

Mine La Motte has been before alluded to,* as being the most celebrated metalliferous locality in Missouri. In the last few years, but little appears to have been done there. In 1841, according to Mr. Hodge, 200 persons were still employed in raising and smelting ore from this tract; "skimming off the surface-ores; one set throwing their rubbish over unwrought tracts, which another set will remove again another year, to get at the ore below. No search is made for any other ore than that which runs in horizontal strata through the clay, or through the rock near the surface." The miners know very well that it is useless to search for that which does not exist. The granite is everywhere near the surface, and it would be of no avail to attempt to find anything in that rock. In 1852, the number of men employed was not more than twenty, and, from all appearances, the amount of lead produced could not exceed one or two hundred thousand pounds.

The "Mammoth Lead," a few miles north of Potosi, was visited in 1845 by Mr. Christy,† and is described by him as being an irregularly-formed cavern, in which the lead, with calcareous spar, was found in large masses, mixed with clay and sand.

The lead deposits of Missouri, on the whole, strikingly resemble those of the Upper Mississippi, and the same theoretical observations in regard to the occurrence of the ore will apply to both. As they have been considerably longer worked in the former state, they are now nearer to exhaustion, and there is little reason to believe that they will ever regain the importance which they once had.

The smelting furnaces throughout the West are all on the same plan, the Scotch hearth being in universal use. The

* See p. 310.

† Letters on Geology, p. 43.

blast is usually supplied by a large bellows, worked by a small over-shot water-wheel, the whole arrangement being of the most simple description. The slag is worked over in a round blast furnace. Owing to the great freedom of the galena from foreign metalliferous substances, and the fact that most of it is smelted in small fragments, these arrangements are sufficiently economical and satisfactory in their working; still, a good deal of the lead is thrown away in the slag. It is doubtful whether more than 70 per cent. of metal is obtained from the ore.

In 1819, according to Mr. Schoolcraft, there were 45 lead mines at work in Missouri, of which 39 were in Washington County; and they were estimated to produce together about 3,000,000 lbs. of metal, giving employment to 1100 hands. The culminating period of their prosperity seems to have been from 1830 to 1845, and since that time they have declined very considerably. No reliable statistics are to be obtained of the last few years. From the best information I could collect in 1852, I gathered that the production at that time did not probably exceed 1500 tons.

The annexed table embodies all the statistics which it has been possible to procure of both the Upper Mississippi and Missouri lead mines, the produce being given in tons of metallic lead. Up to 1839, the figures given have been taken from Capt. W. H. Bell's Report on the Mineral Lands of the Upper Mississippi; these appear to be the most reliable, and are stated to have been obtained from the records of Messrs. Collier and Kennet. The statistics of the produce of the Upper Mines, since 1845, were furnished by Captain Beebe, of Galena, who has kept a careful record of the shipments from Galena, Dubuque, &c. The present yield of the Missouri mines can only be estimated, as the data which I have been furnished with do not agree sufficiently to allow them to be made use of. There can be no doubt that it has fallen off considerably, and I have fixed on 1500 tons as the most probable amount, which, indeed, it is more likely to fall below than to exceed. The price of lead for the different years is also added, in order to throw more light on the fluctuations in the production.

| | Upper Mississippi Mines. | Missouri Mines. | Total. | Price per 100 lbs. at St. Louis. | Price per 100 lbs. at Galena. |
|---|---|---|---|---|---|
| 1819, . . . . . | | 1,300 | | | |
| 1823, . . . . . | 150 | | | | |
| 1824, . . . . . | 78 | | | | |
| 1825, . . . . . | 297 | 984 | 1,281 | | |
| 1826, . . . . . | 428 | 1,343 | 1,771 | | |
| 1827, . . . . . | 2,313 | 1,614 | 3,927 | $4 50 | |
| 1828, . . . . . | 4,958 | 2,857 | 7,815 | 3 30 | |
| 1829, . . . . . | 5,957 | 1,867 | 7,824 | 2 00 | |
| 1830, . . . . . | 5,331 | 1,832 | 7,163 | 2 13 | |
| 1831, . . . . . | 5,369 | 1,277 | 6,646 | 3 00 | |
| 1832, . . . . . | 5,401 | 3,487 | 8,888 | 4 25 | |
| 1833, . . . . . | 6,068 | 3,699 | 9,767 | 4 13 | |
| 1834, . . . . . | 7,699 | 2,853 | 10,552 | 4 25 | |
| 1835, . . . . . | 8,469 | 3,227 | 11,696 | 5 00 | |
| 1836, . . . . . | 11,390 | 2,826 | 14,216 | 5 13 | |
| 1837, . . . . . | 9,708 | 2,286 | 11,994 | | |
| 1838, . . . . . | 10,811 | 2,701 | 13,512 | | |
| 1839, . . . . . | 11,976 | 3,563 | 15,539 | 4 38 | |
| 1840, . . . . . | 11,987 | 2,793 | 14,780 | 4 38 | |
| 1841, . . . . . | 14,150 | 3,317 | 18,171 | 3 50 | |
| 1842, . . . . . | 13,992 | 3,348 | 21,586 | | $2 24 |
| 1843, . . . . . | 17,477 | | | | 2 34 |
| 1844, . . . . . | 19,521 | | | | 2 82 |
| 1845, . . . . . | 24,328 | | | | 2 96 |
| 1846, . . . . . | 23,513 | | | | 2 88 |
| 1847, . . . . . | 24,145 | | | | 3 17 |
| 1848, . . . . . | 21,312 | | | | 3 24 |
| 1849, . . . . . | 19,654 | | | | 3 67 |
| 1850, . . . . . | 17,768 | | | | 4 20 |
| 1851, . . . . . | 14,816 | 1,500 ? | | | 4 08 |
| 1852, . . . . . | 12,770 | | | | 4 12 |
| 1853, . . . . . | 13,307 | | | | 5 50 |

Galena has also been obtained in the southern part of Illinois, from the carboniferous formation, but hitherto in small quantity. I understand that the State Geologist considers it probable that the localities will become of some importance. The argentiferous lead ores of Arkansas probably occur in the azoic, but little is known of their value; the veins are said to be too small for successful working.

In closing the chapter devoted to lead, the usual summing up of the statistics of the production of that metal will be given. And the following table presents, at one view, the most reliable information which has been obtained in regard to the different countries of Europe and the United States. Nothing is definitely known of the manufacture of lead in Southern Asia, Africa, or South America. The amounts are given in tons:—

| | Russia. | Sweden. | Great Britain. | Belgium. | Prussia. | Harz. | Saxony. |
|---|---|---|---|---|---|---|---|
| 1810, . . | | | 12,500 | | | | |
| 1815, . . | | | | | | | |
| 1820, . . | | | 31,900 | | | | |
| 1825, . . | | | | | | | |
| 1830, . . | 711 av. | | | | 1,480 av. | ('31) 5,645 | 146 |
| 1835, . . | | 49 | 46,112 | | 1,213 " | 4,967 | 519 |
| 1840, . . | | 38 | | | 1,643 " | | |
| 1845, . . | | | 52,695 | | 1.880 " | | |
| 1846, . . | | | 50,161 | | 2,179 | | |
| 1847, . . | | | 55,703 | | 1,977 | | |
| 1848, . . | | | 54.853 | | 3.344 | 5,600 | 1,041 |
| 1849, . . | | | 58,727 | | 4,222 | | 1,423 |
| 1850, . . | 750 | 196 | 64.752 | | 5,275 | 5,000 | 2,144 |
| 1851, . . | | | 58.701 | | 7,195 | | 1,884 |
| 1852, . . | | | 64,960 | 1,000 | | | |
| 1853, . . | | | 61,000 | | | | |

| | Bavaria. | Nassau. | Austria. | France. | Spain. | Italy. | United States. |
|---|---|---|---|---|---|---|---|
| 1810, . . | | | | | | | |
| 1815, . . | | | | | | | |
| 1820, . . | | | | | | | 1,300 |
| 1825, . . | | | 5,943 av. | ('26) 774 | ('23) 23,000 | | 1,300 |
| 1830, . . | | | 6,442 " | 708 | ('27) 36,000 | | 7,500 |
| 1835, . . | | | 5,623 " | 574 | | | 12,000 |
| 1840, . . | | | 5,851 " | 488 | ('39) 27,500 | | 15.000 |
| 1845, . . | | | 6,417 " | 970 | ('44) 25,000 | | 26,500 |
| 1846, . . | | | | | 24,800 | 250 | 25.000 |
| 1847, . . | | | 5,937 | | 31,000 | | 25,000 |
| 1848, . . | 32 | | 6,737 | | | | 22.500 |
| 1849, . . | | | | | 30,000 | | 21.000 |
| 1850, . . | | 600 | | | | | 19,500 |
| 1851, . . | | | | | | | 16.500 |
| 1852, . . | | | | | | | 14,000 |
| 1853, . . | | | | | | | 15,000 |

The present production of lead throughout the world may be estimated at 133,000 tons, which is divided among the different countries according to the annexed table of actual and relative amounts:—

| | Tons. | Relative amount. |
|---|---|---|
| Russia, . . . . . . . . . . | 800 | ·6 |
| Sweden, . . . . . . . . . | 200 | ·1 |
| Great Britain, . . . . . . . . | 61,000 | 45·9 |
| Belgium, . . . . . . . . . | 1,000 | ·8 |
| Prussia, . . . . . . . . . . | 8,000 | 6· |
| Harz, . . . . . . . . . . | 5,000 | 3·8 |
| Saxony, . . . . . . . . . . | 2,000 | 1·5 |
| Rest of Germany, . . . . . . . . | 1,000 | ·8 |
| Austria, . . . . . . . . . . | 7,000 | 5·2 |
| France, . . . . . . . . . . | 1,500 | 1·1 |
| Spain, . . . . . . . . . . | 30,000 | 22·5 |
| Italy, . . . . . . . . . . | 500 | ·4 |
| United States, . . . . . . . . | 15,000 | 11·3 |
| | 133,000 | 100·0 |

In order to complete our view of the statistics of lead, as connected with this country, another table is appended of our exports and imports of this metal for the fiscal years from 1840 to 1851:—

| Year. | IMPORTED. Pig, Bar, Sheet, Shot, Pipes, Scrap, and Old Lead. | | IMPORTED. Manufactures of Lead. | IMPORTED. Pewter, Old, and Manufactures of. | EXPORTED. Lead. | | EXPORTED. Manufactures of Pewter and Lead. | Excess of Value of Exports or of Imports. |
|---|---|---|---|---|---|---|---|---|
| | Tons. | Value. | Value. | Value. | Tons. | Value. | Value. | |
| 1840, | exp. 90 | exp. $14,635 | $901 | $24,799 | 394 | $39,687 | $15,296 | $43,918 exp. |
| 1841, | " 29 | " 3,464 | 2,287 | 17,221 | 972 | 96,748 | 20,546 | 101,250 " |
| 1842, | 13 | 579 | 236 | 14.265 | 6,496 | 523,428 | 16.789 | 525,137 " |
| 1843, | | exp. 298 | 35 | 2,538 | 6,850 | 492,765 | 7,121 | 497.611 " |
| 1844, | 1 | 55 | | 252 | 8,223 | 595.238 | 10,018 | 604,954 " |
| 1845, | 8 | 325 | | | 4.548 | 342,646 | 14,404 | 356,725 " |
| 1846, | 4 | 142 | | | 7.510 | 614,518 | 10,278 | 624,654 " |
| 1847, | 12 | 3,380 | 2,164 | 1,188 | 1,485 | 124,981 | 13,694 | 131.943 " |
| 1848, | 146 | 6.377 | 854 | 2,216 | 890 | 84,278 | 7,739 | 82.570 " |
| 1849, | 1,085 | 74,002 | 754 | 3,956 | 304 | 30,198 | 13.196 | 34,958 imp. |
| 1850, | 15.908 | 1,125,604 | 304 | 5,224 | 116 | 12,797 | 22,682 | 1,095,653 " |
| 1851, | 17,741 | 1,370,158 | *exp. 266 | 7,612 | 102 | 11,774 | 16.426 | 1,349,304 " |
| 1852, | 15,251 | 1,151,474 | 554 | 1,991 | 334 | 32,725 | 18,469 | 1,102,825 " |

From the above table it will be seen that, for a number of years previous to 1848, we had not only supplied our own consumption of this metal, but had exported an annual amount of about half a million of dollars in value. At present we are importing a larger quantity than we produce,

* The items thus marked give the excess of re-exportation of foreign lead, imported in previous years, over the importations of that year.

although the price of the metal is nearly double what it was when we were exporting most largely. The fact that so great an increase in the price of lead has not stimulated its production in this country, but that, on the contrary, it has continued to decline, is due partly to the superior attractions of the California gold fields, but still more to the exhaustion of the Western lead deposits, which from their nature can never again attain the importance which they once had. We must now look to the mines of the Eastern States, or to other deposits to be discovered in regions still unexplored in the farthest West, to make up the deficiency thus caused.

# CHAPTER IX.

## IRON.

### SECTION I.

#### MINERALOGICAL OCCURRENCE AND GEOLOGICAL POSITION OF THE ORES OF IRON.

MINERALOGICAL OCCURRENCE.—Hitherto, in considering the form in which metallic ores occur, it has been seen that by far the largest portion of them are in the form of sulphurets, the combinations of the metals with this *mineralizer* being most widely distributed; tin is an exception to this rule, its chief ore being an oxide. Iron, also, is another seeming exception, since the valuable ores of this metal are either oxides or oxidized combinations. The sulphuret of iron is more widely diffused, perhaps, than any other existing ore of any metal, being an almost universal accompaniment of every other metalliferous ore; but it is of no value for the manufacture of iron, since the other ores are too abundant to make it worth while to separate the iron from the sulphur, which could only be imperfectly done, while the trace of sulphur remaining would be deleterious to the quality of the iron. Sulphuret of iron may be considered more properly an ore of sulphur than of iron, since it is used for the production of sulphuric acid, and will eventually be an important article of commerce.

Although the ores of iron which are of economical importance are not many in number, yet iron, in some form, is almost universally diffused both through the organic and inorganic world. Not a rock or a stone can be found without at least a trace of this metal. Nothing is visible around us which is wholly without its presence. The history of its discovery and use is lost in the most remote antiquity; but from its affinity for oxygen, and consequent liability to rust and thus lose its form and identity, it could hardly be expected that we should possess tangible evidence of its use in ancient times. It appears from Mr. Layard's researches,* that the Assyrians were well acquainted with the manufacture of iron, and that they employed it, together with bronze, in useful and ornamental works; they also had the art of coating iron with bronze, and objects thus prepared have come down to the present time with the former metal in its metallic state. The great skill now shown by the East Indian native in the metallurgy of iron, and the surprising results he is enabled to accomplish with the rudest and simplest means, lead us to believe that the knowledge and use of this metal may have been much more widely diffused among the ancients than is frequently supposed, or is inferred from the fact that all the tools and weapons of the ancient Greeks and Romans are of bronze. An iron pick, found at Nimroud, is worthy of notice as being of a shape which would make it serviceable at this day.

### NATIVE METAL.

*Native Iron.* The occurrence of iron in its native state is extremely rare, at least when of terrestrial origin. Meteoric iron, an alloy usually of iron and nickel, is not unfrequently found; and there is a mass of this kind in the Yale College Cabinet, which weighs 1635 pounds; it was found on the Red River, in Texas. The only place in this country where native iron, not meteoric, is said to have been found, is in Canaan, Connecticut, and here the precise locality cannot be designated, so that it still remains a matter of uncertainty.

### ORES.

#### COMBINED WITH SULPHUR, ARSENIC, OR PHOSPHORUS.

*Pyrites.* Sulphuret of Iron; contains iron 46·7, sulphur 53·3. This mineral occurs abundantly in rocks of all ages, from the oldest crystalline to the most recent alluvial; often

* Nineveh and Babylon, pages 191 and 670.

in fine crystallizations, which from their yellow color are every day mistaken for gold. This species affords a part of the sulphuric acid and sulphate of iron of commerce, and also some sulphur; but it is of no use as an ore of iron.

*Marcasite*, White Iron Pyrites; has the same composition as the last-mentioned species, but is crystallized in a different form.

*Magnetic Pyrites*. This is another sulphuret of iron, which contains about 40 per cent. of sulphur and 60 per cent. of iron. It is abundant, though not so much so as the common pyrites.

*Leucopyrite*, an arseniuret of iron, and *Mispickel*, a sulpharseniuret of the same metal, are found in numerous localities; the latter frequently contains cobalt enough to be of value as an ore of this metal, but neither this nor the other can be considered ores of iron.

*Schreibersite*, or phosphuret of iron, is found only in meteorites.

COMBINED WITH OXYGEN.

*Specular Iron*, Peroxide of Iron, Micaceous Iron Ore, Red Hematite, Fer Oligiste, Red Ochre. This is an oxide of iron with two atoms of iron and three of oxygen. When pure, it consists of iron 70, and oxygen 30 per cent. It is a widely diffused species, and presents itself in a great variety of forms, and has received a great number of names. These varieties may be classed under two heads, the crystalline and the amorphous. Specular iron, which often occurs in fine crystals, is at one end of the list, and red chalk at the other. Specular iron includes specimens of a perfect metallic lustre; if in fine scales, it is called micaceous iron; the varieties which have only a slightly metallic lustre, with, generally, a fibrous structure, are called hematite; the soft and earthy varieties are called ochre. This ore occurs in immense abundance and purity, and, if these were the only requisites, would furnish the larger part of the iron of commerce; but in respect to a metal whose elaboration requires such an amount of fuel, and whose transportation in proportion to its cost is so expensive, there are many other circumstances to be taken into consideration besides quality and quantity of ore. At Gellivara, in Sweden, an immense mountain of this ore exists, which has never been touched for manufacturing purposes, and which probably will not be for a great while to come.

This ore frequently contains in combination titanic acid, in varying proportions, the peroxide of iron and titanic acid being isomorphous. A variety of names have been given to these combinations. Titanic acid in any considerable quantity renders the substance valueless as an ore.

*Magnetic Iron Ore*, Magnetite, Magnetic Oxide of Iron. A combination of the protoxide and the peroxide, with 72·4 of iron and 27·6 of oxygen. It is the native magnet or loadstone, and is widely diffused in nature, though not so widely as the peroxide. It furnishes an unrivalled ore. From specular it differs in its crystalline form, in being magnetic, and in giving a black powder instead of a red one.

*Franklinite.* This is an ore of iron containing zinc and manganese; it may be considered as magnetic iron ore in which a part of the protoxide is replaced by the protoxides of zinc and manganese, and the peroxide by the oxide of manganese. It may be considered both as an ore of iron and of zinc. It is only found in New Jersey, where it has just begun to be worked for both metals.

*Chromic Iron*, Chromate of Iron. Valuable as an ore of chrome.

*Limonite*, Brown Hematite, Brown Ochre, Bog Iron Ore, Iron-stone, Yellow Clay Iron-stone. Under all these names is understood a hydrated peroxide of iron, which, when chemically pure, contains 85·58 peroxide of iron and 14·42 water. The purer varieties contain from 60 to 62 per cent. of metallic iron; but this ore is almost always mixed with more or less earthy matter. Brown hematite is the name given to the compact and pure varieties, which have often a mammillary or stalactitic structure. The ochrey ores, brown ochre, yellow ochre, and the like, are earthy decomposed varieties; bog iron ore is a porous aggregate, usually occurring in low ground, as a recent deposit from the decomposition of other ores. Yellow clay iron-stone is the same, mixed with argillaceous matter. This ore forms the coloring matter of so many stratified rocks, and is so universally disseminated through the geological formations, that it is more difficult to say where it does not exist, than where it does.

*Göthite*, Lepidocrocite. This is a hydrated oxide of iron, like limonite, but differing in the proportions of its ingredients. It is comparatively rare, and cannot be called an ore.

#### SILICATES.

The number of silicates into which iron enters as an ingredient is great, and, as they have little or no value as ores, they must be passed over here.

#### CARBONATES, PHOSPHATES, ARSENIATES.

Of these combinations the variety is very great, but only one is important as an ore, that is—

*Spathic Iron*, Sparry Iron, Brown-spar, Clay Iron-stone. This is a carbonate of iron, with carbonic acid 37·94 and protoxide of iron 62·06. It is almost never found pure, but contains manganese, and generally more or less alumina, lime, and magnesia. This is, perhaps, the most important ore of iron; not generally in its sparry state, but as a mixture with clay and the hydrated oxide which results from its decomposition, and constituting a part of the great carboniferous formation; hence, occurring with the coal required for its reduction, it becomes of great importance.

The arseniates and phosphates are not ores; but, on the contrary, are highly injurious to the quality of those with which they are found occurring.

GEOLOGICAL POSITION.—The immense number of the deposits of iron ores found in every part of the world, renders it impossible to attempt anything like a detailed description even of the most important of them; volumes would be required for this subject alone. Under the head of each country will therefore be given only such statistics as may enable the reader to form an estimate of the comparative importance of the production of this metal in various regions and under varying circumstances, with allusions to such features in the manufacture as may appear to be of sufficient importance to be specially dwelt upon.

In order, however, to give as complete a picture as possible of this all-important metal, it is proposed to discuss with some detail the geological position and mode of occurrence of its ores, and reference will be made for illustration to the most important iron-producing districts throughout the world.

At first sight, considering the immense number and variety of the deposits of iron ore, it might seem as if they hardly admitted of being classified into distinct divisions; a farther consideration, however, will show that there are two modes of occurrence which are entirely distinct from each other, and whose difference affects the mode of manufacture, the quality of the metal produced, and its quantity. These main divisions may be again divided up into characteristic subgroups, each of which has its special character and economical importance.

In order to exhibit this clearly, the following scheme is presented for a classification of the ores of iron, with reference to their geological position:—

- DIVISION I. Unstratified ores.
  - A. In mountain masses, eruptive or metamorphic, in the azoic system.
  - B. Eruptive masses in the newer formations.
  - C. In the metamorphic and palæozoic, as segregated masses, or in veins.
- DIVISION II. Stratified ores.
  - A. Masses interstratified with the formations, up to the tertiary.
    - a. Not associated with coal.
    - b. Associated with the coal-measures.
  - B. Deposits in the tertiary and alluvial formations.
    - a. Tertiary ores.
    - b. Recent ores, bog ores.

#### UNSTRATIFIED ORES—IN THE AZOIC.

The masses of ore in the azoic, though developed on a larger scale, and made up of purer ores, than those occupying any other geological position, are not economically so important as those which occur in stratified deposits in connection with the coal. The ores found in this group are the oxides, specular ore, and magnetic ore. When associated with foreign matter, this is almost invariably of a silicious nature, quartz in some form; but they are generally quite pure, often approaching a state of chemical purity. They are particularly valuable as being more likely to be free from arsenic, phosphorus, and sulphur, which have an injurious effect on the quality of iron, than any other ores.

The ores of Sweden and Norway, which furnish so large a portion of the iron used for conversion into the finer qualities of steel, belong chiefly to this class of deposits.

The following scheme of their mode of occurrence is given by Durocher, who has published a very detailed and careful description of the metalliferous deposits of Scandinavia.

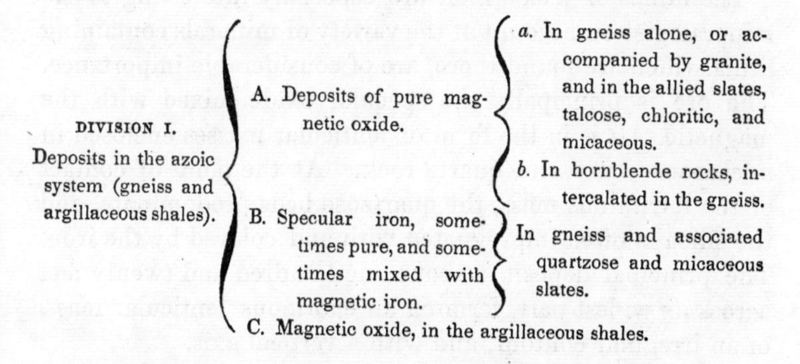

DIVISION I. Deposits in the azoic system (gneiss and argillaceous shales).

- A. Deposits of pure magnetic oxide.
  - *a.* In gneiss alone, or accompanied by granite, and in the allied slates, talcose, chloritic, and micaceous.
  - *b.* In hornblende rocks, intercalated in the gneiss.
- B. Specular iron, sometimes pure, and sometimes mixed with magnetic iron.
  - In gneiss and associated quartzose and micaceous slates.
- C. Magnetic oxide, in the argillaceous shales.

The remaining ores, which are comparatively of little importance, consist of masses of magnetic and rarely of specular ore, near the contact of the palæozoic rocks and the granite, and bog ore, forming deposits in low ground and swamps.

The azoic series in Sweden and Norway is made up principally of a crystalline, granitic gneiss, presenting an almost infinite succession of feldspathic, quartzose, micaceous, and hornblendic laminæ, and often cut through and disturbed by dykes of greenstone and granite. The researches of Murchison and Verneuil show conclusively that these rocks had taken their present form before the deposition of the Lower Silurian strata. There are several localities where the magnetic oxide occurs nearly pure and without gangue. Of this the mine of Bispberg furnishes a good example. It has the form of a lenticular mass, and its longest axis coincides with the direction of the schistose structure of the slates in which it is enclosed. The mines of Danemora are in a ferriferous band of about 600 feet in width and 7000 in length. In the neighborhood, gneiss is the prevailing rock; but in the immediate proximity of the mines, the rock exposed is a grayish limestone, slightly magnesian, accompanied by talcose and chloritic slates, which probably are subordinate to the gneiss. The deposits of iron form imperfectly cylindrical masses, with their axes nearly vertical, and their bases much elongated in the direction of the schistose structure of the rock.

The mines of Utö, which are especially interesting to the mineralogist on account of the variety of minerals containing lithia which are found there, are of considerable importance. The ore is principally the specular oxide mixed with the magnetic. It is in the form of lenticular masses enclosed in micaceous slates and quartz rock. At the point of contact of the ferriferous mass, the quartzose beds predominate, and the silica is often impregnated with and colored by the iron. The principal deposit is about one hundred and twenty feet across its widest part, forming an enormous lenticular mass, of an irregular contour, and with a vertical axis.

At Gellivara the magnetic oxide forms a mountain mass three or four miles in length and a mile and a half in width, a great portion of which is very pure, some parts of it containing specular ore mixed with magnetic. The principal reasons why this enormous mass has not been worked to any considerable extent are its remoteness from navigable waters, and its very high northern latitude (67°).

These deposits are called, in Sweden and Norway, veins, but they differ materially in character from what is generally understood by true veins. With a few exceptions, they appear to have been deposited in the midst of schistose or massive rocks, in forms which approach more nearly to beds, or elongated bands, and irregular masses; and they have evidently not filled previously-existing fissures which cross the strata at an angle, but almost uniformly coincide, in the direction of their greatest elongation, with the strata of the schistose rock.

The micaceous specular ores are generally associated with the quartzose and mica slates, and but rarely with the calcareous rocks. When there is calcareous matter near the junction of the ore and the enclosing rock, there is a great variety of minerals in the gangue, indicating that they were formed under certain conditions by the metamorphic action of the ferriferous mass upon the adjacent rocks. The mine of Hassel, in Norway, offers a good instance of the tendency of the specular ore to associate itself with the quartzose and slaty rocks. The deposit is not a vein, but rather a series of slaty beds, impregnated with peroxide of iron to the amount of twenty or thirty per cent.

There can be no finer instances of the mode of occurrence now under discussion than are to be found in this country. The mountain masses of Missouri have pre-eminently the eruptive character, and are associated with rocks which have always been considered as of unmistakably eruptive origin. The iron region of Lake Superior, which is even more extensive and more abundant in ores than that of Missouri, is another instance of the vast development of these ores in the azoic.

In the State of New York, in the same geological position, we find the same occurrence of the specular and magnetic oxides, and almost rivalling with those of the regions just mentioned in magnitude and importance. Here, however, the evidences of direct eruptive origin are perhaps less conspicuous, and the deposits seem, in many cases at least, to exhibit the appearance of a secondary action having taken place since their original formation. In this region, these ores have in their mode of occurrence the most striking analogy with those of Scandinavia. Like them, they generally coincide in the direction of their greatest development with the line of strike of the rocks in which they are enclosed, forming lenticular or flattened cylinder-shaped masses intercalated in the formation. The enclosing rocks are similar in character to those of Sweden; they are gneiss, quartzose, and hypersthenic rocks. The deposits of these ores will be noticed more particularly farther on, under the head of each state.

Although the iron ores of the azoic have not always had a purely igneous origin, yet even in those cases where they bear the most evident marks of having been deposited in beds parallel with the formation, with the presence of water, we must acknowledge that pre-existing eruptive masses may have furnished the material from which they were derived. That the azoic period was one of long-continued and violent action cannot be doubted, and while the deposition of the stratified beds was going on, volcanic agencies, combined with powerful currents, may have abraded and swept away portions of the erupted ferriferous masses, rearranging their particles and depositing them again in the depressions of the strata. This seems the most probable origin of some of these lenticular beds of ore parallel with the stratification, where it is difficult to conceive of a fissure always coinciding with the line of strike of the formation, and where the mechanical evidences are wanting of the thrusting up of such masses of matter, which we know could not have taken place without many dislocations of the surrounding rocks which would have made themselves very apparent.

The masses of iron ore in the azoic are far more grand in their scale of development than in any other formation, characterizing it everywhere as the age of iron; a fact which has a high degree of significance, when we consider that this is the oldest geological formation, and that we thus, as it were, receive a hint as to the structure of the interior of the earth. In this connection, it will be remembered that the bodies of extra-terrestrial origin which fall upon the earth, or meteorites, are very often found to consist of metallic iron, and we are thus led directly to infer the existence of vast masses of metallic iron within the interior of the globe.

The evidences of eruptive masses of iron ore grow fainter as we ascend in the scale of formations, or recede from the focus of internal heat; but, nevertheless, the ferriferous emanations from volcanoes still in action, as well as the undeniable upheaval of oxydized iron from the interior of the earth during comparatively recent periods, show that there are still supplies of the same material accessible, which are not below the depth at which chemical action is still going on and making itself sensible.

### ERUPTIVE MASSES IN FORMATIONS MORE RECENT THAN THE AZOIC.

The magnetic iron ore hill near Nijny Tagilsk, which is extensively wrought, affords a fine illustration of an eruptive mass in the midst of sedimentary formations. There are numerous points of eruptive rocks, mostly hornblendic greenstone, among stratified masses, which have been highly metamorphosed in their vicinity. The age of the sedimentary beds is referred by Murchison to the Upper Silurian, although the fossils are mostly obliterated by the metamorphic action. These limestones appear to have been rent in twain by a narrow ridge of intrusive greenstone, which rises to the north of Nijny Tagilsk into a high hill (Vissokaya-gora), on the summit and flanks of which iron ore has long been extracted. The chief mass of the ore is seen to occupy the valley on the western side of the hill, where it has been deeply cut into by open quarries, and is found to consist of an enormous body of ore, rudely bedded and traversed by numerous joints, and exposed for a height of a hundred feet and a length of several hundred. In opening out the side of the valley nearest to the hill of greenstone, irregular knobs or points of rocks were met with, on stripping which it was found that the iron ore had accommodated itself to the irregularities of their surface, and that at such points of contact, the ore was not only harder and more crystalline than usual, but also much more magnetic than at a short distance from the greenstone.

The rock associated with the magnetic iron ore of Mount Blagodat, near Kuschwinsk, which has been worked since 1730, is a feldspathic augite porphyry. So far as they have been worked down, the excavations exhibit a continuous mass of fine-grained magnetic iron ore, with flakes of yellow and pink feldspar and brown mica. It is the opinion of Col. Helmersen, who has carefully studied this locality, that these feldspathic iron-stone masses are portions of dykes of eruptive character which have traversed the porphyry, a fragment of that rock even having been found in one of them which rises up from near the base of the hill.

The Katschkanar, one of the loftiest and most rugged summits of the Ural, is made up of igneous rocks (greenstone), having a bedded structure and traversed by regular joints, so as to give it the appearance of a sedimentary rock; it is cut through by courses of magnetic iron ore, and has an abundance of the same substance diffused through it in crystals; but the ore is hard and intractable, and being at one of the most inaccessible points of the Ural, the works which were commenced there are now abandoned.

There can be no doubt, from the consideration of all the phenomena of these localities, that the ores are of purely eruptive origin, and that they have played the same part as the igneous greenstones and porphyries with which they are associated. At Nijny Tagilsk, it is evident that the magnetic iron penetrated the pre-existing greenstone, and flowed, as sub-marine lava or volcanic mud, into the contiguous depressions. This is proved by the fact that the ore expands in width, thickness, and dimensions, as it is traced into the lower parts of the valley, precisely as a lava stream which fills up the sinuosities of the subjacent rock. Such was the opinion of Helmersen, and it was adopted by Murchison, although not agreeing with his preconceived theories.

The iron mines of Elba, which are celebrated alike for the length of time they have been worked and for the purity and beauty of their ores, furnish another interesting example of the class of eruptive masses associated with the more modern rocks. The metalliferous deposits of the island are mostly concentrated towards its eastern extremity, and are associated with serpentine. The sedimentary rocks in that vicinity have been metamorphosed and intermingled with serpentine, so as to give rise to an abundance of beautifully variegated marble. The mass of specular ore worked near Rio is included between the upturned slates which form the flank of the Mountains of St. Catherine. It has all the appearance of having been forced from below upwards through the strata, which are highly metamorphosed at the contact of the ferriferous mass, and into which the metallic emanations have penetrated in every direction. As a proof of this, it may be observed that the attendant minerals vary with the adjacent strata. In the quartzose slates, crystallized quartz predominates; in the calcareous strata, actinolite and yenite have been developed.

The mass of magnetic ore and hematite of Monte Calamita is more extensive than that of Rio, and exhibits even more clearly the phenomena of igneous action. It has uplifted the superincumbent strata, and produced on them all the effects of metamorphism due to igneous action. For instance, the compact limestone which lies adjacent to the ferriferous mass is changed into a saccharoidal dolomite, and along the line of contact silicates of lime, magnesia, and iron have been developed. According to Burat, the whole appearance of the mass is that of an immense wedge driven upwards from below into the calcareous and schistose rocks, producing all the effects to be expected from the intrusion of such a mass by igneous agency.

The geological age of the strata which have been thus metamorphosed by the ferriferous masses, is considered by Burat to be near the Jurassic, but the introduction of the metallic matter itself probably took place at a much later period, perhaps after the deposition of the chalk.

#### SEGREGATED MASSES AND VEINS IN THE PALÆOZOIC ROCKS, GENERALLY METAMORPHIC.

Deposits of this class are widely scattered over the world, and frequently developed on a grand scale, though not so much so as those in the azoic. The more extensively metamorphosed the formation, and the older it is, the more do the deposits of iron ore take on the character of true veins. The spathic ore is one of the most abundant in this position, forming often veins of great extent, and furnishing large quantities of a material eminently calculated for making good iron and steel. The interesting vein of this ore at Roxbury, Connecticut, may be noticed as a good example of the class. Veins and vein-like masses of the same ore, and similar in position, occurring in the valley of the Rhine, furnish the material to the numerous manufactories of steel of that region.

Veins of magnetic iron ore are also frequently found occurring in this position. In this country they are especially numerous. They are generally segregated masses lying in the direction of the stratification. Sometimes they are quite pure, being mixed only with a little silicious matter; at other times they are associated with other metalliferous minerals. Occasionally the iron ore is found, as the depth increases, to be replaced in part by ores of copper. In the Southern States, the occurrence of gold with ores of iron is very frequent, but the latter are too much mixed with pyrites to be of any value apart from the gold which they contain.

In the carboniferous limestone of England there are numerous deposits of hematite of great importance, some of which are intermediate in character between veins and beds, while others appear to occupy previously-formed fissures, and to belong to the class of gash-veins, the ferriferous matter having been deposited in them from above. In general, the kind of ore-deposits now under consideration, when not associated with crystalline rocks, are not distinctly marked in their characters, but seem to form a connecting link between the stratified and unstratified masses.

### STRATIFIED ORES OF IRON.

The stratified ores of iron are distinguished from the unstratified by their earthy character; they are very rarely crystalline in their structure, and consequently are less pure than the crystallized "mountain ores." This lower percentage of iron which they contain is more than compensated, however, by the superior facility of raising the ore, and its easier working in the furnace, aside from another all-important consideration, that of the proximity of coal. Probably five-sixths of the iron of the world is made from ores of this class. They are interstratified and contemporaneous with the rocks in which they occur, and are found in all geological positions, from the very lowest upwards, and are forming even now under our own eyes. For practical purposes, this class of stratified ores may be divided into two groups, which are quite distinct from each other in some important respects: these are, ores in the solid rock, and ores in the superficial formations, or alluvial ores. The first group, that of stratified ores in the solid rock, may be divided into two important sub-groups: 1st. Ores not associated with the coal; 2d. Ores associated with the coal-measures.

The sub-group of stratified ores in non-carboniferous rocks has been hitherto considered of comparatively little importance, but the deposits of this class are of immense extent, and must eventually be worked much more largely than at present. They exist in the Silurian system, where it has not been metamorphosed, in the form of oolitic argillaceous masses. Examples of this class may be seen in numerous localities in the Western United States, where the Silurian and other infra-carboniferous groups are in an unaltered condition. In Europe, as a general thing, these groups have been so much altered that the iron ores which they contain can hardly be recognized in their original character of sedimentary deposits. In the West, these deposits are likely to be of great importance at some future day, since the coal-measures proper are not rich in iron ores.

In the stratified rocks lying immediately below the coal, the deposits of iron ore in England are very extensive; the most important of them are those of Lancashire, Cumberland, and the Forest of Dean, all of which are in the carboniferous or mountain limestone formation. These deposits were worked very extensively before the use of coal was introduced. The ores consist principally of hematites or compact red oxide; those of the Forest of Dean occupy a regular position in the limestone, but are themselves exceedingly irregular, assuming rather the character of a series of chambers than of a regular bed. These chambers are sometimes of great extent, and contain many thousand tons of ore, which is generally raised at a very low cost, since no timbering is required. The iron made from these ores is red-short, and from its superior quality commands a high price. The hematites of Whitehaven and Ulverstone are of great importance, since they exist in large quantity, and are the best ores raised in England. They are, in their mode of ocurrence, of a somewhat singular character, since they occur sometimes in vein-like masses transverse to the stratification, and sometimes in irregular beds, which occasionally attain the thickness of 20 or 30 feet. The ores contain from 65 to 95 per cent. of peroxide of iron, the impurities being chiefly alumina and silica, with a little manganese. Nearly 500,000 tons of this kind of ore were raised in 1851.

Brown hematites are another important class of ores occurring in a similar position. There are large deposits of this ore in the Western States, in strata about the age of the Upper Helderberg Limestone. Some of the lead veins in the carboniferous limestone district of England are associated with very large masses of brown hematite. They usually contain from 20 to 40 per cent. of iron: sometimes they exist as "riders" to the veins, sometimes they form their entire mass, and even in this case attain a thickness of from 20 to 50 yards; sometimes they form distinct and regular beds. In the localities of Alston Moor and its vicinity, the quality of the iron produced does not stand high: it has too great a tendency to cold-shortness. In this country, however, the ores are found of sufficient purity to produce a very excellent quality of iron. They are invariably found to contain manganese, sometimes in considerable quantity.

The Jurassic or Oolitic group furnishes many important deposits of iron ore. Throughout the formations between the Lias and the coal-measures, namely the Triassic and Permian groups, or the Old and New Red Sandstones, peroxide of iron is diffused almost universally as a coloring matter, tinging them red, and the whole amount thus present in these sandstones is enormous, but it is not usually concentrated into workable masses. Burat, in noticing this fact, considers it due to the violent agitation which seems to have prevailed during the formation and deposition of these conglomerates and coarse sandstones. The fact seems to be indisputable, that concentration of the ores of iron has only taken place when the deposits were of a fine material, indicating a long period of repose and tranquil deposition.

The ores in the groups above the carboniferous do not

in this country possess any considerable degree of importance, the formations being too little developed. In Europe, on the other hand, they are not without interest. The Jurassic formation furnishes the larger portion of the ores smelted in Southern France. One of the most interesting deposits of this class is the mine of La Voulte, in the department of the Ardèche. Here are three beds of red oxide of iron, varying from 3 to 15 feet in thickness, and opened on for an extent of two-thirds of a mile; they are interstratified with marls, and contain the same fossils, of the age of the Lias, or perhaps of the Oxford clay. The three subdivisions of the Oolite contain contemporaneous deposits of hematite and brown hematite scattered through them. The ores are sometimes compact, and sometimes have an oolitic or pisolitic structure, the ore taking the same structure as the rock which encloses it. Besides the interstratified deposits, there are large masses of ore filling cavities or long winding fissures, in which they seem to have been deposited by water holding iron in solution.

The ores in this geological position in England have only very recently become of any importance. They are found at the base of the Oolite, as it crops out along an extensive line stretching from the River Tees through the midland counties down to the south coast. But for this discovery, some of the principal iron-works of the North of England must have been closed, as the cheap black-bands of Scotland had destroyed the value of the comparatively expensive argillaceous iron-stones of the Newcastle coal-fields. These ores, on the contrary, cropping out on the surface, could be mined at a very small expense, and new railroads are building to furnish access to them, and they will eventually become of great importance. They are of an extremely varied character, containing hydrated peroxide and carbonate of iron, associated with carbonate of lime, silica, alumina, a little manganese, and traces of phosphoric acid. The amount of metallic iron present in them varies from 20 to 55 per cent.

The beds of argillaceous iron ores in the Lias are worked to some extent near Whitby and Lyme Regis.

The ores of the cretaceous formation are not of much importance. Those of the Greensand and Wealden in England were worked to a considerable extent before the use of coal for smelting. They are generally not rich. In France they are worked on a small scale in the Bas Boulonnais. In the cretaceous, as well as the other sedimentary formations, the ores of iron are chiefly concentrated near the bottom or the top of the group, especially where the beds are thick and homogeneous. If found in the central portions of such beds, it is principally when occupying water-worn cavities and winding fissures. The chalk, notwithstanding its immense thickness, is almost destitute of iron, except occasional nodules of sulphuret, which seem to have been concentrated around organic matter.

### IRON ORES ASSOCIATED WITH THE COAL-MEASURES.

We come now to a class of ores of surpassing importance, and in which England and the United States are pre-eminently rich, ores associated with coal, or in the coal-measures; in both these countries the larger part of the iron smelted is from this source. It is to the abundance of her coal-measure iron-stones that England is indebted for her vastly preponderating production of this metal, and it is thus that she has been able to supply the rapidly-increasing demand for railway iron, which the discovery of a new means of national intercommunication rendered necessary. The coal fields of North and South Wales, North and South Staffordshire, Derbyshire, Yorkshire, and Scotland, while they furnished fuel to smelt the ore, furnished the ore itself and the necessary flux from the same shaft, with hardly any increased expense beyond what it would have cost to raise the coal alone.

There are two distinct classes of ore associated with the coal-measures, the argillaceous iron-stones and the black-band ore. Of these, the latter may be considered as of quite modern use, and is especially developed in Scotland. In this country it is as yet hardly known. The usual chemical form of the ore in the argillaceous iron-stones of the coal-measures is that of a carbonate. Besides the carbonate, they contain often a little peroxide of iron and manganese; these are mixed with varying proportions of clay and sand, or silica and silicate of alumina, carbonates of lime and magnesia, and usually a small amount of sulphuric and a little phosphoric acid. The ores of this country have not generally been carefully enough examined with reference to these ingredients, but they are probably almost always present, although often only in minute traces. The usual form of these ores is that of small nodules, of an earthy texture, having little of a metallic look about them. They are interstratified in beds with the shales and coal. The number of repetitions of iron-stone beds in a single coal-basin is frequently very considerable. In the South Welsh coal-field there are seven distinct districts, which contain respectively 6, 9, 16, 12, 10, 17, 22, beds of iron-stone, and many of these individual series are themselves divided into several distinct courses. These beds of ore usually maintain about the same position in reference to the beds of coal throughout each basin. They are not usually very thick, and this is the greatest drawback to their value. Sometimes they consist of a single layer of spheroidal concretions, or balls of all sizes, up to a ton or more in weight. The amount of iron which they contain is not usually more than from 30 to 33 per cent., though they sometimes rise as high as 40. In England they are rarely used when they fall below 25 per cent. The quality of the iron produced from them is various, according to the locality and composition of the ore, the skill of the iron-master in working them or mixing them with other ores, and the quality of the coal with which they are smelted. In many of the English iron-works where these ores are exclusively used, the produce of iron is of a good quality, neither cold-short nor red-short in undue degree, although there is always a tendency to one or the other of these qualities. In this country the iron produced from them is very various in its quality. Some of it is very inferior.

In France, the iron ores accompanying the coal are limited in quantity, and inferior to those of England, a great drawback to its prosperity. In the coal-field of St. Etienne the iron-stones are hardly worthy of notice. That of St. Aubin is best supplied with iron ores, which occur in nodules disseminated in the slate adjacent to the coal. These two coal-basins are the only ones in France which contain iron enough to be worth working.

The black-band ores are exclusively worked in England; they exist in our own coal-fields, but have not yet begun to be worked to any extent. They differ from the argillaceous ores principally in containing a considerable proportion of carbonaceous matter, sometimes amounting to 20 or 25 per cent. When calcined, they lose a large part of their weight, generally as much as one-half, and are then very rich in iron. Mushet's Black-band contained, according to Dr. Colquhoun's analysis—

| | |
|---|---|
| Carbonic acid, | 35·17 |
| Protoxide of iron, | 53 03 |
| Lime, | 3·33 |
| Magnesia, | 1·77 |
| Silica, | 1·40 |
| Alumina, | ·63 |
| Peroxide of iron, | ·23 |
| Bituminous matter, | 3·03 |
| Moisture and loss, | 1·41 |
| | 100·00 |

The iron produced from this class of ores has a decided tendency to cold-shortness, owing, probably, to the phosphoric acid which they contain. The principal localities where they are found in Great Britain, are in Northumberland, North and South Wales, North Staffordshire, and Scotland; of these, the two latter are much the most important. The principal Scotch black-bands are two, known as Mushet's and Crofthead.

### IRON ORES IN THE TERTIARY AND ALLUVIAL.

The ores in this position are of minor importance; but nevertheless, they form no inconsiderable portion of the iron-making resources of some countries. Lying near the surface, and being easily recognized and cheaply excavated, they are often worked, and sometimes worked out, before attention is directed to the ores which are less superficial in their occurrence. They frequently produce an excellent quality of iron, and work easily in the furnace.

In England, the ores of this class have very little importance. In France, they are more largely developed, and, in the general scarcity of good ores in that country, are worked to a considerable extent. The deposits of Berry, especially those in the valley of the Cher, are in the tertiary, and consist of pisolitic ores, disseminated in beds in the argillaceous slates. Through the northeastern part of France, in the departments where wood is abundant, as in Ardennes and Marne, the extent to which the tertiary and alluvial ores are used is very great. Burat notices the following localities where their working is principally developed: 1. The country between the Sambre and the Moselle, which furnishes ores for the furnaces of Ardennes, the Meuse, and the Moselle. They consist of hydrated oxides, disseminated through sands overlying strata of Jurassic age. The beds of Saint-Pancré and Aumetz furnish ores of superior quality. 2. The Bas-Rhin, the larger portion of whose ores occur in clays and alluvial sands, and consist of impure earthy carbonate, and tertiary deposits which have been washed away from their original position. 3. The superficial ores of the Jura, scattered over the surface and in the depressions of the limestone. These are often found in winding, cave-like excavations in the strata, which communicate with the surface by vertical columns of ore, apparently filling the orifices through which the ore was originally introduced into the cavity. 4. Oolitic and pisolitic ores of alluvial formation in the Nivernais. 5. Hydrated oxides and hematites, in the superficial clays of the Charente, Dordogne, Lot-et-Garonne, Lot, and Tarn-et-Garonne. These are of very considerable importance, being of good quality, and sufficiently pure to be worked in the Catalan forges. The excavations are mostly open to the day, and descend to a depth of between 60 and 70 feet. 6. The sandy alluvia of the Landes contain deposits of hydrated oxide and bog ore, which are worked to a considerable extent. Burat estimates that one-third of the production of iron in France is derived from alluvial ores.

In this country ores of this class are abundant, and are worked in numerous localities, often furnishing iron of a very superior quality. The best deposits are probably in the tertiary; the true bog ores are of less importance. The formation of this latter class is one which is constantly going on where ores of iron are decomposing. The resulting hydrated oxide of iron accumulates in low grounds, and is frequently found spread over them to the depth of from a few inches to several feet. The bog ores have been worked to some extent in the New England States, but are now of little consequence. In some of the Western States they are abundantly distributed, and may eventually furnish an important supply of metal.

---

## SECTION II.

### STATISTICS OF IRON IN FOREIGN COUNTRIES.

Russian Empire.—The erection of the first iron furnace in Russia is said to date back to 1623. The want of means of internal communication is a great drawback to the development of this branch of the mineral industry of the empire. It appears from official data that, taking the mean of the six years previous to 1844, and the six succeeding that year, the annual production was as follows:—

| | 1838–1844. | 1844–1850. |
|---|---|---|
| Pig iron, . . . . | 169,000 tons. | 188,300 tons. |
| Wrought iron, . . | 111,650 " | 124,300 " |

Which shows an increase of only 11½ per cent. in the last period of six years. The dearness of iron in many parts of the interior renders it difficult to carry on the simple operations of agriculture, and there are whole districts where the shoeing of horses and the use of iron instruments for cultivating the ground are unknown. The production of this indispensable metal has not even kept pace with the increase of population.

The present production of the empire may be estimated at 200,000 tons of pig iron, of which about three-fourths is worked into wrought iron. Poland furnishes about one-tenth of the whole production.

SCANDINAVIA.—The very interesting deposits of iron ores in Sweden have been noticed in the preceding section. The quality of the iron made is the best known, and it is principally exported to England and there worked up into steel, mostly in Sheffield and its vicinity. The furnaces which produce iron for exportation use the best qualities of ore, and bestow a special care on its manufacture. Charcoal is exclusively used. The mine of Danemora produces about one-twelfth of the iron of Sweden.

From the latest official returns, it appears that there had hardly been any increase in the production of iron in Sweden between 1840 and 1850. The present production is about 150,000 tons of pig and 100,000 tons of bar iron.

The exportation of wrought iron has been as follows:—

| | Tons. |
|---|---|
| 1834–38, | av. 79,300 |
| 1839–43, | " 89,200 |
| 1844–48, | " 92,000 |
| 1849, | 88,500 |

It is principally sent to four different countries, in the following proportions:—

| | Tons. |
|---|---|
| Great Britain, | 33,300 |
| United States, | 19,850 |
| Denmark, | 8,150 |
| France, | 5,200 |

Besides, a small quantity of cannon, bomb-shells, &c., are sent to Norway, Holland, and Denmark. The number of blast-furnaces in 1850 was 220; of workmen employed in mining the ore, 5241; at the blast-furnaces and foundries, 3096; at the wrought iron establishments, 3983.

Norway produces, from magnetic iron ore, about 5000 tons of pig iron, of which about two-thirds is worked up into wrought iron, and one-third consumed in castings.

GREAT BRITAIN.—Singular as it may appear in respect to a country whose prosperity is based on its coal and iron, and which extracts from the earth more of each of those prime-movers of civilization than all the rest of the world put together, it is only very recently that anything approaching to a complete idea of its resources has been obtained. That a tolerably compendious account can now be given of the iron manufacture of England is due to the Great Exhibition, and the enterprise of a private individual, S. H. Blackwell, Esq., who made a collection of English iron ores, gathered statistics with regard to them, and embodied them in the Official Catalogue, and in a lecture "On the Iron-making Resources of the United Kingdom," delivered before the Society of Arts in London.

From this source, and from a report on the iron trade of Great Britain, made to the British Association, in 1847, by G. R. Porter, whose authority on these subjects is of great weight, the following abstract has been compiled.

The history of the iron manufacture of Great Britain may almost be said to have commenced with the discovery of the use of mineral coal as a fuel for smelting the ores, so great has been its progress since that time. Still, the great abundance of localities in which ore is found gave an importance to the production of iron by charcoal so long as this kind of fuel lasted. In 1615, there were a large number of furnaces at work, producing, as was estimated, nearly 180,000 tons per annum. This was the maximum production of charcoal iron. Owing to the destruction of the forests consequent on so large a demand for charcoal, the produce of iron fell off gradually, and had, in 1740, declined to 17,350 tons, the product of fifty-nine furnaces.

The first attempts to smelt iron with mineral coal are said to have been made as early as 1620; but they do not seem to have been successful, as it was not until 1740 that the use of this fuel can be said to have become introduced to any extent, and for many years from that time it advanced but slowly.

In 1788, the whole quantity of pig iron made in England and Wales is said to have amounted to no more than 61,300 tons, of which quantity 48,200 tons were made with coke of pit-coal, and the remaining 13,100 tons were still made with charcoal. In the same year, the production in Scotland did not exceed 7000 tons. In Ireland charcoal iron was made on a moderate scale during the 17th century, and in 1672 the quantity amounted to 1000 tons; but, as the timber was exhausted, it gradually declined, and in 1788 there was not an iron-work in Ireland.

About this time the iron-masters in Great Britain began to avail themselves of Watt's improvements of the steam-engine; and they were thus enabled greatly and rapidly to increase the productive power of their works, so that, in eight years from 1788, the quantity of British-made iron was nearly doubled. An inquiry made in 1796, consequent upon the proposal of Mr. Pitt, which was afterwards abandoned, to put a tax upon coal at the mouth of the pit, showed the make of British iron to be then—

| | |
|---|---|
| In England and Wales, . . . . . | 108,993 tons. |
| In Scotland, . . . . . . . . | 16,086 |
| | 125,079 |

Ten years later, in 1806, it was proposed to tax the production of iron; and again, on that occasion, the amount was ascertained to be—

| | |
|---|---|
| In England and Wales, . . . . . . | 234,966 tons. |
| In Scotland, . . . . . . . . | 23,240 |
| | 258,206 |

Of this quantity, about 95,000 tons were converted into bars and plates, and the capital invested amounted to £5,000,000.

The next account of this manufacture was prepared by Mr. Francis Finch, and has reference to 1823, when the production was ascertained to be 452,066 tons; in 1830 it amounted, on the same authority, to 678,417 tons.

From this time forward began a new era in the manufacture of iron, caused by the introduction of the hot blast, the discovery of Mr. Neilson, of Glasgow, who patented it in 1829. The plan was adopted in Scotland in 1830, and its effects were most extraordinary, reducing the quantity of coal used in the furnace from 7 tons per ton of iron produced to 2, and, in some instances, $1\frac{1}{2}$ tons, rendering it possible to use the black-band ores exclusively. By this improvement, the quantity of iron made in that district was increased from 20,000 tons in 1820, and 37,500 tons in 1830, to 200,000 in 1839, and in 12 years after to 775,000; as Mr. Blackwell remarks, the most wonderful increase of production in any branch of manufactures which the world has ever seen.

The use of the hot blast gradually spread from Scotland into the other iron-making districts, effecting in all a most important economy of coal and limestone, especially in the anthracite district of South Wales. One of its consequences was the enlargement in size of the blast furnaces, which has been found materially to increase production and economize fuel.

This invention, all-important as it was to the interest of Great Britain, was not accepted without much opposition, and for a long time many of the most eminent engineers positively forbade the use of hot-blast iron for any works in which they were engaged; but at the present time more than nineteen-twentieths of the iron manufactured in Great Britain is made with hot-blast. As Mr. Porter remarks, "but for the introduction of the hot-blast, we should in all likelihood not have witnessed the unequalled development exhibited during the past fifteen years in this, which has now become one of the greatest branches of our national industry. Without this discovery our railroad system could not have marched forward with such giant strides, and in all probability the application of iron to the building of ships—an application from the extension of which, in future years, so many advantages may be made to arise, might have continued unthought of."

The introduction of the black-band ore formed another important addition to the resources of England in the manufacture of iron. This was first discovered, so long ago as 1801, by Mr. David Mushet, but for many years it was exclusively used in the works of Mr. Mushet himself, and there, even, in combination with other ores of argillaceous character. It was not until 1825 that it was first used alone, by the Monkland Company, whose success in the experiment led gradually to its adoption by other establishments, and to the erection of additional works.

Mr. Mushet, in his "Papers on Iron and Steel," thus describes the advantages of this kind of ore: "Instead of 20, 25, or 30 cwts. of limestone, formerly used to make a ton of iron, the black-band now requires only 6, 7, or 8 cwts. to the production of a ton. This arises from the extreme richness of the ore when roasted, and from the small quantity of earthy matter which it contains, which renders the operation of smelting the black-band with hot blast more like the smelting of iron than the smelting of an ore. When properly roasted, its richness ranges from 60 to 70 per cent., so that little more than a ton and a half is required to make a ton of pig iron; and as one ton of coal will smelt one ton of roasted ore, it is evident that, when the black-band is used alone, 35 cwts. of raw coal will suffice to the production of one ton of good gray pig iron." In corroboration of these results, it appears that to make 400,400 tons of iron in Scotland, it required 934,266 tons of coal, or a little over 2 tons 6 cwt. for each ton of iron.

According to M. Le Play, who made a thorough inspection of all the iron-works in Great Britain, the produce in 1836 amounted to 1,000,000 tons, which had increased in 1839 to 1,248,781 tons, on the authority of Mr. Mushet.

The opening of the Liverpool and Manchester Railway, which took place in July 1830, may be said to have commenced a new era for the iron interests of Great Britain, for from that time forward a demand was created for this metal for a hitherto unknown purpose on a scale of immense magnitude. The development of the railway system of England went on with rapid steps, and in 1836 and 1837 Parliament passed 77 railway bills, of which 44 were for new lines, and the capital authorized to be raised amounted to more than £36,000,000. The lines thus sanctioned would demand for their construction over 500,000 tons of iron, and the natural effect of this stimulus was to advance the price of this metal, at the same time largely increasing its production, as we have seen above. Such an immense expansion must naturally be followed by a depression, and a variety of causes tended to produce a commercial stagnation, which pressed heavily upon the iron interest from 1839 to 1844. Prices fell to half what they had been during the height of the railway speculations, and the iron-masters were obliged to reduce their production as much as possible. In 1840, it appeared from Mr. William Jessop's researches that the produce of iron amounted to about 1,400,000 tons, but in 1842 it had fallen off 22 per cent. in the principal districts, and the whole production of the kingdom probably did not exceed 1,000,000 tons, or the same amount which was made in 1836.

This period of depression could not last for many years, and since 1845, when prices rose again rapidly, the production of the English and Scotch furnaces has gone on increasing with but little fluctuation. In 1847 and 1848 the amount had risen to 2,000,000 tons, and in 1851 it made a greater advance than ever before, rising to 2,500,000 tons. Of this enormous quantity South Wales produced upwards of 750,000; Scotland, 775,000; South Staffordshire and Worcestershire, nearly 600,000; and the other districts about 400,000 tons. Of this total it was estimated that about one-third was converted into castings or exported as pigs; and two-thirds manufactured into wrought iron. In 1852, the production was 2,701,000 tons.

It is probable that the amount produced in 1854 will not fall below 3,000,000 tons.

The importance of the iron trade to England may be inferred from the following returns of the exports of iron of all kinds in 1850 and 1851:—

| Year. | Est. No. of Tons. | Declared Value. |
|---|---|---|
| 1850, . . . | 1,122,084 . . . | £9,567,108 |
| 1851, . . . | 1,296,873 . . . | 10,424,137 |

The annexed tabular statement shows the estimates for various periods since the beginning of the present century, according to the best authorities:—

| | Tons. | | Tons. |
|---|---|---|---|
| 1802, | 170,000 | 1839, | 1,250,000 |
| 1806, | 258,000 | 1840, | 1,400,000 |
| 1823, | 452,066 | 1842, | 1,000,000 |
| 1825, | 581,367 | 1847, | 1,999,608 |
| 1828, | 702,584 | 1848, | 2,093,736 |
| 1830, | 678,417 | 1851, | 2,500,000 |
| 1836, | 1,000,000 | 1852, | 2,701,000 |

BELGIUM.—The Belgian iron-works have reached a high degree of perfection in respect to economical employment of materials and skilful arrangement of machinery. The coals of that country are of excellent quality for smelting purposes, especially in the blast-furnace; but in the province of Liège, the chief seat of the iron manufacture, the exploitation of the coal is attended with some difficulties, which render its cost higher than that of the English. The ores are of pretty high percentage, although they require washing, which is very simply and easily effected, and the facilities of transport are admirable.

In 1844 there were 80 blast-furnaces using charcoal in existence, of which 26 were in blast, and 51 using coke, 23 of which were in blast. The charcoal-furnaces are of small dimensions, but produce a good quality of iron, the quantity of which is constantly diminishing, owing to the scarcity of that kind of fuel. The coke-furnaces are generally of large size, worked with a high pressure of blast, and give a high yield of iron, which averages 100 tons per week of white iron, for refining, or 75 tons of gray foundry pig.

In 1844 the produce of pig iron amounted to 106,878 tons, seventeen-twentieths of which was from furnaces using coke. In 1845 there were in operation 44 blast-furnaces using coke, and the total produce of the kingdom was 220,000 tons. The production continued to increase until 1848.

The coke-furnaces are divided into three groups. That of the river Maas is the most extensive, and comprehends, among others, the celebrated establishment of Seraing, where six blast-furnaces of very large dimensions, with puddling-furnaces, &c., are erected, together with immense machine and locomotive building establishments.

The wrought iron is mostly puddled, and amounted in

1844 to 46,913 tons. For the later years statistics cannot be given for the whole country; but in the province of Liège, in 1851, the production of pig iron amounted to 65,000 tons, the capacity of the furnaces being about 100,000. The quantity of wrought iron produced in the same district was in 1849 only 19,000 tons; but it had increased in 1850 to 20,000 or 25,000, and was in 1851 about the same.

On the whole, the Belgian establishments are admirably arranged, and might furnish a much larger quantity of iron than they now produce; but they are hemmed in on every side by almost prohibitory duties, and the greater cost of raising their coal renders it impossible for them to compete with England in supplying the United States with their products.

Prussia.—The manufacture of iron in this country is of great importance. The principal works are in the provinces of Upper Silesia and the Rhine. The government owns and manages some of the most important, which are intended as models for imitation by others, and into which the new improvements are introduced to be tried.

The production of iron had rapidly increased up to 1847, when, owing to various political causes, and to the completion of the great system of Prussian railways, it decreased somewhat for two or three years. In 1851, however, it amounted to 145,000 tons of pig iron, a quantity greater than that produced in 1847 by nearly 10,000 tons.

The following table shows the amount, in tons and percentage, furnished by each of the great mining districts into which Prussia is divided, for the year 1847.

| Mining District. | Pig Iron. | Castings from the furnace. | Total. | Per cent. |
|---|---|---|---|---|
| Brandenburg, . . . . . | | 784 | 784 | 0·6 |
| Silesia, . . . . . . . . | 43,888 | 7,185 | 51,073 | 39·6 |
| Saxony, . . . . . . . . | 4,214 | 527 | 4,741 | 3·7 |
| Westphalia, . . . . . | 2,153 | 7,103 | 9,256 | 7·2 |
| Rhine, . . . . . . . . | 54,326 | 8,604 | 62,930 | 48·9 |
| | 104,581 | 24,203 | 128,784 | 100·0 |

Austria.—The most extensive iron-works of the Austrian Empire are in Styria, the product of whose furnaces is mostly converted into steel of an excellent quality. The ores are chiefly spathic iron. The furnaces used are the German Blauöfen. Carniola and Carinthia also furnish large quantities of iron, a very considerable portion of which is converted into steel: the blast furnaces at Lolling are models of economy in respect to fuel, consuming only from 50 to 70 lbs. of charcoal in the production of 100 lbs. of pig metal.

The following table shows the yield of the different provinces, for the year 1847, in pig iron and castings direct from the blast furnace:—

| Province. | Pig Iron. | Castings from furnace. | Total. | Per cent. |
|---|---|---|---|---|
| | Tons. | Tons. | Tons. | |
| Styria, . . . . . . . . | 46,480 | 1,430 | 47,910 | 24·3 |
| Carinthia and Carniola, . | 36,340 | 1,300 | 37,640 | 19· |
| Tyrol, . . . . . . . . | 3,345 | 600 | 3,945 | 2· |
| Lower Austria, . . . . | 1,570 | | 1,570 | ·8 |
| Salzburg, . . . . . . . | 2,790 | 180 | 2,970 | 1·5 |
| Lombardy, . . . . . . . | 6,320 | 1,040 | 7,360 | 3·7 |
| Bohemia, . . . . . . . | 19,600 | 9,870 | 29,470 | 14·9 |
| Moravia and Silesia, . . | 16,160 | 8,180 | 24,340 | 12·3 |
| Hungary, . . . . . . . | 33,300 | 2,765 | 36,065 | 18·3 |
| Transylvania, . . . . . . | 1,270 | 75 | 1,345 | ·7 |
| Gallicia, . . . . . . . | 3,575 | 1,260 | 4,835 | 2·4 |
| Military frontier, . . . . | 250 | | 250 | ·1 |
| | 171,000 | 26,700 | 197,700 | 100·0 |

Saxony.—This state produced in 1851 nearly 7000 tons of pig iron, of which about 6000 was worked into bar. The quantity has remained nearly stationary for the last few years.

Brunswick.—The iron-works of this state are of some importance. They are estimated as producing about 4000 tons of pig iron annually, of which about half is converted into wrought iron, and the other half employed in castings. The most important furnaces are in the Harz.

Hanover.—The production of this kingdom is about 7000 tons yearly, mostly from iron-works in the Harz district.

Other German States.—Of these the whole production

amounted in 1847 to about 60,000 tons, a small amount compared with the yield of the English furnaces, but sufficient to supply the home consumption. Baden produced in 1847 about 7500 tons of pig iron, and 2000 of castings, and made nearly the same amount of wrought iron from home-manufactured and imported pig iron.

Würtemberg furnishes about 6000 tons. Nassau is rich in deposits of iron ore, and its produce of metal is very considerable, considering the insignificant extent of the Duchy. The amount in 1847 was 15,000 tons of pig, most of which is made into wrought iron. The yield of Bavaria is small for a kingdom of its size, being about the same as that of Nassau. The other smaller states, in regard to their production of iron, are of little importance.

France.—The geological position and the character of the iron ores of this country have already been sufficiently described. It remains to give some statistics of the yield of metal. It may be said with truth, that the circumstances under which iron is manufactured in France are not favorable to the development of this branch of her industry, and it is only by the aid of a high protective tariff that the production has been raised to its present amount, making her, next after Great Britain and the United States, the greatest iron-producing country in the world. Were the markets of France freely opened to Belgian and English iron, her production of this metal would sink to comparative insignificance.

The amount of pig iron produced in 1849, the last year for which accurate returns have been obtained, was 514,172 tons.

Very extensive changes have recently been made in the French tariff, by which a considerable reduction is already effected of the duties previously levied on coal and iron. In 1855 a still farther reduction will take place, when pig iron will pay only £1 15*s.* 4*d.* per ton, instead of £2 16*s.* 8*d.*, which had formerly been charged. Bar iron, formerly admitted at from £6 12*s.* to £8 4*s.* per ton, according to its size, will then be charged only £4 8*s.* to £6 3*s.* These changes will, no doubt, add largely to the consumption of

English and Belgian iron and coal, and give a new impetus to the manufacture of machinery and the use of this metal in France.

SPAIN.—The latest accounts give, as the yield of the Spanish iron-works, 14,000 tons of pig, and 15,500 tons of bar iron, entirely produced by Catalonian furnaces direct from the ore. Portugal produces the insignificant amount of 300 tons, by the same process.

ITALY.—*Tuscany.*—In this country the iron-works are managed with skill, and produce about 7500 tons.

*Sardinia.*—Estimated production 11,000 tons; and the Roman territory, together with the kingdom of the Two Sicilies, may be put at 4000 tons.

SWITZERLAND.—The Swiss iron-works are in the Cantons of Berne, Solothurn, Schaffhausen, St. Gallen, Grisons, and the Valais. There were in 1847 twelve blast-furnaces in operation, producing about 14,500 tons of pig iron, most of which was worked into bar iron for home consumption.

---

## SECTION III.

### DISTRIBUTION OF THE ORES OF IRON IN THIS COUNTRY.

UNDER this head, it will be impossible to describe the different localities of iron ore with anything like the same detail which has been thought necessary in treating of the other metals. The ores of iron are so widely distributed throughout the United States, that, even were it possible to obtain accurate information concerning them in all cases, want of space would prevent its being incorporated into this work. Besides, the mere fact of the existence of deposits of these ores, even although they may be of the greatest purity and existing in the greatest abundance, is not by any means sufficient to enable one to pronounce on their real value. In the manufacture of iron there are many other questions besides quality and quantity of the ore to be taken into consideration, such as the cost and nature of the fuel to be employed, the facilities for procuring the proper flux, and the accessibility of the locality. And besides these, there is the great question of the cost of labor, which, as connected with protection and tariffs, makes the whole subject one rather to be studied as a politico-economical than as a geological one. It is sufficient for the geologist, as such, to establish the fact that we have unrivalled natural facilities for the manufacture of iron; the more or less speedy development of these immense resources depends on circumstances, the discussion of which could not properly be included in a work like the present.

In order to give as comprehensive a view as possible of our deposits of iron ore, each state will be taken up separately, and the principal localities mentioned, with more or less detail, as it may be deemed best; and at the close of the section a short general review of the whole, with statistical illustrations, will be added.

MAINE.—The ores of iron in this state are either veins and vein-like masses in the metamorphic palæozoic, or bog and other surface ores. The most extensive deposit known at present is on the Aroostook, about fifty miles above its mouth. It is in the calcareous slates, which are much metamorphosed. The bed is fully thirty-six feet thick, and consists of red hematite, containing considerable manganese. No doubt there are other similar beds of ore in this region, but their situation is so remote that they cannot, at present, be considered as of any value.

Another such bed occurs in New Brunswick, near Woodstock, and not far from the state line. It lies on the St. John's River, and so is much more likely to become of value than the ore of the Aroostook. In 1848 a furnace was building to work this ore, which could be delivered at forty cents per ton. Charcoal is, of course, abundant in that region.

There are numerous veins of magnetic iron ore along the coast, but none of them are sufficiently wide to make their working a matter of profit. There are some localities of iron ores in the interior of the state, especially of bog ore. A furnace was worked on a small scale at Shapleigh, and another fifty miles back from Bangor, on the Piscataqua. Both these have been stopped for some time. At the latter locality, the ore was an ochrey hydrated oxide, with considerable sulphur. It was found covering the surface of a large dry knoll, lying just beneath the soil, and occurred in the form of limbs, branches, and leaves of trees, having replaced the organic particles entirely.

I do not know of any blast-furnace now at work in the state; but there is one given in the Census of 1850 as in operation.

New Hampshire.—The deposits of iron ore in this state are more abundant than in Maine, but they are not much more favorably situated with respect to facility of access. The iron-furnace at Franconia was for a long time the only one in operation in the state; indeed, I do not know of any other one ever having been built.

The ore is the magnetic oxide, forming a vein three to four feet wide, running north 30° east, and dipping nearly vertically. It has been unskilfully opened, and worked as an open cut for a depth of 150 feet. The best of the ore is nearly pure magnetic oxide, containing from two to three per cent. of silica, and a small amount of titanic acid.

The furnace was erected in 1811, and was producing, a few years since, when in blast, about two and a half tons of iron per day, with a consumption of 160 bushels of charcoal to the ton.

There is a very extensive deposit of specular iron ore in Piermont, near Haverhill. It occurs in heavy beds associated with white quartz, and of every degree of intermixture with that mineral. An abundance may be obtained of an ore yielding from 50 to 60 per cent. of metallic iron. The only impurity in it is quartz, with a little titanic acid. This appears, from its position, and the character and abundance of the ore, to be one of the most valuable localities in the state.

In the town of Bartlett, on Baldface Mountain, there are powerful veins of magnetic iron ore of good quality; but their remote and elevated position renders it unlikely that they will be worked at present.

Localities of bog ore are numerous throughout the state;

they could be profitably worked only on a small scale, and for the supply of the immediate vicinity.

VERMONT.—There were, in 1849, ten blast-furnaces in this state, with a capacity of production of about 8,000 tons per annum, but then producing hardly more than half that amount.

The number of the deposits of ore is very considerable, and some of them are capable of furnishing an excellent quality of iron.

Bog ore is abundantly distributed throughout the state, but is of little importance, compared with the ores of tertiary age which are found along the western base of the Green Mountains. Of these the most interesting deposits are at Brandon and Chittenden. They consist of irregular beds of brown hematite and yellow ochre, associated with manganese and a great variety of differently-colored clays and beds of gravel, and also, at Brandon, with an extensive deposit of lignite, containing fossil fruits.

MASSACHUSETTS.—The first ores smelted in this state were the bog ores, and the first furnace was erected in Pembroke, Plymouth County, in 1702.* In the early part of the present century there were ten blast furnaces in operation in that county, working bog ores, but they have all stopped, both for want of ore and of fuel. The quantity of iron manufactured in these furnaces is estimated at 1500 tons per annum. Cannon were cast at Bridgewater during the Revolutionary war.

The veins of ore in the metamorphic rocks of the western part of the state appear to be of little value, owing to the expense of mining in the hard rock. One of the best known is at Hawley, in Franklin County. There are two beds here, included in mica slate, and at a distance of ten feet from each other,† one of which is made up of magnetic iron ore, and the other of a beautifully micaceous specular ore. Its thickness is only about 2½ feet, and of course too little to be of any value at present.

The only ores of any considerable value in this state are

* J. T. Hodge, Am. R. R. Journal, No. 683. † Ibid. No. 684.

those of tertiary age in Western Massachusetts. They have been and are extensively worked. The operations of the furnaces in this region have been carefully studied by Mr. Hodge, from whose accounts of them the following information is mostly derived.* The ores are chiefly brown hematite, in a great variety of forms, and usually containing considerable manganese and a little phosphoric acid, besides some earthy silicious matter. Zinc is also present to some extent, and the oxide of this metal is found lining the walls of the furnaces. Black oxide of manganese also occurs with these ores, but is usually too much mixed with silicious matter to be of any value as an article of commerce. Magnesian limestone is generally found near them, being a nearly pure dolomite, with a small percentage of silica. It makes the most suitable flux. Quartz rock and fire clay, suitable for use in and about the furnaces, are also abundant in the neighborhood of these deposits.

The fuel used in the furnaces is exclusively charcoal, which is furnished at from 5 to 7½ cents per bushel, delivered at the works. The wood is charred in large kilns built of brick, producing about 3000 bushels at a charge.

In 1849 there were 7 furnaces in Berkshire County, all using hot blast, and all but one driven by water-power.

The North Adams Furnace is supplied from the Kingsley ore-bed, which furnishes a very silicious ore, and from the Anthony bed, at South Adams.

The Lanesboro Furnace is supplied from the Sherman and Newton ore-beds with an excellent brown hematite. At the former the ore lies on the east side of a low ridge of limestone, in layers interstratified with clays and ochres, and it is very abundant. The Newton bed is described by Mr. Hodge as being situated, in a nearly vertical position, in a high ridge of mica and talcose slates.

The Richmond Furnace and the Lenox Furnace are supplied chiefly from the West Stockbridge ore-beds.

The West Stockbridge ore-bed supplies the furnaces of the "Stockbridge Iron Company," as well as the two last-

* See Am. R. R. Jour., No. 684.

mentioned. It lies along the southeast side of one of the limestone ranges so common in this region. It has been opened for a third of a mile in length, in two portions, one of which is called the Leet-bed, and the other the Chauncey Leet-bed. In the former, the ore is found in alternating strata with ochres and clays. The layers of solid ore sometimes exceed 10 feet in thickness, and parallel layers are separated from each other by a variable bed of earthy matters. The ore is of good quality, of a chocolate-brown color, passing into black, and containing more or less manganese, and a little phosphoric acid. The Chauncey Leet-bed has been worked considerably; its extent is not known, but is very great. The iron produced from it is of excellent quality.

The Vandeusenville Furnace is in the town of Great Barrington, and is also supplied with ore from West Stockbridge.

The capacity of these furnaces is about 12,000 tons per annum.

CONNECTICUT. — There are a few localities in this state where magnetic ore has been found in veins in the metamorphic rocks, but they have been, thus far, little worked.

The great spathic-iron vein in Roxbury also lies unworked, so far as I am informed. This vein is described by Prof. Shepard as being from 6 to 8 feet in width, and consisting of pure carbonate of iron mixed with white quartz. The locality is said to be well situated in every respect, and the ore abundant.

The only iron-beds which are of any importance in this state are in its northwestern corner, and belong to the district just described in Massachusetts; the ores are similar in character and in geological position.

The furnaces are in the towns of Salisbury, Canaan, Cornwall, Sharon, and Kent, and the adjacent establishments in Dutchess County, New York, belong to the same group.

In 1849, Mr. Hodge gives the number of furnaces in Litchfield County, Connecticut, using hematite ores, at 16, and their capacity of production is estimated at 12,000 tons per annum.

All these are supplied from one source, the Ore Hill Mine

of Salisbury. This is a vast deposit of ochres, clays, and hematite, belonging to the tertiary formation. The ore lies in irregular-shaped masses, the whole arrangement here, as in the other localities in this position, being one very difficult to decipher. Hence their geological place was for a long time misunderstood, and they were all referred to the drift, previous to the discovery of the lignite and fossil fruit of Brandon, Vermont. At the Salisbury mine, the ore is of a character which furnishes forge pig of the finest quality. The best ore is a fibrous and massive hematite, and, from the circumstance that it is furnished to the different companies at a uniform price, each one being allowed to select the ore where it chooses and of such quality as it prefers, the workings are very irregular, and only the best kinds have been taken. Mr. Hodge estimates that, in the fifty years previous to 1850, from 250,000 to 300,000 tons of ore had been taken from this hill. The best quality of iron is made with cold blast, and its actual cost, delivered on the railroad, is no less than $30 per ton. This, however, is considered as superior to any other iron manufactured in the country, for articles requiring great strength and tenacity. The iron made by hot blast is not so good; but the furnaces running with cold blast produce only half the quantity of those working with hot blast, and use fifty bushels more of charcoal to the ton.

The Kent ore-bed was formerly of considerable importance, being similar to that of Salisbury; it has, however, been pretty much worked out.

The cost of ore and coal was, in 1849, for the furnaces using the ore of the Salisbury bed, from $20 to $23 per ton of iron produced.

New York.—The deposits of iron ore in this state are on the most extensive scale and of the most varied character.

In the eastern part of the state, in Columbia and Dutchess Counties, hematite ores occur in extensive and valuable deposits, being similar in character and adjacent to those just described as occurring in Massachusetts and Connecticut. There were, in 1849, seven furnaces using these ores. According to Professor Beck, there are several important beds.

That of Fishkill is mentioned as occurring in a hill, near the junction of mica slate with the gray and white limestone. The Clove ore-bed is an extensive deposit of brown hematite in the southwestern part of the town of Union-Vale. The Foss ore-bed is of less importance, but similar to the one last mentioned. The Amenia bed is one of the most important in this region, and has yielded a large amount of ore, and, according to Professor Mather, will continue to do so for many years to come. A layer of broken rocks and gravel covers the ore to a depth of from five to twenty feet, and the ore-bed beneath had in 1838 been worked to a depth of forty-five feet without its bottom having been found. The amount furnished by this bed is estimated at 5000 tons per annum, yielding fifty per cent. of pig iron. Prescott's ore-bed is in Columbia County, in the town of Hillsdale. It seems to be similar in position, as well as in the quality of the ore furnished by it, to the one just described. The quantity is also very large, as the bed has been penetrated to a depth of thirty-two feet, and the ore found to improve in quality, and to continue still farther.

The belt of metamorphic rocks which passes through the southeastern corner of the state, and is developed principally in Putnam, Orange, and Westchester Counties, contains very numerous deposits of magnetic iron ore, but which appear to have been, at least until quite recently, very much neglected. Prof. Beck remarks of the ores of Orange County, that "it is doubtful whether the quantity (of the magnetic oxide of iron) which exists here does not exceed that found in an equal area in any part of the world." In 1849, there were only two furnaces using these ores, although another was built, and ready to commence operations when the price of iron should rise. It is difficult to see what drawback there can be to the future prosperity of this region, situated, as it is, so near to the metropolis of the country. Among the important deposits of ore, the following are mentioned by Prof. Beck: Stirling Mountain, in the town of Monroe; the ore is granular magnetic oxide, containing some pyrites. This deposit was opened in 1750, and a blast furnace built near it during the next year. The

Belcher Mine, $1\frac{1}{2}$ miles southwest of the Stirling Mine, has been worked over a width of 115 feet, without finding the wall-rock. Crossway Mine; a bed fourteen feet thick, which has been mined to some extent, making a moderately good red-short iron. Paterson Mine; the ore from this locality contains considerable silica; the iron produced from it was of good quality, the ore yielding about 56 per cent. in the furnace. Forshee Mine yields a very good ore, which is abundant. Forest of Dean Mine; six miles west-northwest of Fort Montgomery; according to Mr. Hodge, a vein of magnetic ore, from 10 to 16 feet wide, of good quality and well situated for working.

It would be impossible to enumerate all the localities known to exist in this part; and it is not easy to understand why they have not been more extensively worked. Recently, it is said, operations have been commenced here on a somewhat more extended scale than previously, but I have not the particulars of what is now doing.

The great azoic region of New York, which occupies the larger portion of the northern part of the state, extending from Lake Champlain to Lake Ontario, is exceedingly rich in the specular and magnetic ores of iron. It would be vain to attempt to state all the localities where these deposits exist, but some of the most important may be specified. The counties where they are most developed, or best known, are Essex and Clinton. St. Lawrence also contains some very valuable mines. Those of Franklin and Jefferson Counties are of less importance. The iron ores of this region have been described by Prof. Emmons and Prof. Beck in their official reports, and also, partially, by Mr. Hodge.

The ores of Clinton County are among the most celebrated in the country. They are all in masses intercalated in the gneissoidal rocks, and coinciding with them in their line of strike, but not always in their apparent planes of dip. The Arnold Veins are best known. They are four in number; of which one, the "Old Blue Vein," is considered of great value on account of the purity of its ores. This varies from two to eight feet in width, and in 1842 had been worked to a depth of 260 feet, and over a length of about 80 rods. The ore of this vein, according to Prof. Beck's analysis, consists of nearly pure magnetic oxide of iron, with a small percentage of silicious matter. The four veins are parallel with each other, and have all been heaved simultaneously by trap-dykes which intersect them.

The Palmer Vein is a bunch of magnetic iron ore 35 feet wide, without any distinct walls, but gradually passing into the adjacent rock. The ore is much mixed with quartz, and requires washing before being worked. The Cook Veins are four or more in number, of which the widest is 14 feet across. They dip nearly perpendicularly, with the rock. The ore is said by Prof. Emmons to make iron of the first quality for toughness. The principal vein has been traced for 1½ miles. The Winter bed appears in the form of a thick plate, overspreading several square yards of the rock with which it is associated. According to Dr. Emmons, "it appears as if deposited horizontally on the rocks, like an overflowing melted mass of lava."

In 1849, the largest bloomery in the country was engaged in working these ores. It had 21 fires, and one oven for reheating the blooms. The bar iron produced is said to be much valued for nails. Mr. Hodge remarks of the forges in this and the adjoining county, that their business is very extensive, requiring about 50,000 tons of ore a year to supply their demand. The blooms and bar iron made by them will bear the high cost of transportation, which pig iron would not do. Besides, it is generally believed that these ores work better in the bloomery fire than in the blast furnace.

The ores of Essex County are not less valuable than those of Clinton. Port Henry is the head-quarters of the iron manufacture of this district. Among the most important iron-producing localities are the Cheever Mine, the Sanford Vein, the Penfield Vein, &c. The Cheever Veins are two in number, one of which is six feet, and the other from eight to ten in thickness. The ore is a very pure magnetic oxide, of rather coarse grain. The cost of mining and conveying the ore to the furnaces on the Lake, in 1849, was about 80 cents. A deep adit has been driven in from near the Lake, which will afford access to many hundred thousand tons of ore.

The Sanford Vein is about four miles northwest of Port Henry. This ore is remarkable for its coarsely-crystalline texture, and for containing considerable phosphate of lime mixed with it. Its mass is very large, so that it may be easily and cheaply mined. It works easily in the blast furnace, but is not adapted for the bloomery.

The other and more remote parts of the county contain numerous deposits of ore, many of which are of great extent, but they have been but little worked. The cost of manufacturing pig iron in this region, in 1849, was about $20 per ton; the cost of the ore being estimated at $2 per ton, and of charcoal at 6 cents per bushel, which was the price usually paid.

The ores of St. Lawrence County are chiefly the specular. One of the most important localities is the Parish Mine in Gouverneur. From an analysis of this ore by Prof. Beck, it appears that it is almost pure peroxide of iron, containing only two or three per cent. of silicious matter. The deposit of ore is represented by him as a flat bed, between the gneiss and the Potsdam Sandstone; but according to Prof. Emmons, it appears to have the character of an eruptive mass, which has lifted up the strata of the sandstone, and caused them to dip each way from the igneous nucleus. The Tate and Polley Veins are also mentioned by the State Geologist as being of some importance. Veins of magnetic ore occur in the gneiss of this county; they are, however, of less consequence than the deposits of specular ore. Bog ores are also abundantly distributed through the northern counties.

In the western part of the state, there are deposits of iron ore of considerable value in strata of the Clinton group. They form a band of small thickness, but lie near the surface, so as to be easily mined. The ore is a compact and earthy peroxide, often fossiliferous; and there were, in 1849, five furnaces using it.

New Jersey.—The very important iron-producing district of New Jersey is a continuation of that previously noticed as occurring in Dutchess, Orange, and Putnam Counties, in New York. The rocks in which the ores of New Jersey occur are gneiss and hornblende slates, having a steep inclination, and a direction nearly northeast and southwest. The ore is the magnetic, and it occurs in heavy bands, parallel with, and having the same dip as the enclosing rocks. The most extensively-worked beds are in Morris County, but the ores appear to be abundantly distributed through the whole of that part of the state which is underlaid by the rocks denominated "primary" in the State Geological Report. The following localities are mentioned by the State Geologist as of importance. In Pompton there are several veins, which occur in the continuation of Stirling Mountain of New York. Two parallel veins have been extensively opened, and one of them traced for three miles. They occur in a granitic gneiss, abounding in hornblende and destitute of mica. The width of the ore in the eastern vein seems to vary from 6 to 15 feet, and its quality is excellent, answering either for the bloomery or the forge. The western vein furnishes an ore which makes a very cold-short iron. In the range of country extending for several miles northeast of Succasunny, in Morris County, there are deposits of a very fine magnetic ore, of great value. Prof. Rogers is of opinion that there are probably at least two parallel veins, as the openings, although not continuous, are distributed along two parallel lines. There are mines opened on these lines of ore at intervals for a distance of ten miles. The thickness of the ore varies from a few feet to fifteen. The most easterly bed furnishes the best ore. Scott's Mountain, in Warren, is another locality abounding in iron ores, whose mode of occurrence is similar to that just described.

From a very complete account of the iron manufactures of Morris County, published recently in the New York Tribune, it appears that there were in that district, in 1853, about 50 forges in operation, with 90 fires, each fire producing about 75 tons of blooms and bar iron annually, with a consumption of 42,000 bushels of charcoal. There are extensive rolling mills at Dover, Rockaway, Powerville, Boonton, and Charlottenburgh. The Boonton works are among the most extensive and best-managed in the country. The blast furnace produced, in 119 weeks from Feb. 13, 1851, 11,755 tons of pig iron, working with anthracite coal, of which it consumed 20,995 tons. At the same establishment, the following was the amount of business done from October 1, 1852, to May 1, 1853, a period of seven months: Pig iron puddled, 3774 tons; nail-plate rolled, 3009 tons; spike-rods rolled, 885 tons. During the same period, 836 tons of rail-road spikes, and 5617 of cut nails were produced. The five rolling mills above mentioned employ 500 hands, and work up 16,000 tons of iron, with a consumption of 9000 tons of anthracite, and produce, in the form of bar iron, hoops, nails, spikes, &c., 13,780 tons, worth in market about $1,000,000.

PENNSYLVANIA.—The immense resources of this state in coal and iron can hardly be more than hinted at in the space allowed by the plan of this work. Indeed, until the completion of the geological survey, which has been going on at intervals since 1836, and the publication of its results, it will be difficult to give an account of the localities of iron ore which shall be anything more than fragmentary, like the materials from which it is drawn. Mr. R. C. Taylor, in his valuable work,* has given a connected account of the coal-fields of the state, and some statistics of the iron manufacture; the two are so intimately connected that they can hardly be separated in their development.

Mr. Taylor has computed that in Pennsylvania about 15,000 square miles are occupied by the coal-measures; and when the abundance of the iron ores which these and the strata below them contain is taken into consideration, some idea may be formed of the capacity of production of the state.

In one respect Pennsylvania has a great advantage over the other states underlaid by the coal measures. She is the almost exclusive possessor of the anthracite of the country, whose introduction into the iron-making business has proved to be a great step in advance, and has given an immense development to her production.

* Statistics of Coal (Phil. 1848), p. 72.

The iron-works of the state are grouped into two divisions, separated by the Alleghany Mountains. In Eastern Pennsylvania, both anthracite and charcoal are used, with hot and cold blast. In Western Pennsylvania, charcoal and cold blast are principally employed.

According to the State Geologist, it appears that the ores used in Pennsylvania belong to three distinct species: magnetic iron ore, brown oxide of iron, and the compact carbonate; the two latter kinds being much the most extensively diffused. The magnetic ores occur only in the southeastern division of the state, in the older metamorphic rocks, or adjacent to the trap-dykes of the middle secondary region. The carbonate abounds in the anthracite and bituminous coal-measures, where it exists in many of the basins in inexhaustible quantities. The brown hematites occur in greater or less abundance in all the formations within the state, not only with many of the older secondary strata of the Appalachian region, but also in the metamorphic rocks and the coal-measures. The analyses of the magnetic ores show from 63 to 65 per cent. of metallic iron, the chief impurity being silica. The brown hematites of the rocks of Lower Silurian age contain usually from 45 to 55 per cent., and the carbonates from 33 to 45 per cent. of iron.

The furnaces in Eastern Pennsylvania use principally hematite and magnetic iron ore; those of the western portion of the state are supplied almost entirely with the argillaceous carbonate, only a few of them using hematite alone, or mixed with other ores.

The following statistics from the "Documents relating to the Manufacture of Iron in Pennsylvania," published in 1850 by a committee of those interested in the coal and iron trade, will show the extent of the business at that time.

Of the sixty-two counties in the state, iron-works had been established in forty-five; and in nine of the seventeen in which there were none, there was believed to be an abundance of coal and iron, the only drawback being the want of means of communication.

The annexed table exhibits the principal facts in regard to the Pennsylvania iron-works producing iron from the ore:—

| | No. | Capital invested. | Capacity. | Make in 1847. | Make in 1849. |
|---|---|---|---|---|---|
| Blast furnaces using anthracite, | 57 | $3,221,000 | 221,400 | 151,331 | 109,168 |
| " " bituminous coal, | 7 | 223,000 | 12,600 | 7,800 | 4,900 |
| " " coke, | 4 | 800,000 | 12,000 | 10,000 | |
| " " charcoal with hot blast, | 85 | 3,478,500 | 130,705 | 94,519 | 58,302 |
| " " charcoal with cold blast, | 145 | 5,170,376 | 173,654 | 125,155 | 80,665 |
| Bloomeries, | 6 | 28,700 | 600 | 545 | 335 |
| | 304 | 12,921,576 | 550,959 | 389,350 | 253,370 |

The following statement shows the most important facts with regard to the conversion of cast into wrought iron:—

| | No. | Capital invested. | Forge-fires. | Puddling furnaces. | Capacity. | Make in 1847. | Make in 1849. |
|---|---|---|---|---|---|---|---|
| Charcoal furnaces, | 121 | $2,026,300 | 402 | | 50,250 | 39,967 | 28,495 |
| Rolling mills, | 79 | 5,554,200 | | 436 | 174,400 | 163,760 | 108,358 |
| | 200 | $7,580,500 | 402 | 436 | 224,650 | 203,727 | 136,853 |

A small amount of iron is also converted into steel, the number of establishments being, in 1849, thirteen, and their product 6,078 tons.

From the same source, the total number of all the iron-works in the state is given as 504:—

| | |
|---|---|
| Capital invested in lands, buildings, and machinery, | $20,502,076 |
| Number of men employed, | 30,103 |
| Number of horses employed, | 13,562 |
| Number of men otherwise dependent on the iron-works for their support, | 11,513 |

The consumption of fuel was, in 1847, as follows:—

| | |
|---|---|
| Anthracite, 483,000 tons, at $3, | $1,449,000 |
| Bituminous coal, 9,007,600 bushels, at $0 05, | 450,380 |
| Wood, 1,490,252 cords, at $2 (= $0 05 per bushel for charcoal), | 2,980,504 |
| | $4,879,884 |

The progress of the iron manufacture in this state may be seen from the following table of the amount of pig iron made at different periods since 1828. It is compiled from the Census returns, R. C. Taylor's work on Coal, and the Report of the Pennsylvania Committee above referred to:—

| | Pig iron made. Tons. | | Pig iron made. Tons. |
|---|---|---|---|
| 1828, | 24,822* | 1844, | 246,000 |
| 1830, | 31,056* | 1846, | 368,056 |
| 1840, | 98,395 | 1847, | 388,805 |
| 1842, | 151,885 | 1849, | 253,000 |
| 1843, | 190,000 | 1850, | 285,702 |

Since 1850, the increase of production has been very considerable, but the statistics cannot be given for any later period.

MARYLAND.—In this state the production of iron has been extensively and successfully carried on. As far back as 1756, there were eight furnaces and nine forges in operation. The ores used are from the tertiary and the coal-measures principally. The greater part of the furnaces in the state are situated near tide-water, and are supplied partly from beds of clay iron-stone, occurring in the tertiary belt which occupies the eastern portion of the state. This ore is an argillaceous carbonate, similar to that found in the coal-measures. A considerable quantity of hematite is obtained, from deposits which seem to be analogous to those described as occurring in connection with clays and ochres in the western part of Massachusetts, and in Vermont and Connecticut. Mr. Hodge† describes a locality at Beaver Dam, three miles from the railroad at Cockeysville, where the ore lies under a thickness of 15 feet of clays, in masses closely packed together, some of which weigh several tons. In the gneissoidal belt which traverses this state the specular and magnetic ores are occasionally found, but they do not appear to have been worked to any considerable extent. The iron-stones of the coal-measures are the most valuable deposits in the state, and are similar in character to those usually found in this position. The superficial coal area of the state, according to Mr. Taylor, occupies only 550 square miles,

* These years are considered by Mr. Taylor as very much underrated.

† Am. R. R. Journal, No. 703.

and is principally confined to Alleghany County, where the great works of Lonaconing and Mount Savage are in active operation. Dr. Higgins, the State Chemist, divides the ores occurring in this county into four classes: fossil ore, red hematite, brown hematite, and clay iron-stone. He remarks that there are five distinct beds of the last-named variety in the coal-fields of Youghiogeny River, the thinnest being about one foot, and the thickest five feet in thickness. Their analysis shows them to contain from 27 to 43 per cent. of metallic iron. The quantity and quality both of the ores of iron and the associated coal, is such as to make this a region likely to become of great importance for the manufacture of this metal. The hematites of this section of the state are worked to a considerable extent on Dan's Mountain. They have supplied most of the ore to the furnaces of the district. The analyses show them to be of good quality, and to contain from 50 to 60 per cent. of iron.

In 1853, there were in the state 31 blast furnaces, with a capacity of 70,500 tons per annum. The Census of 1850 gives the number in operation at that time as 18, and the produce of pig iron as 43,641 tons.

VIRGINIA.—This state abounds in iron ores as well as coal, and the manufacture is of great and growing importance; but, for want of full reports of the geological survey, it seems hardly possible to give even a synoptical view of its resources in this branch of its mineral wealth.

It appears that, as early as 1732, there were four furnaces in operation in Virginia, and that these were among the first works of this kind erected in North America.

According to the report of the State Geologist on the reconnaissance of Virginia, there are veins of magnetic iron ore and hematite in the older crystalline rocks; and, as they occur in the vicinity of the Richmond bituminous coal-field, they are likely to be of much value. Throughout the Southwest Mountain and its spurs, specular and magnetic iron ores are also found, similar to the deposits of the same in New Jersey. In the Valley of Virginia hematites occur in abundance. The western part of the state is underlaid by a portion of the great Appalachian coal-field, the area occupied by which amounts to about 20,000 square miles. As elsewhere, this coal is associated with iron-stone; and hardly any portion of the United States is more highly favored as respects the location and extent of its mineral wealth than this. The development of these resources has thus far been but limited, although there are considerable establishments for manufacturing iron at Wheeling. According to the Census of 1850, there were 29 blast furnaces in the state, producing 22,163 tons of pig, and 39 establishments furnishing 15,328 tons of wrought iron.

North Carolina.—The manufacture of iron has made but little progress in this state. The Census returns of 1850 show the amount produced within its limits to have been only 400 tons. The geological survey now in progress has not yet thrown any light on its resources in this respect. It appears, however, that the coal-field of Deep River contains little iron ore of value. The metamorphic rocks of the state undoubtedly contain veins of magnetic and specular ores; but, up to this time, the means of communication with the interior, where they should occur, are too limited to render their existence a matter of much consequence.

South Carolina.—According to M. Tuomey, the ores of iron are sufficiently abundant in this state; but if the Census returns may be believed, there was no iron made in 1850. At the date of the Geological Report, there were eight or ten blast furnaces; but the information with regard to their working is too indefinite to be used. The quantity of iron made would seem to be very small, and the business to be wastefully conducted.

The ores are described as being the magnetic and specular, and brown hematite. The two former are chiefly confined to a narrow belt of slates in York, Union, and Spartanburg Districts, extending for a distance of six or eight miles in a direction north 50° east. The magnetic oxide occurs in a belt of talcose slate, in a series of bands or veins parallel with the stratification of the enclosing rock, sometimes swelling out to a thickness of fifteen or twenty feet, at others diminishing to a foot, or even less. At the surface the ore is generally much mixed with the slates, and has an unpromising appearance; but at a little depth it improves in quality. The specular ores occur in a belt of mica slates, immediately overlying the talcose slates which contain the magnetic oxide. This belt extends from the North Carolina line to Gelkey's Mountain, in Union District. These deposits, according to M. Tuomey's account, can be of little value, since they gradually change into the sulphuret on being worked downwards. Brown hematite occurs in a greater number of localities than any other ore, but nothing is known of the value of the deposits.

GEORGIA.—But little can be said of the iron deposits of this state. In 1849 there were two blast furnaces on the Etowah River, a few miles from Cartersville, on the Western and Atlantic Railroad, turning out, when in blast, about 6 tons of pig iron per day. In the Census returns for 1850, three furnaces are set down as in operation in the state; their yield is given at 900 tons of pig iron, and $28,000 in value of other products.

ALABAMA.—Brown hematite and the specular oxide of iron are said by the State Geologist to exist in abundance in the older Silurian strata of the northern portion of the state. The manufacture of this metal, however, is very little developed. There were, in 1849, eight bloomeries and two blast furnaces in operation. The principal works were in Benton County, where the furnace was producing about three tons per day.

TENNESSEE.—The production of pig iron in this state is given by the Census for 1850 as 30,420 tons, showing it to be one of the most important iron-producing states of the Union. But little is known in detail of the geology of the Tennessee coal region, the geological reports heretofore published containing scarcely any information of any practical value. A new survey has been recently organized, and it is to be hoped that its results will aid in developing the mineral resources of the state. The great Appalachian coal-field extends through Tennessee, and there are numerous furnaces scattered over it. According to local authorities, there were in 1849, in Dickson County, 10 blast furnaces, of which 7 were in blast, and 5 forges, all but one of them in operation.

In Montgomery County, there were 5 blast furnaces, of which 4 in blast, 3 forges, and one rolling mill; in the other counties of Middle Tennessee, there were 14 blast furnaces and 9 bloomeries. In West Tennessee, there were 5 blast furnaces and forges; in East Tennessee, 12 furnaces, and 70 bloomeries, furnaces, and rolling mills. The total number of furnaces in the state, according to this authority, is 47, and of bloomeries, &c., 92. Tennessee was, in 1840, the third iron-producing state of the Union; but her progress has not kept pace with that of the other states, and she is now probably the fifth in rank.

KENTUCKY.—This state is underlaid by portions of two distinct coal-fields; its eastern end belongs to the great Appalachian, and its western to the Illinois coal-basin. If we may believe Professor Mather, the resources of Kentucky must indeed be great, so far as coal and iron are concerned; he estimates the extent of the coal region at 12,000 square miles, of which seven thousand contain workable coal beds; and states that the bituminous coal-seams are everywhere accompanied by beds of iron-stone, averaging one yard in thickness over the whole extent. Allowing for considerable exaggeration in these estimates, it will be seen that the state possesses abundant resources for the production of this metal.

By the returns of the last Census, there were 21 blast furnaces in operation here, placing the state about sixth in rank in her production of iron. Later and more reliable statistics are wholly wanting.

OHIO.—The progress of the iron manufacture in this state has lately been very rapid. The use of charcoal seems to be still very general, although mineral coal is abundant and of good quality. From the State Geological Reports, it appears that both coal and iron occur together, and are easily mined, throughout the southeastern part of the state, especially in Hocking and Muskingum Counties. One of the most important points for the manufacture of this metal is at Ironton, in Lawrence County, where there were, in 1853, ten blast furnaces, which produced during that year about 20,000 tons of pig metal. The whole amount produced in that county is estimated at 28,000 tons.

According to the last Census, there were 35 blast furnaces in operation, making 52,658 tons of iron yearly. These figures must be far below the amount now produced.

MICHIGAN.—*Lower Peninsula.*—But little has been done, as yet, towards developing the iron ores of the southern portion of this state. The great coal-field also remains quite unexplored. According to R. C. Taylor, the beds of coal, as observed by him on the Shiawassee River, are accompanied by courses of excellent argillaceous carbonate of iron. Bog iron ore is also quite abundantly distributed through the state.

*Upper Peninsula, Lake Superior Iron Region.*—The very interesting deposits of iron ores on and near Lake Superior, like those of Sweden and Missouri, are in the azoic, and they form literally mountain masses, sufficient to furnish an unlimited quantity of the purest and finest ore.*

The distance of these deposits from the Lake, at the nearest point, is about twelve miles; and a railroad is now in process of construction which will render them quite accessible. From this point, the ores of iron are found at intervals in a belt of slates from six to twenty-five miles wide, extending for a distance of 150 miles or more westward into the State of Wisconsin. The ore is mainly the peroxide, or specular ore, sometimes nearly chemically pure, although generally containing a small quantity of silicious matter. At some of the localities a little magnetic oxide is mixed with it, either in fine crystals or intimately incorporated with the mass. It contains hardly a trace of sulphur, phosphorus, or titanic acid, and, as might be expected, makes a remarkably tough and fibrous iron. The whole region where these ores occur is densely wooded, so that charcoal may be furnished for a long time at a moderate price. At present only those deposits of ore which are nearest to navigable water can be considered as of much value, since they can supply an almost unlimited demand. No mining is required, as the ore lies

* See Foster and Whitney's Report, Part II. p. 50.

in knobs and ridges at a considerable elevation above the general level, and needs only to be blasted off, or worked in a quarry, like any other rock.

Up to this time but little has been done towards developing these deposits. Two bloomeries have been built in the region, one at Marquette, on the Lake shore, and another near one of the principal iron-knobs, ten miles from the Lake. From both these establishments about 800 tons of blooms were shipped in 1853. Upon the completion of the canal around the falls at Saut Ste. Marie, by which the necessity of transshipment will be avoided, a considerable quantity of these ores will undoubtedly be shipped to ports on the Lower Lakes, to be worked, with other ores or by themselves, in furnaces erected at convenient points and in the vicinity of coal.

Indiana and Illinois.—Nearly the whole of Illinois and a considerable part of Indiana is underlaid by coal-measures, and valuable beds of argillaceous iron ore are said to exist in them. Thus far, however, they appear to have been but little developed, as there were in 1850, by the Census returns, only four blast furnaces in operation in both these states. The vast system of railroads now constructing must open access to all parts of the region; and if the deposits of iron are as valuable as they have been represented to be, these states must soon take a high place in the list of the iron-manufacturing states of the Union. The geological survey now in progress in Illinois will, no doubt, throw more light on the occurrence of the useful ores and minerals; at present but little is definitely known respecting them. The coal and iron ores of the Big Muddy River region, in the southern part of the state, are stated to be of superior quality.

Missouri.—The resources of this state for the manufacture of iron are very great. Hardly any localities of ore in the country have a higher interest than those in the azoic of Missouri; among which the mountains of peroxide of iron called the Iron Mountain and the Pilot Knob are most conspicuous, and are everywhere known. There are also numerous beds of brown hematite in the various divisions of the palæozoic strata, many of which are capable of furnishing large quantities of good ore, and are already worked to some extent.

The mode of occurrence of the eruptive ores in the azoic has already been described, and the Iron Mountain alluded to as a remarkable instance of a mass of this character. It is a flattened dome-shaped elevation, of about 200 feet above its base, and forms the western extremity of a ridge of reddish feldspathic porphyry, which rises one or two hundred feet higher than the knob of iron ore, and stretches to the east for a mile or two. The surface of the Iron Mountain is entirely covered with loose pieces of ore, which become more and more conspicuous toward the summit, on account of the small quantity of soil and vegetation covering them, as well as from the fact that the masses of ore themselves grow larger and more angular. The summit is covered with moss-grown blocks, some of which are many tons in weight, piled together in the greatest confusion. Nowhere about the mountain can the rock or ore be seen *in place*. On the west end of the hill a considerable excavation has been made for the purpose of getting out the ore. A vertical cut was carried to the depth of 8 feet, and a shaft sunk 7 feet farther; at the bottom, a bed of red clay, destitute of boulders, was struck and penetrated for one foot only, without reaching the solid ore. It appears, therefore, that the bed of loose masses covering the side of the mountain is at this point at least 15 feet thick; it is here made up entirely of small, somewhat rounded pieces of ore, packed together without any other substance than a little bright-red ferruginous clay between them. The ore requires no selecting or washing, as there are no foreign boulders or stones mixed with it. It is dug out close to the furnace, and of course can be furnished in unlimited quantity. It is a nearly pure peroxide of iron, containing only a small percentage of silica.

In 1852, there were two blast furnaces built and in operation at the Iron Mountain, making together 9 tons of pig iron per day. Flux is abundant at a distance of half a mile, costing 25 cents at the tunnel-head per ton of iron produced. Charcoal, the only fuel used, costs 3½ cents a bushel; the company owning the woodland, and paying that sum for burning and hauling to the furnace. About 110 bushels of coal are required to make one ton of iron. The furnaces are 36 feet high, and one is 8, the other 7 feet across the boshes. The ore costs 80 cents a ton, mined, roasted, broken up, and delivered at the tunnel-head.

The only difficulty in the way of success at this point seemed to be the distance from a market, transportation to the Mississippi costing at that time $7 50 per ton, and not being practicable except during a part of the year. A plank-road was then constructing to Ste. Genevieve, which would reduce the expense about one-half. Since that time a rail-road has been commenced to connect St. Louis with the iron region, and when this enterprise has been completed, there can be no doubt that the ore will be carried to the river in large quantities.

The Pilot Knob differs considerably from the Iron Mountain in character. It is much higher, by estimate 650 feet above its base, and is mainly composed of a dark silicious rock, distinctly bedded, and dipping to the south at an angle of 25° or 30°. For about two-thirds of the distance to the summit, the quartz rock predominates; above that, the iron is found in heavy beds, alternating with silicious matter. Some of these beds are very wide, and made up of nearly pure micaceous and specular ore. The richest ores show a very evident slaty structure, differing in this respect entirely from those of the Iron Mountain, which are compact and without any noticeable cleavage. The summit of the Pilot Knob is ragged and bare, except where covered by moss, and forms a conspicuous object in the distance; hence the name.

There are numerous other localities in the vicinity, which, although not so well known as those just described, furnish an ore at least equal, if not superior, to any obtained from the Pilot Knob itself. The ore of Shepherd's Mountain is the magnetic, and is much valued. The Bogy Bank furnishes a fine ore, containing occasional druses of quartz. The Russell Bank is a fine-grained peroxide, very pure, and making excellent iron.

There was at the Pilot Knob in 1852 a bloomery with six fires, built in 1848, and a blast furnace. Both together make about 4000 tons of iron a year. The cost of making blooms was stated to me as $30 per ton. The ore cost 20 cents a ton delivered at the roasting-heap; that of Shepherd's Mountain cost 55 cents. The woodland is owned by the company; 35 cents a cord is paid for cutting, and charged to the coalers, who are paid from 2¾ to 3 cents a bushel for the coal delivered at the furnace. There are about 300 persons employed at this place and at the Iron Mountain. The abundance and purity of the ore in the vicinity can hardly be surpassed; and it will eventually be carried in large quantities to the Mississippi and mixed with other ores to be smelted by hard coal. At present, charcoal is abundant and cheap.

There are several furnaces on the Maramec River, working mostly brown hematite ores, but I have no information as to their operations.

At Birmingham, on the Mississippi River, about 120 miles below St. Louis, are valuable deposits of brown hematite; the ore occurs in abundance, and is of good quality. Nothing had been done in 1852 towards developing this important locality.

There seem to be but few, if any, beds of iron ore in the coal-measures of this state, although the explorations have thus far been very slight.

Iowa.—But little is known of the character and extent of the iron ores contained in the Iowa coal formation. The beds of coal are thin, and appear to be hardly workable. Dr. Owen speaks of continuous beds of iron-stones of various qualities, several inches in thickness, in the middle division of the coal-field, but no precise information is given with regard to them. The inference may be drawn, however, that they are hardly likely to be of much value.

Wisconsin.—The iron ores of the Lake Superior region, already noticed as occurring within the limits of the State of Michigan, continue into Wisconsin, under the same circumstances and with the same characters as in the former state. According to Mr. Whittlesey, however, the position of the nearest workable beds thus far discovered is such that they would be of no value at present. The distance from Lake Superior of the best exposures of ores is from eighteen to twenty-eight miles, through an entirely uninhabited country. The abundance of the deposits of iron ores through the whole of this region is so great, that for a long time to come only those which are most favorably situated with respect to navigable water will be worked.

The most promising locality of ore in the state, thus far known, is in Dodge County, at the so-called "Iron Ridge." It is a peroxide, having an oolitic structure, and forming a heavy bed in rocks apparently of the age of the Clinton group of New York. The quality seems to be good and the quantity abundant. A rolling mill and blast furnace have recently been erected here, being the only ones in operation in the state, so far as I have learned.

The iron ores of the Mississippi Valley must eventually become of great importance, since the rapid growth of the population in numbers and wealth indicates an immense future consumption of this metal, which can be nowhere so economically manufactured as here. The extension of a network of railroads through the vast region west of the Mississippi will create a demand for iron, which will not fail to lead to the development of the rich mineral resources of the northwest.

CANADA.—The British Provinces of North America abound in various ores, although their mineral resources have hardly yet begun to be developed. An interesting collection of their ores of iron was exhibited by Mr. Logan, the Provincial Geologist, at the Great Exhibition in 1852. From his account of them, it appears that the magnetic and specular oxides are most abundantly distributed throughout the Provinces. They occur chiefly in a formation consisting of gneiss interstratified with important bands of a highly crystallized limestone, which sweeps through the Province from Lake Huron to Labrador, and connects near the Thousand Islands with the great azoic district of New York, already noticed as so rich in iron ores.

The ore of Canada, as in New York, forms immense beds, interstratified with the gneiss, and dipping at a high angle. In the township of Marmora there is a bed 100 feet in thickness. In Madoc there is one which has been traced for several miles and found to have a breadth of 25 feet. This locality has been worked to some extent. At Myer's Lake, in South Sherbrooke, there is a sixty-foot bed. In South Crosby a mass of ore 200 feet in thickness is also mentioned.

The ores from these localities are usually the magnetic, containing from 60 to 70 per cent. of metallic iron; and, being found in the midst of a region abounding in wood and water-power, they cannot fail to become at some future day of great value. Specular iron ore is also abundant. At Macnab it occurs in a bed 25 feet thick, and most favorably situated in every respect.

In New Brunswick and Nova Scotia the richness and abundance of the ores is also remarkable, and fuel in the form of wood and coal is extensively distributed.

An account, imperfect, it is true, has thus been given of the distribution of the ores of iron through the United States. The magnitude of the subject prevents its being treated with any detail, and unfortunately the materials are wanting to enable us to exhibit the manufacture in its present state of development. The localities of the latter are so widely distributed over a vast extent of country that it is beyond the power of any individual to visit them all, or to collect complete statistics of them. For a general summing up we are obliged to depend on the returns of the Census of 1850, which, unreliable as they confessedly are, may yet be taken as a general guide. And for the purpose of convenient reference, the two annexed tables are reprinted, in a slightly modified form, from Mr. Kennedy's Abstract of the Seventh Census. The first gives detailed statistics of the production of pig iron in the different states of the Union; the second does the same with respect to the manufacture of wrought iron. A third, which has not been deemed of sufficient importance to be presented here in full, exhibits the condition of the manufacture of castings: it shows that there were in operation, in twenty-nine states, 1391 establishments, employing 23,489 persons, working up 356,969 tons of pig and old iron and 9850 tons of ore, and producing 322,745 tons of castings, which were valued, along with other products, at $25,108,155.

## TABULAR STATEMENT OF THE MANUFACTURE OF PIG IRON IN THE UNITED STATES.

| STATES. | Number of establishments in operation. | Capital invested. | Tons of ore used. | Tons of mineral coal. | Bushels of coke and charcoal. | Value of raw material, fuel, &c. | Number of hands employed. | Average wages per month. | Tons of pig iron made. | Value of other products. | Value of entire products. |
|---|---|---|---|---|---|---|---|---|---|---|---|
| Maine, . . . . . . . . | 1 | $214,000 | 2,907 | | 213,970 | $14,939 | 71 | $22 00 | 1,484 | | $36,616 |
| New Hampshire, . . . . | 1 | 2,000 | 500 | | 50,000 | 4,900 | 10 | 18 00 | 200 | | 6,000 |
| Vermont, . . . . . . . | 3 | 62,500 | 7,676 | 150 | 326,437 | 40,175 | 100 | 22 08 | 3,200 | | 68,000 |
| Massachusetts, . . . . | 6 | 469,000 | 27,909 | | 1,855,000 | 185,741 | 263 | 27 52 | 12,287 | | 295,123 |
| Connecticut, . . . . . | 13 | 225,600 | 35,450 | | 2,870,000 | 289,225 | 148 | 26 80 | 13,420 | $20,000 | 415,600 |
| New York, . . . . . | 18 | 605,000 | 46,385 | 20 | 3,000,074 | 321,027 | 505 | 25 00 | 23,022 | 12,800 | 597,920 |
| New Jersey, . . . . . | 10 | 967,000 | 51,266 | 20,865 | 1,621,000 | 332,707 | 600 | 21 20 | 24,031 | | 560,544 |
| Pennsylvania, . . . . | 180 | 8,570,425 | 877,283 | 316,060 | 27,505,186 | 3,732,427 | 9,294 | 21 65 | 285,702 | 40,000 | 6,071,513 |
| Maryland, . . . . . | 18 | 1,420,000 | 99,866 | 14,088 | 3,707,500 | 560,725 | 1,370 | 20 14 | 43,641 | 96,000 | 1,056,400 |
| Virginia, . . . . . . . | 29 | 513,800 | 67,319 | 39,982 | 1,311,000 | 158,307 | 1,129 | 12 76 | 22,163 | | 521,924 |
| North Carolina, . . . . | 2 | 25,000 | 900 | | 150,000 | 27,900 | 31 | 8 00 | 400 | | 12,500 |
| Georgia, . . . . . . | 3 | 26,000 | 5,189 | | 430,000 | 25,840 | 138 | 17 44 | 900 | 28,000 | 57,300 |
| Alabama, . . . . . . | 3 | 11,000 | 1,838 | | 145,000 | 6,770 | 40 | 17 50 | 522 | 5,000 | 22,500 |
| Tennessee, . . . . . . | 23 | 1,021,400 | 88,810 | 177,167 | 160,000 | 254,900 | 1,822 | 12 81 | 30,420 | 41,900 | 676,100 |
| Kentucky, . . . . . . | 21 | 924,700 | 72,010 | | 4,576,269 | 260,152 | 1,855 | 20 23 | 24,245 | 10,000 | 604,037 |
| Ohio, . . . . . . . | 35 | 1,503,000 | 140,610 | 21,730 | 5,428,800 | 630,037 | 2,415 | 24 48 | 52,658 | | 1,255,850 |
| Michigan, . . . . . . | 1 | 15,000 | 2,700 | | 185,000 | 14,000 | 25 | 35 00 | 660 | 6,000 | 21,000 |
| Indiana, . . . . . . | 2 | 72,000 | 5,200 | | 310,000 | 24,400 | 88 | 26 00 | 1,850 | | 58,000 |
| Illinois, . . . . . . | 2 | 65,000 | 5,500 | | 170,000 | 15,500 | 150 | 22 06 | 2,700 | | 70,200 |
| Missouri, . . . . . . | 5 | 619,000 | 37,000 | 55,180 | | 97,367 | 334 | 24 28 | 19,250 | | 314,600 |
| Wisconsin, . . . . . . | 1 | 15,000 | 3,000 | | 150,000 | 8,250 | 60 | 30 00 | 1,000 | | 27,000 |
| Total, . . . . . . | 377 | 17,346,425 | 1,579,309 | 645,242 | 54,165,236 | 7,005,289 | 20,448 | | 564,755 | 259,700 | 12,748,777 |

TABULAR STATEMENT OF THE MANUFACTURE OF WROUGHT IRON IN THE UNITED STATES.

| STATES. | Number of establishments in operation. | Capital invested. | Tons of pig metal. | Tons of blooms used. | Tons of ore used. | Tons of mineral coal. | Bushels of coke and charcoal. | Value of raw material used. | Number of hands employed. | Average wages per month. | Tons of wrought iron made. | Value of other products. | Value of entire products. |
|---|---|---|---|---|---|---|---|---|---|---|---|---|---|
| New Hampshire, . . | 2 | $4,000 | 145 | | | | 50,000 | $5,600 | 6 | $32 00 | 110 | | $10,400 |
| Vermont, . . . | 8 | 62,700 | 750 | 525 | 2,625 | | 337,000 | 66,194 | 57 | 31 05 | 2,045 | | 163,986 |
| Massachusetts, . | 6 | 610,300 | 7,030 | | | 11,022 | 78,500 | 221,194 | 260 | 22 50 | 6,720 | | 428,320 |
| Rhode Island, . . | 1 | 208,000 | 3,000 | | | 6,000 | | 111,750 | 220 | 26 00 | 2,650 | | 222,400 |
| Connecticut, . . | 18 | 529,500 | 7,081 | 1,644 | | 5,062 | 783,600 | 358,780 | 374 | 31 59 | 6,325 | $5,000 | 667,560 |
| New York, . . . | 60 | 1,131,300 | 8,530 | | 44,642 | 13,908 | 5,554,150 | 838,314 | 1,037 | 26 00 | 13,636 | 195,000 | 1,423,968 |
| New Jersey, . . | 53 | 1,016,843 | 10,430 | | 14,549 | 4,507 | 1,994,180 | 320,950 | 593 | 27 78 | 8,162 | | 629,273 |
| Pennsylvania, . . | 131 | 7,620,066 | 163,702 | 20,405 | | 325,967 | 3,939,998 | 5,488,391 | 6,771 | 27 68 | 182,506 | 219,500 | 8,902,907 |
| Delaware, . . . | 2 | 15,000 | 510 | 60 | | | 228,000 | 19,500 | 50 | 24 19 | 550 | | 55,000 |
| Maryland, . . . | 17 | 780,650 | 10,172 | 3,389 | | 10,455 | 246,000 | 439,511 | 568 | 23 33 | 10,000 | | 771,431 |
| Virginia, . . . | 39 | 791,211 | 17,296 | 2,500 | | 66,515 | 103,000 | 591,448 | 1,295 | 23 62 | 15,328 | | 1,254,995 |
| North Carolina, . | 19 | 103,000 | | | 4,650 | | 357,900 | 28,114 | 187 | 10 37 | 850 | | 66,980 |
| Georgia, . . . | 3 | 9,200 | 100 | | | | 76,600 | 5,986 | 27 | 11 35 | 90 | | 15,384 |
| Alabama, . . . | 1 | 2,500 | 120 | | | | 30,000 | 3,000 | 14 | 20 00 | 100 | | 7,500 |
| Tennessee, . . . | 42 | 755,050 | 11,696 | 325 | 9,151 | 62,038 | | 385,616 | 786 | 15 20 | 10,348 | 38,800 | 670,618 |
| Kentucky, . . . | 4 | 176,000 | 2,000 | 1,600 | | | 280,000 | 180,800 | 183 | 32 06 | 3,070 | | 299,700 |
| Ohio, . . . . . | 11 | 620,800 | 13,675 | 2,900 | | 22,755 | 466,900 | 604,493 | 708 | 33 61 | 14,416 | | 1,076,192 |
| Indiana, . . . . | 3 | 17,000 | 50 | | 3,150 | | 85,000 | 4,425 | 24 | 27 45 | 175 | | 11,760 |
| Missouri, . . . | 2 | 42,100 | 1,204 | | | 9,834 | | 24,509 | 101 | 30 00 | 963 | | 68,700 |
| Total, . . . | 422 | 14,495,220 | 251,491 | 33,344 | 78,787 | 538,063 | 14,510,828 | 9,698,109 | 13,257 | | 278,044 | 458,300 | 16,747,074 |

The following is a comparison of the principal results of the Census of 1840, and that of 1850:—*

| | Census of 1840. | Census of 1850. |
|---|---|---|
| Number of blast furnaces in the United States, . | 804 | 377 |
| Tons of pig iron produced, . . . . . . | 286,903 | 564,755 |
| Rolling mills, bloomeries, and forges, . . . . | 795 | 422 |
| Tons of wrought iron produced, . . . . . | 197,233 | 278,044 |

By the Census of 1850, it appears that there were, of the thirty-one states, ten in which no blast furnaces were in operation, and twelve which were without any works for the manufacture of wrought iron. Of these non-producing states, almost all are known to contain abundant deposits of iron ore, and several of them possess extensive coal-fields, so that the time cannot be far distant when some of them will take their places among the first in this branch of industry.

Closely connected with the richness and abundance of our iron deposits, is the extent of our coal-fields, in respect to which we take precedence of all other nations. In Mr. Taylor's work, the coal area known to exist in the United States is set down as 133,132 square miles. This, however, does not include the Iowa and Missouri, and the Arkansas coal-fields. It is true that little is yet known of their actual value, so far as workable thickness and quality of the coal are concerned, but the results of Dr. Owen's survey have given us a better knowledge of the extent of territory underlaid by the coal formation, and it appears that there cannot be less than 50,000 square miles occupied by the coal-measures to the west of the Mississippi, within the boundaries of the organized states: this, added to Mr. Taylor's amount, after deducting 6000, the number allowed by him for Missouri, would give in round numbers 177,000 square miles of coal area within the borders of the thirty-one states. Considering the immense development of our coal-fields, and the quantity of our iron ores of every possible variety, it is difficult to set any limits to our possible production of iron.

* The discrepancies of these two statements are self-evident. The Census of 1840 was notoriously inaccurate and unreliable; that of 1850 is probably somewhat more to be depended on.

The facilities of internal communication by canal or steamboat and railroad, already ample, are daily increasing, and new works are constructing to bring every section of the country into connection with the nearest and most valuable coal and iron districts. It may, indeed, be said with truth, that in our capacity for producing this most important metal we are unrivalled.

The estimates which have been made at various times of our production are as follows:—

| | Tons. | | Tons. |
|---|---|---|---|
| 1810, | 54,000 | 1842, | 215,000 |
| 1828, | 130,000 | 1845, | 486,000 |
| 1829, | 142,000 | 1846, | 765,000 |
| 1830, | 165,000 | 1847, | 800,000 |
| 1831, | 191,500 | 1849, | 800,000 |
| 1832, | 200,000 | 1850, | 600,000 |
| 1840, | 347,000 | | |

The figures for the years from 1810 to 1832 are compiled from the published statements of H. C. Carey and R. C. Taylor. That of 1840 is the result of an inquiry made by the Commission of the Home League in New York. It is larger by 61,000 tons than the amount given by the Census, but is probably nearer the truth. For 1846, the estimate was made by the Secretary of the Treasury, from returns collected officially. The Commission of the Pennsylvania iron masters give 800,000 tons as the production of 1849.

Within the last year or two the manufacture, which had been in a very depressed condition, has greatly increased, and bids fair to equal nearly 1,000,000 tons during the present year.

In closing this chapter, the statistics of the produce of iron throughout the world will be given, in the same form as heretofore. An inconsiderable amount of this metal is manufactured in some of the Asiatic and South American countries, but principally for local consumption, and as no detailed information can be given with regard to the quantity, they have been omitted from the table.

| Year. | Russia. | Sweden. | Norway. | Great Britain. | Belgium. | Prussia. | Saxony. |
|---|---|---|---|---|---|---|---|
| 1800, . . | | | | 180,000 | | | |
| 1810, . . | | | | 294,642 | | | |
| 1820, . . | | | | 368,000 | | | |
| 1825, . . | | | | 581,367 | | | |
| 1830, . . | 179,850 | | | 678,417 | | 44,409 av. | |
| 1835, . . | | 104,785 | | 1,000,000 | | 58,926 " | 4.371 |
| 1840, . . | 168,975 av. | 133,675 | | 1,396,400 | 106.878 | 98,452 " | |
| 1845, . . | | | 8,119 | 1,512,500 | 220,000 | 102,174 " | |
| 1846, . . | | | | | | 115,216 | |
| 1847, . . | | | | 1,998,558 | | 135,731 | |
| 1848, . . | | | | 2,093,736 | | 125.916 | 7.140 |
| 1849, . . | 188,337 av. | 122,925 | | | | 115,254 | 6,591 |
| 1850, . . | | 147,295 | 5,000 | 2,380,000 | | 132,873 | |
| 1851, . . | | | | 2,500,000 | | 145,519 | 6,979 |
| 1852, . . | | | | 2,701,000 | | | |

| Year. | German States. | Switzerland. | Austria. | Italy. | Spain. | France. | United States. |
|---|---|---|---|---|---|---|---|
| 1810, . . | | | | | | | 54,000 |
| 1825, . . | | | 75,783 av. | | | 170,000 | |
| 1830, . . | | | 88,979 " | | | 262,165 | 165,000 |
| 1835, . . | | | 96,138 " | | | 289,985 | |
| 1840, . . | | | 132,597 " | | | 342.278 | 315,000 |
| 1845, . . | | | 172,385 " | | 29,034 | 441,876 | 502,000 |
| 1846, . . | | | | | | 531.750 | 765,000 |
| 1847, . . | 70,000 | | 197,713 | | 40,053 | 522,386 | 800,000 |
| 1848, . . | | | 201,350 | | | | |
| 1849, . . | | | | | 29,715 | 514,172 | 650,000 |
| 1850, . . | | 14.500 | | 22.500 | | | 600.000 |

The present production of this metal in Europe and the United States may be estimated as follows:—

| | Tons. | Relative Amount. |
|---|---|---|
| Russian Empire, . . . . . . . . | 200,000 | 3·4 |
| Sweden and Norway, . . . . . . . | 155,000 | 2·7 |
| Great Britain, . . . . . . . . . | 3,000,000 | 51·6 |
| Belgium, . . . . . . . . . . | 300,000 | 5·2 |
| Prussia, . . . . . . . . . . | 150,000 | 2·6 |
| Saxony, . . . . . . . . . . | 7,000 | ·1 |
| Austrian Empire, . . . . . . . . | 225,000 | 3·9 |
| Rest of Germany, . . . . . . . . | 100,000 | 1·7 |
| Switzerland, . . . . . . . . . | 15,000 | ·2 |
| France, . . . . . . . . . . | 600,000 | 10·3 |
| Spain, . . . . . . . . . . | 40,000 | ·7 |
| Italy, . . . . . . . . . . | 25,000 | ·4 |
| United States, . . . . . . . . . | 1,000,000 | 17·2 |
| | 5,817,000 | 100·0 |

If we adopt Heron de Villefosse's estimate of the production of iron throughout the world in 1808, namely 740,000 tons, we find that the manufacture of this metal has increased nearly eight-fold in the last half century. Great Britain alone produces above one-half of the iron of the world, and next in the list stands our own country, furnishing a little more than one-sixth of the whole amount.

Great as is our production of iron, it by no means suffices for our consumption, as the annexed table of imports for the years 1849–52 will show.

| | Pig, Old and Scrap, Castings. | | Bar and Wrought. | | Steel. | | Manufactures. | Total Value. |
|---|---|---|---|---|---|---|---|---|
| | Tons. | Value. | Tons. | Value. | Tons. | Value. | Value. | |
| 1849, | 115,028 | $1,564,951 | 196,343 | $7,162,640 | 6,349 | $1,172,094 | $5,010,676 | $14,910,361 |
| 1850, | 86,034 | 1,189,469 | 279.875 | 9.042,433 | 6,116 | 1,292,060 | 6,000,497 | 17.524,459 |
| 1851, | 75,837 | 919,526 | 298.024 | 9,228,702 | 7,803 | 1,531,692 | 7,058,182 | 18.738,102 |
| 1852, | 99,654 | 1,052,635 | 356,383 | 10,923,654 | 9,027 | 1,672,030 | 6,846,767 | 20,495,086 |

In order to show from what countries our demand for iron is supplied, the following detailed statement of our imports of this metal for the year ending June 30, 1852, the latest for which the returns have been published, is appended.

| Imported from | Kind. | Tons. | Value. | Value. |
|---|---|---|---|---|
| Great Britain, | pig, | 91,149 | $927,055 | |
| " " | old and scrap, | 6,049 | 81,554 | |
| " " | castings, | 351 | 18.114 | |
| " " | bar, | 318,236 | 8,967,669 | |
| " " | wrought (rods, hoop, sheet), | 19,125 | 798,140 | |
| " " | steel, | 8,550 | 1.629,222 | |
| " " | manufactures, | | 6,065,918 | |
| | | | | 18,487,672 |
| Sweden and Norway, | bar, | 14,104 | 751,050 | |
| " " | steel, | 365 | 22,624 | |
| | | | | 773,674 |
| Russia, | bar, | 142 | 7,984 | |
| " | sheet, | 2,315 | 312,106 | |
| | | | | 320,090 |
| Belgium, | manufactures, | | 424,049 | |
| " | other, | | 698 | |
| | | | | 424,747 |
| France, | manufactures, | | 240,790 | |
| " | other, | | 2,637 | |
| | | | | 243,427 |
| Other countries, | | | | 411,982 |
| | Total, | | | $20,661,592 |

In this statement the unimportant amounts re-exported have been taken no note of, which explains the discrepancy between the sum given here and that in the table above It will be seen that by far the larger portion of this metal which we import comes to us from England. Sweden and Norway furnish a comparatively small quantity of the finer kinds of bar iron, and Russia supplies us with the peculiar sheet iron which she alone has the secret of manufacturing. From Belgium and France we draw a small amount of manufactures of iron and steel, but no unmanufactured iron. From all other sources our supply is entirely insignificant.

That our commerce in iron, however, is not entirely confined to importation, will be made evident by the next-following table, which exhibits the amount and value of the iron and manufactures of iron, the production of this country, exported during the same period of four years. Though the items are small as compared with those of our importation, they are not insignificant, and bid fair steadily to increase in importance.

| | Pig. | Castings. | Nails. | | Wrought. | Manufactures. | Total Value. |
|---|---|---|---|---|---|---|---|
| | Tons. | Value. | Tons. | Value. | Tons. | Value. | |
| 1849, . . | 83 | $60.175 | 1,400 | $149,358 | 251 | $886.639 | $1,110,500 |
| 1850, . . | 25 | 79,318 | 1,703 | 154,210 | 388 | 1,677,792 | 1,931,500 |
| 1851, . . | 351 | 164,425 | 2,366 | 215,652 | 215 | 1,875.621 | 2,273,500 |
| 1852, . . | 105 | 191.388 | 1,620 | 118,624 | 8,316 | 1,993,807 | 2,720,000 |

Nothing is better calculated to impress us with the immense material progress of the United States than the view thus afforded of our consumption of iron, and the rapidity of its increase. In 1852 we imported nearly half a million tons of iron and steel in their various forms; our consumption must have amounted in 1849 to over 1,100,000 tons, and has been increasing since that with rapidity, so that it cannot now fall much below 1,500,000 tons.

# CHAPTER X.

## METALS NOT USED IN THEIR SIMPLE METALLIC FORM.

FOR convenience of description, the metals included under this head will be arranged in the following order:—

I. Metals used chiefly in alloy with other metals. BISMUTH, ANTIMONY, NICKEL.

II. Metals used chiefly in a non-metallic form. COBALT, ARSENIC, MANGANESE, CHROMIUM, TITANIUM, MOLYBDENUM, URANIUM, TUNGSTEN.

As a matter of course, these metals are of very minor importance, compared with those which have been hitherto described; but as they are all more or less objects of commerce, and all occur in this country, it would not be proper to pass them by unnoticed in this work. Their production is quite limited in quantity, and dependent on the demand, and statistics in regard to them cannot usually be given.

### BISMUTH.

Bismuth is a metal which possesses two qualities rendering it of no value by itself in the metallic state; it is brittle, and fuses at a very low temperature, 476°, F. It has a grayish-white color, with a slight tinge of red, and a crystalline fracture. No metal shows a greater tendency to take the crystalline form; beautiful crystals may be obtained by fusing a quantity in a Hessian crucible, allowing it to cool very gradually, and then piercing a hole in the solidified crust and pouring off the still liquid metal, when the interior will be found lined with crystals. The bismuth of commerce is never entirely pure: it contains silver, sometimes in considerable quantity, iron, lead, arsenic, and other substances.

*Native Bismuth.* The native metal is found chiefly in

foliated masses and scales; also in reticulated and arborescent forms. The bismuth of commerce is chiefly obtained from this source. Its principal localities are in the Saxon and Bohemian mines of the Erzgebirge.

*Bismuthine.* A sulphuret of bismuth containing 18·49 of sulphur, and 81·51 of the metal. It is found in Cornwall and Cumberland, Eng., at Altenberg and Schneeberg, in Saxony, and elsewhere, but not in sufficient abundance to be considered valuable as an ore.

*Bismuth Ochre.* An oxide of the metal, which occurs in small quantity, in a pulverulent form. It accompanies other ores of bismuth at the Saxon mines.

*Bismuth Blende.* This is, mainly, a silicate of bismuth.

*Bismutite.* A carbonate of bismuth; this and the last-mentioned ore occur at Schneeberg, and are the products of decomposition of the native metal.

Bismuth is also found, in nature, in combination with copper and sulphur, as *cupreous bismuth;* with lead, copper, and sulphur in the mineral known as *aciculite* or needle ore, from its delicate acicular crystallizations; with tellurium in the mineral *tetradymite* and *Bornite.*

All the ores of bismuth occur in small quantity, although found in numerous localities. Nothing can be more simple than the metallurgic treatment of this metal; it is separated from the ore by fusion in iron tubes laid obliquely across the furnace, and from which it flows down into iron pots.

The metal bismuth is of quite modern use; it was first recognized as a distinct metal about the middle of the sixteenth century, by the distinguished mineralogist and miner, George Bauer, commonly called Agricola. Its introduction into commerce and the arts dates from the latter part of the last century. The principal consumption of it is in some kinds of type and stereotype metal, and to impart fusibility to various alloys of lead and tin. Its only use in the non-metallic form is as a cosmetic; pearl powder, or the subnitrate of this metal being an article of commerce.

The principal supply of bismuth is from Saxony, where it occurs with ores of cobalt, nickel, silver, and other metals. At the furnace in the Schneeberg district, the production in

1851 was 20,250 lbs., of which 17,850 lbs. were sold for about $8,250. The nickel works of Oberschlema produced during the same year 4700 lbs. of this metal.

In this country, bismuth has been found in several localities, although only in small quantity. At Lane's Mine, in Monroe, Connecticut, the native metal occurs associated with wolfram, tungstate of lime, galena, and blende. It has also been found at Brewer's Mine, in Chesterfield District, South Carolina. The sulphuret is said to have been observed at Haddam, Connecticut. Carbonate of bismuth occurs in some of the South Carolina gold mines in small quantity, and the telluret has been observed in Virginia and North Carolina.

## ANTIMONY.

Antimony is a metal of considerable importance in the arts, although never used in its simple metallic state. It is of a silvery-white color, with a faint tinge of blue, has a brilliant lustre and a coarsely crystalline fracture, and is very brittle. It fuses at a temperature a little above that of zinc; if strongly heated, it boils and is volatilized. From its property of rendering the softer metals harder and more brittle when alloyed with them, even in small quantity, it is used in a variety of combinations.

*Native Antimony.* The native metal is not of frequent occurrence. It was first discovered in the Sala lead mines of Sweden, and has also been found in the Harz, at Przibram in Bohemia, and other places.

*Antimony Glance*, Sulphuret of Antimony. This ore contains 27·11 of sulphur and 72·89 of antimony. It is the ore from which nearly all the antimony of commerce is obtained.

*Berthierite* is a mixture of the sulphurets of iron and antimony. The metal produced from this ore is of inferior quality.

*Arsenical Antimony.* An arseniuret of antimony; a rare ore.

Two combinations of antimony with oxygen occur sparingly in nature. They are, *White Antimony*, which contains one equivalent of antimony to three of oxygen, or 15·68 of oxygen to 84·32 of metal, and *Stiblite*, an oxide with four atoms of oxygen to one of metal, together with a small percentage of water.

*Red Antimony* is a combination of the oxide and sulphuret of antimony.

A great number of other minerals contain a considerable proportion of this metal in the form of antimonio-sulphurets. Some of them are enumerated under lead. There are also compounds of antimony and nickel, but none are of importance as ores of antimony.

The supply of antimony comes chiefly from Borneo, where the quantity of the ore is said to be unlimited. About 1400 tons of this metal are said by M'Culloch to be annually exported from that island to Singapore, where it is sold at from 16*s.* to 20*s.* per ton.

In this country the ores of antimony appear to be quite rare. It is said to occur in small quantity at Carmel, Penobscot County, Maine, and at Cornish and Lyme in New Hampshire. For the present, at least, no country can compete with Borneo in supplying the demand for it.

Prussia and Austria, however, furnish a small quantity; the former country produced, in 1849, 73,500 lbs., and the latter, in 1848, 539,000 lbs.

The alloys of antimony used in the arts are numerous and valuable. The most common is type-metal, which is a mixture of lead with from 17 to 20 per cent. of antimony. The proportion of the latter metal is diminished as the size of the type increases. Tin, bismuth, or copper, are occasionally added in small quantity. Britannia-ware is a variable compound of copper, tin, brass, and antimony. Antimony is also in common use as an ingredient in the alloy known as "anti-friction metal," for machinery bearings, pillow-blocks, railway locomotive axle-boxes, and the like. These are mostly alloys of antimony and tin. An English locomotive-bearing metal yielded, on analysis, 26 per cent. of antimony, 72 of tin, and 2 of copper. A patent alloy for sheathing consists of lead with about three per cent. of antimony. It does not seem to have come into use. Antimony is said to harden and improve the quality of pewter, but is not generally used in that very common alloy.

The use of antimony in medicine is well known.

## NICKEL.

The occurrence of metallic nickel in nature is confined exclusively to bodies of extra-terrestrial origin, commonly called meteoric iron. Such masses are not unfrequently found, and consist of an alloy of iron and nickel, the latter metal forming usually from five to ten per cent. of the whole. This metal, as produced artificially, is hard, takes a high polish, is quite ductile, and has a specific gravity of from 8·2 to 8·6. Its color is white, with a light shade of gray.

The following minerals contain it in combination.

*Millerite, Capillary Pyrites.* A sulphuret of nickel, containing sulphur 35·63 and nickel 64·37. It occurs usually in delicate capillary crystallizations. Not sufficiently abundant to be valuable as an ore.

*Copper Nickel, Kupfer Nickel.* An arseniuret of nickel, which contains arsenic 55·98 and nickel 44·02. This is the most important ore of this metal. It frequently contains a little iron, lead, and sulphur, also cobalt.

*Breithauptite,* Antimoniuret of Nickel; very rare.

*Rammelsbergite,* Arseniuret of Nickel; occurs occasionally with other ores of this metal.

*Nickel Glance.* A number of different minerals are included under this name; they consist mainly of arsenic, sulphur, nickel, and iron. This is one of the ores most usually found in sufficient abundance to be worked.

*Ullmannite.* A sulpharseniuret of nickel, containing about 25 per cent. of the latter. It is found in some quantity in the German mines.

*Placodine,* Arseniuret of Nickel; rare.

*Bismuth Nickel;* a rare combination.

*Emerald Nickel* is a carbonate of this metal, and *Nickel Green* an arseniate; both of them are products of decomposition of the usual ores.

The two metals cobalt and nickel have a remarkable similarity to each other, and this resemblance extends to their combinations; they are almost invariably found in company.

Nickel is not only obtained from its ores proper, but is also separated from the peculiar arsenical product of the metallurgic treatment of nickeliferous ores of lead, copper, and cobalt, which is called *sperse*. It is only since the discovery of the alloy called in this country German-silver, and in Germany new-silver, or argentan, that this metal has become an object of much importance. The separation of it from its ores is a complicated and difficult process, of which many of the details are kept secret by the manufacturers. The best German-silver is an alloy of 8 parts of copper, 3 of nickel, and 3½ of zinc. The larger the quantity of nickel, the harder and whiter the alloy. The proportions used differ greatly, and there are various names for the different products. The use of nickel was probably introduced from China; their *white copper* contains 31·6 nickel, 40·4 copper, 25·4 zinc, and 2·6 iron.

A very large quantity of nickel is manufactured in Birmingham; the ores there worked are obtained principally from Norway and Hungary. In the former country, a mine has been worked at Espedalen for some time, by an English house, and, in two years, 370 tons of ore have been sent to England. On the Continent, the manufacture of nickel is carried on in Saxony and Prussia. In Saxony, the amount produced in 1851 was 20,540 lbs.; in Prussia, it varied, between 1847 to 1849, from 4500 to 9000 lbs.

The principal locality of the ores of nickel and cobalt in this country is at Chatham, Connecticut, where they are found in veins traversing gneiss and mica slate. The minerals occurring there are copper, nickel, smaltine, and Rammelsbergite. A company has been recently organized to open a mine at this place. In their printed report, it is stated that the washed ores yield from 13 to 18 per cent. of a mixture of the oxides of cobalt and nickel, in nearly equal quantities. The ore as taken from the lode is stated to contain 2·2 per cent. of the mixed oxides.

The existence of these ores at this point has been long known, and a vein has been wrought here at different times.

Governor Winthrop is believed to have been the discoverer of the locality. Later, in 1787, a quantity of the cobalt ore was taken out by a Mr. Erkeleus, and still later, in 1818, a considerable amount was expended here by Mr. Seth Hunt, and more recently by Prof. Shepherd, the State Geologist. In his report, the vein is described as being probably about one foot wide, and consisting of an aggregate of quartz, garnet, and hornblende. The principal ore was smaltine, and it was accompanied by copper, nickel, blende, galena, and a little copper pyrites. At the time this mine was most worked, nickel was very little known, and had hardly any value. If the ore is really rich in nickel, it may be of value, as this metal is now much used, and would be more so if it could be furnished in such abundance as to reduce the price. A small quantity of the ores of nickel has been obtained in connection with those of cobalt, at Mine La Motte; the supply from this locality is very limited.

## COBALT.

Cobalt does not occur in the native state, but is usually found in combination with arsenic or sulphur, and sometimes with antimony and bismuth. It is never used in the arts in its metallic state; but as prepared artificially, the metal is a hard and slightly malleable substance, of a reddish gray color, and a specific gravity of about 8·5. The only use to which cobalt has been applied in the arts, as yet, depends on the property which its oxide has of imparting a beautiful and permanent blue to glass.

The principal combinations in which cobalt occurs are as follows:—

*Syepoorite*, Subsulphuret of Cobalt; containing sulphur 35·36, and cobalt 64·64; it occurs only in India, where it is used by the natives.

*Smaltine*, Gray Cobalt; an arseniuret of cobalt, containing, when pure, arsenic 71·81, and cobalt 28·19; but a portion of the cobalt is generally replaced by nickel, and always by a portion of iron. This is the chief ore of cobalt, and it occurs in numerous localities.

*Cobaltine*, Cobalt Glance. This is a sulpharseniuret of

cobalt, containing sulphur, 19·35, arsenic, 45·18, and cobalt, 35·47; it almost always has more or less iron in its composition. This is another important ore which occurs at most of the cobalt localities.

*Glaucodote.* This is an ore containing a sulpharseniuret of cobalt and iron; it occurs in Chili, in the province of Huasco.

*Linnæite*, Cobalt Pyrites. A sulphuret of cobalt, containing sulphur, 44·98, and cobalt, 55·02. It is frequently associated with the other cobalt ores. *Skuterudite* is another sulphuret, with 79·26 of arsenic, and 20·74 of cobalt.

*Cobalt Vitriol*, Sulphate of Cobalt; found as a product of the decomposition of other cobalt ores, in small quantity.

*Cobalt Bloom.* A hydrated arseniate of cobalt, containing 38·43 arsenic acid, 37·55 oxide of cobalt, and 24·02 water. This is another product of the decomposition of the sulpharseniurets.

Before the introduction of artificial ultramarine, the manufacture of smalt and zaffre was of very considerable importance. It was most extensively carried on at Modum in Norway, and in Saxony.

The first discovery of cobalt in the parish of Modum was made in 1772, on the estate of Skuterud, which was purchased by the king, and works were established there in 1783, under German management.* From 1827 to 1840, the manufacture was carried on here by a private company, and was well-managed and productive, until the introduction of artificial ultramarine completely ruined the business.

The ores of this locality occur in fahlbands in the gneiss, a mode of occurrence characteristic of the sulphurets in Scandinavia. There are two groups of mines, but both are in the same metalliferous belt. The chief ore is cobaltine or cobalt glance, which frequently occurs crystallized, and cobaltiferous mispickel, which sometimes contains as much as 10 per cent. of cobalt. The ores of copper also abound, especially copper pyrites and variegated copper ore. Besides this, there is a great variety of other interesting minerals.

* Böbert, Karsten and Dechen's Arch. xxi. 207.

The dressing of the ores is attended with great difficulty, since it requires an experienced eye to detect the minute particles of ore scattered through the quartz veinstone. Some of the lowest quality yield only 1 or 1¼ per cent. clear ore, or ⅝ to ¾ per cent. of regulus. The richest and purest ores, after calcination, are smelted with quartz and potash in variable proportions, in order to form zaffre and smalt, which are nothing but potash glass colored by cobalt, and ground fine. They differ only in the degree of fineness of the powder, smalt being coarser and lighter colored. The poorer ores are subjected to a much more complicated process, the object of which is to concentrate the metal, and procure it in the form of a regulus, which is then treated in nearly the same manner as the rich ores. A large quantity of arsenic is one of the results of the smelting processes, and is a great annoyance, since it is of no value, and is in no way to be got rid of. The different shades of smalt are distinguished by letters.

These mines are said to have made $500,000 profit in the twenty years of their prosperity. In the year 1838–9, smalt and zaffre to the value of 136,547 silver species was prepared for sale. But such was the effect of the introduction of artificial ultramarine, that in 1847 the sales only reached the amount of about $35,000. The demand for smalt and zaffre in England is still said to be considerable.

The works and all the property were sold at auction in July 1849, and purchased by an English firm for $130,000. The stock of oxide of cobalt sold for a little over $2 per pound. What disposition has since been made of the works has not yet transpired.

The cobalt works on the continent are also in a very depressed state. In the Saxon cobalt works not a pound was sold during the year 1851.

Chatham, in Connecticut, which has already been noticed, is one of the principal localities of this metal in the United States. The Patapsco Mine, in Maryland, is another locality where unsuccessful attempts have been made to work cobaltiferous ores. It is evident that in the present state of the cobalt business, the demand for this metal being so limited, it will be impossible for manufacturers, on any considerable scale, to compete with the European establishments already built, and supplied with an abundance of ore, where the labor is much cheaper and the skill greater than here. If the cobalt ores contain nickel, they may be valuable, as the demand for that metal is considerable. A small quantity of the cobalt and nickel ore of Mine La Motte has been worked in Philadelphia, but with what success I am unable to state.

## ARSENIC.

This metal occurs in the native state. It has a tin-white color, which tarnishes to dark-gray, and an uneven and granular fracture: its specific gravity is between 5·6 and 5·9. It is found in the argentiferous veins of the Erzgebirge in considerable quantity; also in the Harz, and in the mines of Transylvania and the Banat.

The form in which arsenic is usually seen is the powder called white arsenic; this is arsenious acid, a combination of the metal with oxygen, containing 24·24 of oxygen, and 75·76 of arsenic. It occurs in nature, but not abundantly, being an occasional product of the decomposition of arsenical ores.

The most abundant arsenical combinations are: *realgar*, or red orpiment, a sulphuret, which contains 70·03 per cent. of the metal, and 29·97 of sulphur. It occurs, sometimes finely crystallized, with ores of silver and lead, especially in Hungary, Transylvania, and in the Erzgebirge.

*Orpiment* is the yellow sulphuret of arsenic, containing 39 of sulphur, and 61 of arsenic. It occurs in considerable quantity in Koordistan, also in the Hungarian and Transylvanian mines.

Arsenic is found in combination with the other metals in great abundance, but is most frequently associated with iron. The mineral called *Leucopyrite* is an arseniuret of iron, containing 72·82 of arsenic and 27·18 of iron. *Mispickel*, an arseniuret and sulphuret of iron, contains arsenic 46·01, sulphur 19·64, and iron 34·35. It is found in great abundance in the older crystalline rocks, associated with ores of silver, lead, and especially tin.

Arsenic is used in a variety of forms and for many different purposes. In combination with potash it is consumed in large quantities in calico-printing establishments. It is an ingredient in several pigments.

It is frequently associated with the ores of cobalt and nickel, and hence is sometimes prepared at the same establishments where those substances are manufactured. Saxony and Silesia furnish the principal portion of the arsenical salts used in the arts.

The fabrication of these substances has to be conducted with great skill and caution, to prevent the workmen from being injured by inhaling the poisonous vapors; and it is only where the treatment of a variety of the cobalt, nickel, and arsenical ores is carried on in the large way, that the business can be made at all profitable.

In this country the arsenical combinations of iron are very abundant, but they are not at present of any value whatever

## MANGANESE.

This substance in the metallic state is hardly known. It is gray and brittle, and resembles cast iron. It never occurs native; but in combination with oxygen, and occasionally sulphur, it is almost as universally diffused as iron itself, with which metal it seems to be intimately related, since it is rare to find an iron which does not contain at least a trace of manganese.

The sulphurets of manganese are comparatively rare, and of no value as ores. The oxides are numerous, and furnish the manganese of commerce.

*Pyrolusite* contains manganese 63·4 and oxygen 36·6. This is the principal ore, and is found in great abundance in many localities.

*Hausmannite* is an oxide containing manganese 72·4 and oxygen 27·6. It is not an ore, and is quite rare.

*Braunite;* another oxide, containing manganese 69·75 and oxygen 30·25. It is of no value as an ore.

*Manganite* is a hydrated oxide which contains about ten per cent. of water.

*Psilomelane* is a mixture of the superoxide with various

earthy bases, and usually contains a little water; it is one of the most common ores of manganese, and occurs with pyrolusite.

*Wad*, or bog manganese, consists mainly of the oxides of manganese and water, with some oxide of iron, and a varying amount of earthy substances. It is a secondary product of the decomposition of other manganesian and ferriferous ores. Cobalt is frequently found associated with this substance. In this form it occurs at Mine La Motte. The bog manganese is of little value for any purpose. The silicates of manganese are of no value at present.

The manganese of Cornwall and Devon was formerly of a good deal of importance, and the value raised was estimated by De la Beche in 1839 at £40,000 per annum. The chief mines were near Tavistock and Launceston. The deposits are mostly superficial.

At present, Nassau supplies almost all the manganese used in the arts. The deposits there are very extensive, and the quality excellent. In 1852, 3291 tons of this substance were brought to Liverpool from that country. Its average price was £7 per ton.

Bog manganese is very abundant in this country, but there are few localities where any considerable quantity of a really valuable article is found. The most important localities are in Vermont, where the gray ore has been mined to some extent.

Manganese is used very extensively in the arts to decompose muriatic acid, and thus furnish chlorine for various purposes, especially for manufacturing bleaching powders. It is also used to correct the green tinge of glass containing too much iron. The peroxide of manganese disengages oxygen when heated, and is an important source from which that gas is obtained for various purposes.

## CHROMIUM.

The only combination of this metal which is sufficiently abundant to be considered as an ore, is chromic iron or chromate of iron, as it is usually called. It is a compound of the oxides of chromium and iron. The metal itself is hardly known. It is not found in the native state.

The bichromate of potash is much used in calico printing, and the chromate of lead is one of the most beautiful and valuable yellow pigments. The oxide of chromium is also used in the arts, and furnishes a fine green for porcelain painting.

The ore from which these various salts of the metal are prepared is quite abundantly distributed in this country. The principal locality is the Bare Hills, near Baltimore, where it has been mined in considerable quantity in serpentine. It also occurs in several other places in Maryland; in Chester and Lancaster Counties in Pennsylvania; at Hoboken, New Jersey; at Jay, Newfane, Troy, and Westville, in Vermont; and at Chester and Blanford in Massachusetts.

At Jay, Vermont, the quantity is said by the State Geologist to be large. There is also a vein about one mile south-east of the farm of Mr. Price, on the east side of the Missico River. An analysis of the Jay ore by T. S. Hunt gave:—

| | |
|---|---|
| Green Oxide of Chromium, | 49·90 |
| Protoxide of Iron, | 48·96 |
| Alumina, Silica, and Magnesia, | 4·14 |
| | 100·00 |

## TITANIUM.

Titanium occurs principally in the form of rutile, or titanic acid, a combination of 38·86 per cent of oxygen, with 61·14 of the metal. This mineral is not rare, but is not often found in any considerable quantity. Titaniferous iron, a mixture in varying proportions of the oxides of titanium and iron, is a very common substance. The combinations of titanium have been, as yet, but little used in the arts. Oxide of titanium is employed to give a flesh color to artificial teeth, and in porcelain painting to a very limited extent. The demand for it is so small that it can hardly be said to be an article of commerce.

## MOLYBDENUM.

Molybdenum occurs in nature principally in the form of

the sulphuret, or *molybdenite*, a substance considerably resembling graphite in its external characters. It has hardly been introduced into use in the arts; but one of its combinations is of considerable value as a reagent for the analytical chemist.

### URANIUM.

This metal does not occur in the native state. It is most commonly found in the form of *pitch-blende*, an oxide, which generally contains more or less iron, lead, and other foreign substances. The other combinations in which it exists are somewhat numerous, but occur only in minute quantities.

The only use to which uranium has yet been put, is to furnish a fine black for porcelain painting.

### TUNGSTEN.

Tungsten is interesting from the circumstance, that in various forms of combination it frequently accompanies the ores of tin. It has not yet, however, been made subservient to the purposes of the arts.

# CHAPTER XI.

## GENERAL SUMMARY.

In the preceding chapters of this work, each metal has been taken up, and after a discussion of the mineralogical and geological occurrence of its ores, their distribution throughout the world has been briefly described, the idea of geographical and political divisions being subordinated to that of the metal under consideration.

There is another mode of treating the subject of the metallic resources of the world, which might have been adopted, according to which each country should have been taken up successively, its mines described together, and its statistics embodied in one table. Such a method would not have been without its advantages, but the object of the work being pre-eminently to illustrate our own resources by comparison with those of other countries, the system which has been followed seemed preferable. By combining, however, the statistics of each country given under the head of the several metals, as complete an idea of its resources may be obtained as is possible to be given by figures alone.

Before closing, therefore, it has seemed advisable to embody in a tabular form the general results which have been arrived at, so that the reader may obtain, at one glance, a view of the comparative capacity of production of each country, for each metal, and for all the metals together. For this purpose, two tables have been prepared; in the first, the estimated amount in weight of the metals annually produced throughout the world at the present time is given; in the second, the same estimates are repeated, the weights being converted into values.

## TABULAR STATEMENT OF THE ESTIMATED AMOUNT OF METALS PRODUCED THROUGHOUT THE WORLD IN THE YEAR 1854.

| | GOLD. | SILVER. | MERCURY. | TIN. | COPPER. | ZINC. | LEAD. | IRON. |
|---|---|---|---|---|---|---|---|---|
| | lbs. troy. | lbs. troy | lbs. av. | Tons. | Tons. | Tons. | Tons. | Tons. |
| Russian Empire, | 60,000 | 58,000 | | | 6,500 | 4,000 | 800 | 200,000 |
| Sweden, | 2 | 3,500 | | | 1,500 | 40 | 200 | 150,000 |
| Norway, | | 17,000 | | | 550 | | | 5,000 |
| Great Britain, | 100 | 70,000 | | 7,000 | 14,500 | 1,000 | 61,000 | 3,000,000 |
| Belgium, | | | | | | 16,000 | 1,000 | 300,000 |
| Prussia, | | 30,000 | | | 1,500 | 33,000 | 8,000 | 150,000 |
| Harz, | 6 | 30,000 | | | 150 | 10 | 5,000 | |
| Saxony, | | 60,000 | | 100 | 50 | | 2,000 | 7,000 |
| Rest of Germany, | | 3,000 | | | | | 1,000 | 100,000 |
| Austrian Empire, | 5,700 | 90,000 | 500,000 | 50 | 3,300 | 1,500 | 7,000 | 225,000 |
| Switzerland, | | | | | | | | 15,000 |
| France, | | 5,000 | | | | | 1,500 | 600,000 |
| Spain, | 42 | 125,000 | 2,500,000 | 10 | 500 | | 30,000 | 40,000 |
| Italy, | | | | | 250 | | 500 | 25,000 |
| Africa, | 4,000 | | | | 600 | | | |
| South Asia and East Indies, | 25,000 | | | 5,000 | 3,000 | | | |
| Australia and Oceanica, | 150,000 | 8,000 | | | 3,500 | | | |
| Chili, | 3,000 | 250,000 | | | 14,000 | | | |
| Bolivia, | 1,200 | 130,000 | | } 1,500 | } | | | |
| Peru, | 1,900 | 300,000 | 200,000 | } | } 1,500 | | | |
| Equador, New Grenada, &c., | 15,000 | 13,000 | | | } | | | |
| Brazil, | 6,000 | 700 | | | | | | |
| Mexico, | 10,000 | 1,750,000 | | | | | | |
| Cuba, | | | | | 2,000 | | | |
| United States, | 200,000 | 22,000 | 1,000,000 | | 3,500 | 5,000 | 15,000 | 1,000,000 |
| | 481,950 | 2,965,200 | 4,200,000 | 13,660 | 56,900 | 60,550 | 133,000 | 5,817,000 |

## ESTIMATED VALUE OF METALS PRODUCED IN 1854.

| | GOLD. | SILVER. | MERCURY. | TIN. | COPPER. | ZINC. | LEAD. | IRON. |
|---|---|---|---|---|---|---|---|---|
| Russian Empire, | $14,880,000 | $928,000 | | | $3,900,000 | $440,000 | $92,000 | $5,000,000 |
| Sweden, | 496 | 56,000 | | | 900,000 | 4,400 | 23,000 | 3,750,000 |
| Norway, | | 272,000 | | | 330,000 | | | 125,000 |
| Great Britain, | 24,800 | 1,120,000 | | $4,200,000 | 8,700,000 | 110,000 | 7,015,000 | 75,000,000 |
| Belgium, | | | | | | 1,760,000 | 115,000 | 7,500,000 |
| Prussia, | | 480,000 | | | 900,000 | 3,630,000 | 920,000 | 3,750,000 |
| Harz, | 1,488 | 480,000 | | | 90,000 | 1,100 | 575,000 | |
| Saxony, | | 960,000 | | 60,000 | 30,000 | | 230,000 | 175,000 |
| Rest of Germany, | | 48,000 | | | | | 115,000 | 2,500,000 |
| Austrian Empire, | 1,413,600 | 1,440,000 | $250,000 | 30,000 | 1,930,000 | 165,000 | 805,000 | 5,625,000 |
| Switzerland, | | | | | | | | 375,000 |
| France, | | 80,000 | | | | | 172,500 | 15,000,000 |
| Spain, | 10,416 | 2,000,000 | 1,250,000 | 6,000 | 300,000 | | 3,450,000 | 1,000,000 |
| Italy, | | | | | 150,000 | | 57,500 | 625,000 |
| Africa, | 992,000 | | | | 360,000 | | | |
| South Asia and East Indies, | 6,200,000 | | | 3,000,000 | 1,800,000 | | | |
| Australia and Oceanica, | 37,200,000 | 128,000 | | | 2,100,000 | | | |
| Chili, | 744,000 | 4,000,000 | | | 8,400,000 | | | |
| Bolivia, | 297,600 | 2,080,000 | | 900,000 | | | | |
| Peru, | 471,200 | 4,800,000 | 100,000 | | 900,000 | | | |
| Equador, New Grenada, &c., | 3,720,000 | 208,000 | | | | | | |
| Brazil, | 1,488,000 | 11,200 | | | | | | |
| Mexico, | 2,480,000 | 28,000,000 | | | | | | |
| Cuba, | | | | | 1,200,000 | | | |
| United States, | 49,600,000 | 352,000 | 500,000 | | 2,100,000 | 550,000 | 1,725,000 | 25,000,000 |
| | 119,523,600 | 47,443,200 | 2,100,000 | 8,196,000 | 34,140,000 | 6,660,500 | 15,295,000 | 145,425,000 |

The estimates here presented are based on the assumption that the present year, for which they are made, would be one of normal production, the fluctuations being due to circumstances naturally connected with the development of this branch of industry, and not to extraordinary political changes: such, for instance, as might be a general European war, which should draw off any considerable portion of the miners, or check the exportation and consumption of the metals.

The value of the estimates for various countries is very different. In some of the well-regulated states of Europe, where the mining interest is under the especial protection of the government, the most minute record of its progress is kept; and as there is little fluctuation of production from year to year, the amounts given are undoubtedly very near approximations to the truth. For Sweden, Norway, Prussia, Saxony, Austria, and France, the official returns are very full, although not always procurable in this country down to a very recent date, nor, indeed, always published at once. As to Great Britain, the greatest metal-producing country of the world, its production of some of the metals was, until recently, very imperfectly known. The public ticketings of copper ores in Cornwall and at Swansea have furnished the means of determining pretty accurately the quantity of copper annually raised; it is chiefly owing to the exertions of Mr. R. Hunt, Keeper of Mining Records, that a fair estimate of the production of the other metals can be formed. In regard to Spain, our information is mostly derived from the accounts of French and German mining engineers, who have travelled or held official stations in that country; as no statements are published by the government, and the production is somewhat fluctuating, the estimates are to be taken as approximations only.

Passing from Europe to Asia and South America, the difficulty of presenting figures which can lay any claim to accuracy is greatly increased. Of only one country, Chili, have we anything approaching to a complete picture of its metallic resources. Something is known, however, of the production of silver and gold in other South American countries likewise; and as the metals of less value are almost entirely neglected, the want of information respecting them is of less consequence.

To come nearer home, we find as great difficulties to contend with in this country as anywhere else in forming estimates. We have no office where such information is recorded, as in England, and the government has no official connection whatever with the development of our mining interests. Nor has the importance of recording the facts belonging to this branch of our industry been generally acknowledged.

It has been exceedingly difficult to obtain desired information without personally visiting each locality, and this, of course, has not in the majority of instances been practicable. With regard to the two metals of predominating importance, gold and iron, the data on which our estimates are based have already been given with sufficient detail. In the case of the other metals, also, the amounts stated will not probably be found to vary much from the truth, as they are the results of extensive inquiries.

In constructing the second table, that of values, the prices of the metals have been taken pretty nearly at an average of English prices for the last few months. It must be remarked, however, that these are only the values of the raw material, which may bear but a small proportion to its worth when manufactured and brought into the shape in which it is exported.

The increase which may be thus given to the value of iron, for instance, is out of all proportion to that which it originally had as pig iron, in which character it appears in the tables above. This distinction must be borne in mind when drawing comparisons between the production of different states. If one furnishes only gold, which demands no labor or expenditure of capital to bring it into the form in which it is to be consumed, such a production will add vastly less to the real wealth of the country than would an amount of a cheaper metal much less in value originally, but which has been wrought into a form by which its price has been raised to ten or perhaps even a hundred fold that of the raw material.

In order still farther to illustrate the subject of the comparative value of the metallic productions of different countries, the following statement is appended, from which may be seen the ratio of their production, as compared, first, with that of this country taken as the unit, and, secondly, with that of Great Britain.

| | Value of metals produced. | Ratio of production to that of United States. | Great Britain. |
|---|---|---|---|
| United States, . . . . . . . . | $79,827,000 | 1· | 5-6 |
| Great Britain, . . . . . . . . . . | 96,169,800 | 1·205 | 1 |
| Australia, . . . . . . . . . . | 39,428,000 | ·494 | 5-12 |
| Mexico, . . . . . . . . . . . | 30,480,000 | ·382 | 1-3 |
| Russian Empire, . . . . . . . | 25,240,000 | ·316 | 4-15 |
| France, . . . . . . . . . . . | 15,252,500 | ·191 | 1-6 |
| Chili, . . . . . . . . . . . . | 13,144,000 | ·165 | 2-15 |
| Rest of South America, . . . . . | 16,176,000 | ·203 | 1-6 |
| Austrian Empire, . . . . . . . | 11,708,600 | ·147 | 1-8 |
| Prussia, . . . . . . . . . . . | 9,680,000 | ·121 | 1-10 |
| Belgium, . . . . . . . . . . | 9,375,000 | ·118 | 1-10 |
| Spain, . . . . . . . . . . | 8,016,416 | ·100 | 1-12 |
| Sweden and Norway, . . . . . . | 5,460,896 | ·068 | 1-17 |
| Saxony, . . . . . . . . . . | 1,455,000 | ·018 | 1-67 |
| Harz, . . . . . . . . . . . | 1,147,588 | ·014 | 1-86 |
| Italy, . . . . . . . . . . . | 832,500 | ·010 | 1-120 |
| Switzerland, . . . . . . . . . | 375,000 | ·005 | 1-240 |

The great importance of our own metallic resources will be at once apparent from an inspection of the above table. It will be seen that we are second only to Great Britain in our production, as we are also in our consumption, of the metals. The two great Anglo-Saxon countries stand far before all others; and Australia, a colony of England of but a few years' growth, is the next competitor on the list. As our production of gold, which now forms so important an item of our metallic wealth, falls off, as it assuredly will, the deficiency may be more than made up by the development of our resources for the production of the other metals.

DATE D'